STATISTICS
IN
ACTION
A CANADIAN OUTLOOK

STATISTICS *IN* ACTION

A CANADIAN OUTLOOK

EDITED BY
JERALD F. LAWLESS

CRC Press is an imprint of the
Taylor & Francis Group, an **informa** business
A CHAPMAN & HALL BOOK

Statistical Society of Canada
Société statistique du Canada

CRC Press
Taylor & Francis Group
6000 Broken Sound Parkway NW, Suite 300
Boca Raton, FL 33487-2742

First issued in paperback 2019

© 2014 by Taylor & Francis Group, LLC
CRC Press is an imprint of Taylor & Francis Group, an Informa business

No claim to original U.S. Government works

ISBN-13: 978-1-4822-3623-1 (hbk)
ISBN-13: 978-0-367-37884-4 (pbk)

Library of Congress Cataloging-in-Publication Data

Statistics in action : a Canadian outlook / edited by Jerald F. Lawless.
    pages cm
  Includes bibliographical references and index.
  ISBN 978-1-4822-3623-1 (hardback)
  1. Statistics--Canada--History 2. Statisticians--Canada. I. Lawless, Jerald F., 1944-

QA276.15.S74 2014
519.50971--dc23   2013048946

Visit the Taylor & Francis Web site at
http://www.taylorandfrancis.com

and the CRC Press Web site at
http://www.crcpress.com

Contents

Foreword

The year 2013 was proclaimed as the International Year of Statistics (`http://www.statistics2013.org`). To many, statistics is about computing averages and proportions on relatively small sets and tables of numbers. But nowadays data from a variety of sources routinely come in many shapes and forms, and in larger quantities than ever before. "Big data" consists of numbers, vectors, curves, text, images, etc.

Extracting useful information from the surrounding noise is what statistical science is all about. Statistical methodology is ever more sophisticated — and exciting — and its application to a vast number of areas, from health science and genetics to fisheries, climate change, and finance, has far-reaching consequences in our lives. Moreover, unbeknownst to many, Canada has been for a long time — and remains — a leader in statistical research.

To help both general readers and users of statistics better appreciate its scope and importance, the Statistical Society of Canada has commissioned this book of expository chapters, written by leading Canadian researchers and showing "statistics in action." What you have in hand is the result of this effort. It is hoped that it will give you a glimpse of the excitement, but also of the importance, of statistics at the beginning of the twenty-first century.

Having a good idea is one thing, but leading it to fruition requires more. We are indebted to Jerry Lawless for taking the lead and editing this book. We are also grateful to all the authors who have worked hard at making their research accessible to a much larger public than usual. Thanks are due to Brian Allen for his help in advertising and marketing the final product, and to Rob Calver, senior editor at Taylor & Francis, for his support. Finally, we would like to express our sincere thanks to Christian Genest, who served as managing editor for this volume, and who expertly dealt with the production of the final manuscript, a task which took him many, many hours.

In the spirit of the International Year of Statistics, this book is intended to help increase public awareness of the importance of statistics in all aspects of human life and society. In support of this objective, its contents are being made freely available at `http://www.ssc.ca/statistics-in-action`, with the publisher's authorization. Color versions of some of the figures can also be found there. I wish you all a stimulating time reading this book!

Christian Léger, PhD, PStat
2012–13 President of the Statistical Society of Canada

About the SSC

The Statistical Society of Canada (in French: Société statistique du Canada) is the main professional organization for statisticians and related scientists in Canada. Founded in Montréal in 1972, the SSC currently has over 1100 members working in academia, government, and the private sector.

The mission of the SSC is to promote the use and development of statistics and probability. Its objectives are to make the general public aware of the value of statistical thought, the importance of this science, and the contribution of statisticians to Canadian society; to ensure that decisions that could have a major impact on Canadian society are based on relevant data, interpreted properly using statistics; to promote the pursuit of excellence in training and statistical practice in Canada; to encourage improvements in statistical methodology; to maintain a sense of belonging within the profession, and to promote dialogue among theoreticians and practitioners of statistics.

In partial fulfillment of its mission, the SSC each year organizes a scientific meeting attended by over 500 participants. It maintains a website at http://www.ssc.ca and publishes a bilingual quarterly newsletter called *SSC Liaison*. Its peer-reviewed scientific journal, *The Canadian Journal of Statistics* (in French: *La revue canadienne de statistique*), is widely recognized for its quality. The Society each year bestows a number of awards recognizing excellence in the profession, the most prestigious of which is the SSC Gold Medal for research. In addition, the Society offers two levels of accreditation (AStat and PStat) for professional statisticians who practice in Canada.

Statistical education is an important aspect of SSC activities. The Canadian component of the International Census at School project is perhaps the Society's most visible initiative in this regard. The present book, which also has an educational purpose, was commissioned by its board of directors to mark the International Year of Statistics 2013. The SSC acknowledges the substantial support of the Institut des sciences mathématiques du Québec (ISM) and its director, Christian Genest, in producing this document.

Preface

Statistics plays an essential role in diverse fields of human endeavor, including physical science and technology, medicine, public health, the social and behavioral sciences, economics and business. It is the unacknowledged workhorse of scientific analysis and prediction aimed at everything from organ transplants to climate science to financial products. At the same time, the emerging field of "big data" — the vast troves of data generated by advancing technology — suggests future directions of a field that continues to develop.

Yet, the breadth and influence of statistics are largely unknown among the general public. The expository chapters in this volume describe some of the contributions of Canadian statisticians and illustrate the breadth and impact of the field. My hope is that the picture of statistics "in action" they provide will stimulate readers from many backgrounds. Those wishing to find out more about statistics (or statistical science) might consult *Statistics in the 21st Century* (Chapman & Hall/CRC, 2002) and a volume commissioned by the Committee of Presidents of Statistical Societies (COPSS) for its 50th anniversary and the International Year of Statistics, *Past, Present, and Future of Statistical Science* (Chapman & Hall/CRC, 2014).

In the first chapter, Bellhouse and Fienberg review the development of statistics as a discipline in Canada. They note the close relationship with statistics groups in the United States, and the extent to which early generations of Canadian statisticians trained there. The following chapter by Beaumont, Fortier, Gambino, Hidiroglou, and Lavallée describes some of the major contributions to survey methodology made at Statistics Canada, one of the world's premier official statistics agencies.

The next five chapters discuss types of statistical methodology that find application in a wide variety of fields. Ramsay and Hermanussen consider the area known as functional data analysis, and illustrate its use in understanding growth in children. Genest and Nešlehová discuss how to model simultaneous variation in two or more variables and illustrate this methodology in a risk management context involving losses associated with fire insurance claims. Tibshirani describes the "lasso" and related methods for finding important variables among large numbers of factors, with an application to genetics. Rosenthal explains the Metropolis algorithm, a so-called Markov Chain Monte Carlo (MCMC) method originally used in theoretical physics; such algorithms are crucial in statistical applications ranging from artificial intelligence to genetics. Finally, Bingham, Ranjan, and Welch describe computer experiments in which large scale computer models are used to study complex phenomena such as volcanic eruptions or climatic variation.

The next seven chapters describe applications of statistics in medicine and public health. Two chapters, by Bull, Graham, and Greenwood and by Craiu and Sun, describe how statistical methods are used to look for genetic factors associated with human diseases or traits. Gustafson explains how Bayesian statistical methods can help deal with measurement problems associated with observational health sciences data. Cook shows how statistical models are used to understand data from clinical trials involving cancer patients. Asgharian, Wolfson and Wolfson discuss methods for studying the time that people live after onset of dementia. Schaubel and Kalbfleisch use statistics to address an important issue in kidney transplantation concerning the use of lower versus higher quality organs. Steiner describes statistical methods for monitoring patients following medical treatment, and their use in improving health care.

Chapters 15 and 16 describe two topics of considerable current relevance: the application of statistical methods in finance and in e-commerce, respectively. Rémillard discusses how statistical models are used to price financial products and to assess the associated risks. Zhu explains some of the methods used to decide what products or services are presented to Internet users.

The remaining chapters deal with various scientific areas of vital importance. Chapters by Cowen, Challenger, and Schwarz and by Rivest and Baillargeon describe "mark-recapture" methods for the study of populations that are difficult to access. They illustrate how the methods can be used to estimate the size of populations as diverse as salmon stocks, injection drug users and illegal immigrants. Mills Flemming and Field show how statistical models are used in marine ecology to study the abundance of fish stocks and endangered species. Zwiers, Hegerl, Zhang, and Wen discuss how evidence for climate change is assessed, and how statistics is used to study the impact of human factors. Finally, Kouadio and Newlands describe how satellite data and other new forms of information are combined with statistical models to provide forecasts of crop yields and other agricultural features.

Canadian statisticians have made major contributions to many other areas of statistical theory and methods, as well as to the foundations of statistical inference. In the selection of areas represented here we have tried to emphasize topics of major current importance and to consider what might lie ahead. One strand running through these chapters is the importance of computing to statistical modeling and analysis. A topic of great current interest is so-called "big data," which refers to the volumes of information generated, for example, by the Internet, financial systems and other automated processes. Issues involved with huge amounts of data are discussed in several of the chapters in this book, and we can expect more and more attention to this area in future.

I want to thank Christian Léger, 2012–13 president of the Statistical Society of Canada, for suggesting that a volume of this type be prepared. I also want to acknowledge the help I had reviewing chapters from numerous readers and from the associate editors: David Bellhouse, Hugh Chipman, Richard Cook, Christian Genest, Paul Gustafson, Agnes Herzberg, Nancy Reid, and Doug Wiens. Brian Allen provided good advice and promotion for the volume. Finally, I am most grateful to the managing editor, Christian Genest, who has expertly compiled this volume.

Jerald F. Lawless, PhD, FRSC
Editor

About the Editor

Jerald F. Lawless is distinguished professor emeritus in the Department of Statistics and Actuarial Science at the University of Waterloo, where he was a faculty member from 1972 to 2007. His research interests include biostatistics, survival and event history analysis, reliability, and general methodology. He is the author of three books and numerous research articles. He has also consulted and collaborated widely in fields that include manufacturing, medicine, risk and safety analysis, social sciences, and law. He is a past editor of *Technometrics* and a former president of the Statistical Society of Canada (SSC). He has received the SSC Gold Medal for research and the Shewhart Medal of the American Society for Quality. He is an elected fellow of the Royal Society of Canada, a fellow of the American Statistical Association, and a fellow of the Institute of Mathematical Statistics.

Contributors

Masoud Asgharian
McGill University
Montréal, QC

Sophie Baillargeon
Université Laval
Québec, QC

Jean-François Beaumont
Statistics Canada
Ottawa, ON

David R. Bellhouse
Western University
London, ON

Derek Bingham
Simon Fraser University
Burnaby, BC

Shelley B. Bull
Lunenfeld–Tanenbaum
   Research Institute and
   University of Toronto
Toronto, ON

Wendell O. Challenger
Simon Fraser University
Burnaby, BC

Richard J. Cook
University of Waterloo
Waterloo, ON

Laura L. E. Cowen
University of Victoria
Victoria, BC

Radu V. Craiu
University of Toronto
Toronto, ON

Christopher A. Field
Dalhousie University
Halifax, NS

Stephen E. Fienberg
Carnegie Mellon University
Pittsburgh, PA

Susie Fortier
Statistics Canada
Ottawa, ON

Jack Gambino
Statistics Canada
Ottawa, ON

Christian Genest
McGill University
Montréal, QC

Jinko Graham
Simon Fraser University
Burnaby, BC

Celia M. T. Greenwood
McGill University
Montréal, QC

Paul Gustafson
University of British Columbia
Vancouver, BC

Gabriele C. Hegerl
University of Edinburgh
Edinburgh, Scotland

Michael Hermanussen
Christian-Albrechts-Universität
Kiel, Germany

Mike Hidiroglou
Statistics Canada
Ottawa, ON

John D. Kalbfleisch
University of Michigan
Ann Arbor, MI

Louis Kouadio
Agriculture and Agri-Food Canada
Lethbridge, AB

Pierre Lavallée
Statistics Canada
Ottawa, ON

Joanna Mills Flemming
Dalhousie University
Halifax, NS

Johanna G. Nešlehová
McGill University
Montréal, QC

Nathaniel Newlands
Agriculture and Agri-Food Canada
Lethbridge, AB

James O. Ramsay
McGill University
Montréal, QC

Pritam Ranjan
Acadia University
Wolfville, NS

Bruno Rémillard
HEC Montréal
Montréal, QC

Louis-Paul Rivest
Université Laval
Québec, QC

Jeffrey S. Rosenthal
University of Toronto
Toronto, ON

Douglas E. Schaubel
University of Michigan
Ann Arbor, MI

Carl J. Schwarz
Simon Fraser University
Burnaby, BC

Stefan H. Steiner
University of Waterloo
Waterloo, ON

Lei Sun
University of Toronto
Toronto, ON

Robert J. Tibshirani
Stanford University
Stanford, CA

William J. Welch
University of British Columbia
Vancouver, BC

Qiuzi Wen
Environment Canada
Downsview, ON

Christina Wolfson
McGill University
Montréal, QC

David B. Wolfson
McGill University
Montréal, QC

Xuebin Zhang
Environment Canada
Downsview, ON

Mu Zhu
University of Waterloo
Waterloo, ON

Francis W. Zwiers
University of Victoria
Victoria, BC

1

Canadians Studying Abroad and the Development of Statistics in Canada

David R. Bellhouse

Western University, London, ON

Stephen E. Fienberg

Carnegie Mellon University, Pittsburgh, PA

"Canada is the Scotland of America, and a proof is before you. Scotland gave a King to England in 1603; to-day Canada gives a President to the American Statistical Association." (Coats, 1939)

1.1   Introduction

The American Statistical Association (ASA) was created in 1839, and for many years was one of the key professional associations for Canadian statisticians. For much of its early history and into the 1930s, the ASA and its *Journal* were dominated by a focus on economic and other forms of official statistics. Thus it was fitting that Robert H. Coats, the Dominion Statistician and an economist by training (BA 1896 Toronto), became the first Canadian president of the ASA in 1938. By the time of the creation of two statistical societies in Canada in the 1970s, later merged to form the Statistical Society of Canada (SSC), and a separate *Canadian Journal of Statistics*, academic statistics in Canada had come into its own. Bellhouse and Genest (1999) describe some of these activities. Nonetheless, the influence of the US on statistics in Canada has been deep and long-lasting. Well into the 1960s and beyond, many Canadian statisticians received their graduate training in the US and returned for the greater benefit of Canada, while others stayed in the US, enhancing Canada's reputation for the quality of its exports. Here we examine how statistics grew as a discipline in Canada up to about 1970 through the training of statisticians and the external influence, mainly American, on that growth. Those who returned developed or enhanced statistics groups and academic programs at Canadian universities.

To recognize the early "pioneers of statistics in Canada," the SSC created the award of Honorary Member in 1980 and conferred it on five statisticians (Maag, 1980). The next year the Society conferred three more honorary memberships (Fellegi, 1983), again on statistical pioneers. Since the theme of the paper is statistical training, we note that of the eight honorary members in the early years of the award, four had substantial teaching careers at Canadian universities: Daniel DeLury at the University of Toronto, George Edgett at Queen's University, Cyril Goulden at University of Manitoba, and Ernest Keeping at University of Alberta. Of these four, only DeLury received all his degrees in Canada (BA 1929; PhD 1936 Toronto). Edgett (BA 1923; MA 1926 Mount Allison) obtained his PhD from the University of Illinois in 1936 and Goulden (BSA 1921; MSA 1923 Saskatchewan) from Minnesota in 1925. Keeping initially studied in England (BSc 1916 London) and later studied with Harold Hotelling at the University of North Carolina. This trend began before these pioneers and continued for many years afterward.

In the published descriptions of the work of these early pioneers, some claims were made about their contributions to the development of courses in statistics at their universities: DeLury "developed widely-used lecture notes for statistics courses"; Edgett joined Queen's in 1930 and taught "the first statistics course in a Mathematics Department in Canada" beginning in 1933; Keeping "developed an advanced level course in statistics, probably the first such course given in Canada" in the early 1930s. No mention was made of Goulden's teaching career at University of Manitoba; see Bellhouse (1992) for a description. One might be tempted to conclude that these four were the first to develop statistics courses at Canadian universities, although Watts (1984) cites evidence of a few statistics courses offered in the 1910s, mainly in economics departments and business schools. We have found that statistics courses were developed in mathematics departments in Canada in the 1920s or before. Like the "statistical pioneers," these early instructors of statistics were typically trained at the undergraduate level in Canada and at the graduate level in the US. We describe some of these developments.

1.2   Statistics' Rise as a Modern Field in North America

During the Victorian era the field of statistics in both the UK and the US was defined by the collection and tabulation of data about social and economic issues. It is the mathematization of statistics that defines the field today. This occurred in late Victorian times initially with Francis Galton and Francis Y. Edgeworth (Stigler, 1986). In the 1890s, Karl Pearson developed a theory of mathematical statistics (Magnello, 2009). His work was motivated by

Charles Darwin's study of biological variation and evolution, and influenced by W. F. R. Weldon's data and scientific questions. Shortly thereafter Pearson established the Drapers' Biometric Laboratory at University College London in 1903, which in 1911 became the first department of statistics, although it only began to offer a degree in mathematical statistics in 1915. In 1901, with Weldon and Galton, Pearson also founded *Biometrika*, the first journal primarily devoted to the development of statistical methodology and theory. At the time the publications of the Royal Statistical Society (RSS) and the ASA were focused primarily on economic and vital statistics, and contained only limited amounts of methodology and essentially no theory until well into the twentieth century. The founding of *Biometrika* in many ways marks the start of the modern era of statistics; see, e.g., Stigler (1986), Porter (1988), Fienberg (1992), and Hald (1998).

Early in the twentieth century, some American researchers, but apparently no Canadians, came to see the applicability of the new mathematical statistics to their work and came to visit Pearson or to study with him at University College London. They came from a variety of disciplines: biology and evolution, economics, and psychology. Pearson's influence was also felt on American mathematics, albeit indirectly, as some studied his work and began to publish their own research in mathematical statistics. Until 1918, no Canadian mathematicians seem to have carried out any research in this area.

At the end of World War I, American mathematicians began to awaken to this newly emerging field. In his presidential address to a meeting of mathematical societies in the US, Edward Huntington of Harvard University addressed the topic of statistics (Huntington, 1919). He noted the paucity of papers on statistics read before the American Mathematical Society in the previous five years and that only a dozen professional mathematicians were members of the ASA. He encouraged more mathematicians to embrace work in statistical theory and then went on to give an overview of correlation and regression. An address to an American society would have had resonance in Canada. There was no Canadian Mathematical Society and many Canadian professional mathematics contacts were through the American societies. Consequently, no mathematical statistics courses, or at most very few, would be expected to appear at Canadian universities until about 1920. In England, important developments occurred throughout this period with contributions by towering statistical figures such as Ronald A. Fisher, Jerzy Neyman, Egon Pearson, and Harold Jeffreys who helped to create the foundations of statistics as we know it today.

In the US, the earliest department with statistics in its name was the Department of Biometry and Vital Statistics at Johns Hopkins University, created in 1918. The Statistical Laboratory at Iowa State College followed in 1933, and the Department of Statistics at George Washington University in 1935. Other departments often taught courses on probability and statistics, especially mathematics departments, with such notable intellectual leaders as Edwin Olds at Carnegie Tech, Harold Hotelling at Columbia, Henry Reitz and Allen Craig at the University of Iowa, Harry Carver at the University of Michigan, and Samuel Wilks at Princeton University. They also provided much of the impetus behind the creation of the Institute of Mathematical Statistics (IMS) and *The Annals of Mathematical Statistics*, created by Carver and edited later by Wilks when it became the official journal of the IMS. The volume edited by Agresti and Meng (2013) documents the rise of academic departments of statistics in the US, during this and later periods.

In a somewhat parallel development, sample survey theory also emerged in the twentieth century; see Fienberg and Tanur (2001). In the 1930s, spurred on by Neyman's contributions in particular, statisticians at the US Bureau of the Census developed the first large-scale probability sampling approaches to measure unemployment, and for auxiliary information, as part of the 1940 decennial census. During this period, the Dominion Bureau of Statistics (DBS), now Statistics Canada, founded and led by Coats, focused on more traditional official statistical activities. Its move towards a probability-based labor force survey came in 1947, in an effort led by Nathan Keyfitz who went on to a second prolific career as a mathematical demographer in both Canada and the US, following his PhD work at the University of Chicago in the 1950s.

1.3 Statistical Education in Canada between the Wars

Relative to the US, the growth of statistics as an academic discipline in Canada was slow. In his address to the ASA, Coats (1939) commented on Helen Walker's study of statistics curricula (Walker, 1929) in the US. He remarked that, with respect to Canada,

> "I grieve to say she would have found no chair of statistics, and that five of our twenty-two universities do not know the word in their curricula."

Further, the first department in Canada to have "statistics" in its name was the Department of Actuarial Science and Statistics in 1950 at the University of Manitoba (Rankin, 2011); in the US it was 1918 at Johns Hopkins. Similar to most of the US, statistics in Canada grew in mathematics departments. Coats' lament is related, in part, to the size of the mathematics departments in Canada. Typically, these departments were small in size, sometimes comprising as few as two members and usually not more than about six. Toronto was the exception with faculty numbering into the double digits. This situation continued with only slight growth into the early 1950s.

Information on early statistics courses offered in mathematics departments can be obtained from university calendars of the time. The entries often, but not always, include the name of the instructor and the textbook used. We have used these calendars as a check on the claims about Canada's statistical pioneers. Edgett, for example, shows up in the Queen's calendars as a lecturer in the 1931–32 academic year in which he taught "Mathematical Theory of Statistics," two years earlier than the date claimed in Maag (1980). Over the next two years, Edgett taught the courses "Statistics and Probability" (1932–33) followed by "Probability and Statistics" (1933–34). No textbooks for the courses were mentioned. In addition to the courses by Edgett at Queen's, we report on statistical courses from a selection of calendars that we have been able to examine.

At the University of Toronto, the first statistics course in the Department of Mathematics was offered at the graduate level in the 1917–18 academic year; the same course was offered yearly until about 1936. Initially, the course had the title "Actuarial science: Frequency curves and correlation," showing Pearson's impact on statistical education at this time. In the late 1920s the title changed to "Frequency curves and correlation, measurement of groups and series." At the undergraduate level, a course "Mathematics of Statistics" was offered in 1924 at the pass level rather than at the honors level. The graduate course was taught continuously by Michael Mackenzie; presumably he taught the undergraduate course as well. No textbook was given for the undergraduate course until 1935 when Yule's *Introduction to the Theory of Statistics* was put on the syllabus. Mackenzie (BA 1887 Toronto) obtained his MA in mathematics from Cambridge University. He became interested in actuarial science and taught the subject for many years at Toronto. He joined the Institute of Actuaries in 1899 and became a fellow of the Institute in 1907. Prior to Daniel DeLury's arrival in the Department of Mathematics at Toronto in 1937, there were a couple of other developments in the statistics curriculum. In 1930 the mathematician Jacques Chapelon introduced a graduate course "The Mathematical Theory of Probability and Statistics" and the next year added another graduate course "The Fundamental Problems of Mathematical Statistics." A few years later, the "Fundamental Problems" course was taken over by Norris Sheppard, another actuary teaching in the department. Toronto's early experience in offering mathematical statistics courses is a harbinger for other universities that we were able to examine. Mathematicians who were interested in actuarial science and the mathematics of finance were often the developers and teachers of mathematical statistics courses. They obtained their highest degree abroad.

In the 1920s the University of Manitoba had a downtown campus where the Department of Mathematics and Astronomy was located. At first, courses with probability content were driven by actuarial science. In 1912 Professor Neil B. MacLean, whose interests were in geometry, offered a course in probability and life contingencies. Later, Lloyd Warren took over the actuarial courses and program, and in 1938 set up the new Department of Actuarial Science. In 1923, when still a member of the Department of Mathematics and Astronomy, Warren offered a graduate course in mathematical statistics with Yule's book as the text. Like Michael Mackenzie, Warren was a Canadian who took his graduate degree in mathematics abroad, obtaining his PhD from the University of Chicago with a thesis topic far removed from actuarial science. Manitoba had one thing that the University of Toronto did not have: an agricultural school which was located on a campus in Fort Garry where now almost the entire university sits. A course on the collection and analysis of data was offered in 1927 in the economics curriculum of the school and was followed the next year by a course in "advanced statistics." A course on agronomic experimental methods using "biometrical methods of analyzing data" was offered in the Agronomy Department in 1928 by Gordon McRostie. When Cyril Goulden arrived in 1931, the course was named "Statistical Methods in Agronomic Research" and used R. A. Fisher's *Statistical Methods for Research Workers* as the text. Goulden, a Canadian who took his PhD at the University of Minnesota in the genetics of wheat, spent a year at Rothamsted Experimental Station studying with Fisher. Goulden's course is the earliest known example of Fisher's influence on a statistics course offered in Canada. When the Department of Actuarial Science was founded in 1938, Goulden was transferred to the new department and the course was offered there under the name "Statistical Methods for Research Workers." All statistics courses previously offered in mathematics were moved to the new department. In 1944, the department offered five courses in statistics, including Goulden's course and another in mathematical statistics which used Wolfenden's *Fundamental Principles of Mathematical Statistics*, a text intended for actuarial students. To reflect the growth of the statistics curriculum, the department's name was changed to Actuarial Mathematics and Statistics in 1950.

We compare Manitoba's situation to the University of Saskatchewan, which, like Manitoba, has a mathematics department and an agricultural school. Their first course in statistics was offered in the College of Agriculture as early as 1938. James Harrington, a Canadian crop scientist who took his PhD at Minnesota like Goulden, offered a course "Biometry and Field Plot Techniques." The text for the course in 1939 was Goulden's *Methods of Statistical Analysis*. The Department of Mathematics followed in 1940 with a course entitled "Introduction to the Mathematics of Statistics" with Kenny's *Mathematics of Statistics* as the text. It is uncertain who taught this course. There were again some parallels to Manitoba and Toronto. A few years prior to 1940, the Department of Mathematics offered a course in the theory of investments which included topics in probability, life annuities and life insurance. In 1945 Wacław Kozakiewicz joined the department and became the primary instructor for statistics. Following his arrival from Poland after World War II, Kozakiewicz worked initially at the DBS where he gave courses to the DBS personnel in pure and applied mathematics. In 1949 he was lured away from Saskatchewan by the Université de Montréal. This university was beginning to offer a program in mathematical statistics and wanted a mathematician trained in the French tradition to teach statistics (W.L.G.W., 1959).

The initial development of a statistics curriculum at University of Western Ontario is derivative from Manitoba. Harold Kingston, another Canadian who obtained his PhD in mathematics from the University of Chicago, joined Lloyd Warren on faculty at Manitoba. In 1921 Kingston was appointed professor and chair of the Department of Mathematics and Astronomy at the University of Western Ontario. When Kingston arrived, there were only two or three faculty members in the department. The year following his appointment, Kingston began an actuarial program at Western and set up a course entitled "Introduction to Statistics." While statistics grew at Manitoba, it remained static at Western for many years. The introductory course was dropped after one year and another statistics course did not appear on the books until 1927. This time it was offered at the third year with Irving Gavett's *First Course in Statistical Method* as the text. It was mathematical but not at the level of Yule's introductory book. A follow-up course was added ten years later. The courses remained essentially service courses to the actuarial program well into the 1940s. The situation changed in the late 1940s after Randal Cole was hired in the department. In 1949 the text was changed to Hoel's *Introduction to Mathematical Statistics* and a graduate course in advanced statistics was offered. Cole's research area initially was differential equations. He did publish two papers in statistics, one in 1944 on biometry and the other in 1951 on order statistics.

The presence of an actuarial science program did not necessarily go hand-in-hand with the development of a statistics curriculum in a mathematics department. McGill was one university where there was difficulty in laying down early roots for the teaching of statistics in the Department of Mathematics. Herbert Tate arrived in the Department of Mathematics at McGill in 1921 after graduating with an MA in mathematics from the University of Dublin. The same year he became a fellow of the Royal Statistical Society. He taught courses related to actuarial science; one of the earliest courses in that area was "Probability, Finite Differences and Statistics" offered in 1924. Although Tate taught several courses related to actuarial science and wrote two books on the mathematics of finance, actuarial courses were offered only sporadically in the Department of Mathematics. What may have hampered the development of statistics courses in the Department of Mathematics is that the Department of Commerce also offered statistics and actuarial science courses nearly every year over the years 1922 to 1930. In 1922, Bowley's *Elements of Statistics* was one of the textbooks for the statistics course in commerce. The next year Yule's *Introduction to the Theory of Statistics* was added to the list. Tate also taught courses in the Department of Commerce from the mid-1920s onward. In 1927, and continuing for several years, a service course in elementary statistics was offered in the Department of Mathematics, first taught by another mathematician, T. H. Matthews. There was no further development of statistics courses in the Department of Mathematics even into the mid-1940s.

Of all the early pioneers, honorary member of the SSC or not, Daniel DeLury (BA Toronto 1929, PhD Toronto 1936) was the only one educated solely in Canada. DeLury taught first at the University of Saskatchewan and then at the University of Toronto, where he first appears as a lecturer in the Toronto calendar for 1938. He later went to the US and served on the Faculty of Virginia Polytechnic Institute from 1945 to 1947 before returning to the Ontario Research Foundation in Toronto and later back to the University of Toronto, whose mathematics department he chaired.

1.4   World War II as a Watershed

In many ways, World War II was a watershed for statistics in the US as well as the UK. In both countries statisticians became part of the war effort as members of scientific research teams (Wallis, 1980; Barnard and Plackett, 1985; Fienberg, 1985, 2006). These concentrated activities led to major theoretical breakthroughs including the creation of sequential analysis (by Abraham Wald in the US and George Barnard in the UK), statistical decision theory and methods for quality control, and Bayesian tools used in code breaking efforts. Statisticians involved in these efforts went on to establish major departments in the US (Agresti and Meng, 2013) and active statistics groups and programs in various UK universities.

It is less well documented, but Canadian scientists were also involved in the war effort, particularly in the areas of operations research and quality control. There were about sixty Canadian scientists working in operations research scattered over the Canadian Army, the Royal Canadian Navy (RCN) and the Royal Canadian Air Force (RCAF); see Morton (1956). Among this group was John W. Hopkins, who completed an MSc in plant biochemistry at the University of Alberta in 1931, followed by a PhD under R. A. Fisher at University College London in 1934. Hopkins led a statistics laboratory at the National Research Council and during the war was seconded to the Eastern Air Command and Canadian Heavy Bomber Group (Fellegi, 1983). In quality control Isobel Loutit, who was working as a school teacher up to the end of 1941, responded to a Canadian government advertisement for women in science and mathematics to take jobs in industry. She initially worked for the Inspection Board of the United Kingdom and Canada, and then transferred to Northern Electric where she worked testing the quality of a mechanical aiming device for artillery (Bellhouse, 2002). This effort did not have the same effect on the growth of academic statistics programs as was the case in the US.

Another notable Canadian researcher closely linked to statistics was Howard B. Newcombe, who studied at Acadia and then in Trinidad before completing his PhD in genetics at McGill in 1939. Following service in the Royal Navy during World War II, he spent a year at Cold Spring Harbor in the US and then joined the Canadian Atomic Energy Project at Chalk River, Ontario. It was there that in addition to his genetics research he pioneered the development of record linkage techniques that influence government and academic statisticians in both Canada and the US.

A number of Canadians, who became statisticians at various universities, finished high school or an undergraduate degree and then served in the military during the war. This group includes Charles W. Dunnett (army), James L. McGregor (army), John R. McGregor (RCAF), Gilbert I. Paul (RCAF), Norman Shklov (army) and Ralph Wormleighton (army, medical corps).

1.5 A Second Early Generation of Canadian Statisticians

The next generation of Canadian statisticians received their undergraduate training during or just after World War II. They began their academic careers in a variety of disciplines, mainly mathematics, at various universities across the country. Given that there was very limited opportunity for doctoral studies at Canadian universities at this time, with one exception, they went abroad to complete their studies, mostly to the US but some to the UK. Again with one exception, they embraced statistics in their graduate programs. Most returned to Canada; the attraction of the US was stronger for later generations. Here we list some of the most notable of them in approximate chronological order of first degrees:

- Norman Shklov (BA 1940 Manitoba; MA 1949 Toronto). After PhD work at the University of Indiana, he served on the faculty of the University of Saskatchewan and later the University of Windsor.

- Charles W. Dunnett (BA 1942 McMaster; MA 1946 Toronto). He worked as a biostatistician in Canada and the US, receiving a PhD from the University of Edinburgh in 1960. He returned to McMaster in the 1970s (Macdonald, 1988).

- Ralph Bradley (BA 1944; MA 1946 Queen's). After a PhD in statistics from the University of North Carolina, he had a distinguished statistical career in the US at Virginia Polytech, Florida State, and Georgia.

- Colin Blyth (BA 1944 Queen's). After his PhD from the University of California at Berkeley in 1950, he taught at Illinois and returned to Queen's as a faculty member in the 1970s.

- Ralph Stanton (BA 1944 Western Ontario; MA 1945, PhD 1948 Toronto). He was more mathematician and computer scientist but exerted enormous influence on the development of statistics in Canada while on the faculties of Toronto, Waterloo, Manitoba, and Windsor.

- Martin Wilk (BEng 1945 McGill). After work at the National Research Council he stumbled into statistics as a graduate student at Iowa State University earning a PhD in 1955. His 25-year US career included Princeton University, Bell Laboratories, Rutgers University, and AT&T. In 1980 he returned to Canada as Chief Statistician (Genest and Brackstone, 2010).

- Jacques Saint-Pierre (MSc, LSc Montréal). He joined the Département de mathématiques at Université de Montréal in 1947 and obtained a PhD in statistics from the University of North Carolina in 1954. Remaining at Montréal for his career, he founded the Centre de statistique in 1957 and the Département d'informatique et de recherche opérationnelle in 1966.

- Donald Fraser (BA 1946, MA 1947 Toronto). He went to Princeton to study mathematics but gravitated into statistics under the influence of Sam Wilks and John Tukey, who were both in the Department of Mathematics. Following the receipt of his PhD in 1949, he returned to Toronto as a faculty member where he has spent the vast majority of his career. He has influenced generations of undergraduates and graduate students, some of whom followed his lead from Toronto to Princeton (DiCiccio and Thompson, 2004).

- James L. McGregor (BA 1949 MA 1951 British Columbia). He completed a PhD in mathematics at Caltech in 1954 and joined the faculty of Stanford University in mathematics, where he wrote a series of papers on stochastic processes in population genetics with Samuel Karlin.

- Ralph Wormleighton (MA Toronto). He went to Princeton where he completed a PhD in mathematics under John Tukey in 1955. He served on the Toronto faculty for 30 years beginning in 1953.

- Gilbert I. Paul (MSc Alberta). With an undergraduate background in plant breeding, he obtained a PhD in statistics and genetics from North Carolina State University in 1956. He taught population genetics at McGill for four years and then statistics at Manitoba for 30 years, where he influenced a stream of students to take advanced degrees in statistics.

- John McGregor (BSc, BEd, MEd Alberta). He went to England to study mathematical statistics at King's College Cambridge, from which he received a PhD in 1959. He returned to the Alberta faculty serving as chair of mathematics and dean of graduate studies before becoming the founding chair of the Department of Statistics and Applied Probability.

1.6 Generation of the 1950s and Early 1960s

In this generation, many Canadians continued to be attracted to graduate programs in statistics in the US. From the brief biographies of statisticians that we have examined from this time, the main attractors in the US were

Harvard, Princeton and Stanford. The major suppliers of students came from across the country: Alberta, British Columbia, Manitoba, McGill, Montréal, Queen's, Saskatchewan and Toronto. Here we document some of the graduates from that era, focusing on Canadians who took their undergraduate degrees in Canada and then went on to do graduate work abroad. Some of those returned to Canada, others did not but had strong Canadian connections, while yet others forged careers solely in the US.

1.6.1 Princeton Connection

One of the early US attractors for Canadian students was Princeton University. Samuel Wilks, John Tukey and William Feller were on staff at Princeton University, joined briefly by George Box and Frank Anscombe. What also helped to attract students, especially from Toronto, was the early connection made by Fraser and later by Wormleighton. Alexander Mood, one of Samuel Wilks' first students, related the following:

> "In his later years [Wilks] maintained that it was impossible for him to persuade enough sufficiently promising college graduates to undertake work in statistics at Princeton and therefore he had to go to Britain and Canada to find good students whose attitudes had not been corrupted by pure mathematicians in the United States." (Mood, 1965)

Many of the early Princeton Canadian students remained in the US after completing their PhD: Arthur Dempster (BA 1952; MA 1953 Toronto) on the faculty of Harvard; David Freedman (BSc McGill 1959) and David Brillinger (BSc Toronto 1959) on the faculty of the University of California at Berkeley; and Morton Brown (BS McGill 1962) at UCLA and Michigan. Although living and working in the US, Brillinger has long maintained strong Canadian connections and support for statistics in Canada; he has been President of the SSC and received the SSC Gold Medal. Several other Canadians who obtained their PhD from Princeton returned to Canada to work in academia or government. Thomas Wonnacott (BA Western Ontario 1957) returned as a faculty member in statistics to the University of Western Ontario where he wrote a popular introductory statistics textbook that influenced the teaching of statistics at many universities in the US and Canada. Morven Gentleman (BSc McGill 1963) has held positions in computer science at Waterloo and Dalhousie. James Ramsay (BEd Alberta 1964) studied psychology at Princeton and took a faculty position in psychometrics at McGill, and later was SSC president and Gold Medal winner. Gordon Sande (BSc Alberta 1964) briefly taught at Chicago and worked for a time at Statistics Canada on confidentiality and data disclosure issues.

Princeton also attracted Canadians as academic visitors, including Irwin Guttman (BSc 1951 McGill) and David Andrews (BSc 1965 Toronto). Guttman completed his PhD at Toronto in 1955 and Andrews in 1968, both under Fraser's supervision. Both were longtime faculty members in statistics at Toronto, although Guttman also taught in the US at Wisconsin and the State University of New York at Buffalo. Both were awarded the SSC Gold Medal for their pioneering work in statistical theory and methods. Len Steinberg, whose PhD was also from Toronto under Fraser in 1967, taught at Princeton before joining the World Bank in Washington, DC. Miklós Csörgő, also an SSC Gold medalist, completed his PhD at McGill and did a post-doc at Princeton before returning to teach at McGill and Carleton.

William "Bill" Williams, was a McMaster undergraduate (BA 1954) and studied statistics at Iowa State (MS 1956; PhD 1958), returning briefly to teach at McMaster before joining Bell Labs, drawn by John Tukey who worked there as well as at Princeton, and later the faculty at Hunter College, City University of New York. Ronald Pyke also studied at McMaster overlapping with Williams, receiving his BA in 1953. He then migrated to the US, completing an MSc in Mathematical Statistics in 1955 and a PhD in Mathematics in 1956 at the University of Washington. After faculty appointments at Stanford and Columbia, he returned to Washington for the remainder of a distinguished career primarily focused on probability.

1.6.2 More Migration from Toronto

Many coming out of the honors courses in mathematics, statistics and chemistry at Toronto in the 1950s and early 1960s, went to US universities other than Princeton. Among the returnees to Canada, Thomas Stroud (BA 1956; MA 1960 Toronto) took his PhD in statistics from Stanford in 1968 and then joined the faculty at Queen's. David W. Bacon (BA 1957 Toronto) joined Queen's after obtaining his MS and PhD degrees in statistics from Wisconsin in 1962 and 1965, respectively. Paul Corey (BSc 1962; MSc 1965 Toronto) obtained his PhD in Biostatistics from Johns Hopkins University in 1974 and has been on the faculty in Public Health at Toronto since the late 1960s. After her PhD at the University of Illinois in 1969, Mary (Beattie) Thompson (BSc 1965; MSc 1966 Toronto) became a member of the statistics faculty at the University of Waterloo. Another SSC Gold Medal winner, she now leads the Canadian Statistical Sciences Institute. Larry Weldon (BSc 1965 Toronto) completed his PhD at Stanford in 1969 and returned to Canada to teach at York, then Toronto, and Dalhousie, and finally at Simon Fraser. Peter Macdonald (BSc 1966; MSc 1967) is the exception to the "study in the US rule." He received his PhD in 1971 from the University of Oxford and then joined the faculty at McMaster University, in Hamilton.

Among the Toronto alumni, some notable statisticians decided to remain in the United States. John Chambers (BSc 1963 Toronto) went to Harvard for a PhD in statistics, which he completed in 1966. He joined the Bell Labs research staff in 1966 where he remained until his retirement in 2005, leading major efforts in statistical computing. He is currently on the faculty of Stanford University. Stephen Fienberg (BSc 1964 Toronto) followed Chambers to Harvard for his PhD in statistics, completed in 1968. He has taught subsequently at the University of Chicago, University of Minnesota, and Carnegie Mellon University, with a brief stint back in Canada at York University as Vice President Academic (1991–1993). Rudolph Beran (BSc 1964 Toronto) went to Johns Hopkins for his PhD, which he received in 1968. He then joined the statistics faculty of the University of California at Berkeley, and then moved to the University of California at Davis in 2001.

Several Canadian-trained undergraduates received PhDs in statistics from the University of Toronto in the late 1960s and migrated to the US as well. Hans Levenbach was born in Indonesia of Dutch parents at the beginning of World War II and came to Canada via the US. He attended Acadia (BSc in Physics and Math 1961), Queen's (MSc Electrical Engineering 1964), and Toronto (PhD 1968). He then joined Bell Labs and later founded an independent consulting company. Gerald van Belle was born in Holland, did all of his degrees at the Toronto (BA 1962, MA 1964, and PhD 1967), and then moved to the US initially to Florida State, and then to the University of Washington, in biostatistics. Andrew Kalotay was born in Hungary and immigrated to Canada following the 1956 revolution. His initial degrees were from Queen's (BA Math 1964, MS Math 1966) and he completed a PhD in statistics from Toronto 1968. He also went to Bell Labs and now focuses on problems of finance, leading his own company.

1.6.3   Manitoba Influence

When Gilbert Paul arrived at Manitoba, he became the head of the statistics group in the Department of Actuarial Mathematics and Statistics. In the early 1960s, he developed a Master's program in statistics and the first class graduated in 1963. Two of the three graduates were Charles Goldsmith and Brian Macpherson, both coming from Manitoba's undergraduate program (Rankin, 2011). Goldsmith went on to take his PhD at Paul's alma mater, North Carolina State. He graduated in 1969 and returned to Canada to a position at McMaster University. MacPherson remained in the department at Manitoba and subsequently obtained his PhD at Iowa State in 1981. Bruce Johnston (BSc Manitoba 1960) also took his PhD at North Carolina State, graduating in 1968. He returned to Canada, initially to Queen's University, and then back to Manitoba. Other Manitoba students were attracted to Harvard. Paul Switzer (BSc 1960 Manitoba) went to Harvard where he completed a PhD under Arthur Dempster, and then joined the faculty at Stanford where he has spent his career. Aaron Tenenbein (BSc 1965 Manitoba) and Allan Donner (BSc 1965; MSc 1967 Manitoba) were classmates and both studied with William Cochran at Harvard. Tenenbein completed his PhD in 1969 and joined the faculty at New York University. Donner completed his PhD in 1971 and returned to Canada to take a position in biostatistics at the University of Western Ontario, where he has become well known for his work in clinical trials. The connection that the Department of Actuarial Mathematics and Statistics had to Manitoba's business school probably led George Alexander Whitmore to go to the University of Minnesota to take his MSc and PhD (1968) in business following his undergraduate work at Manitoba. He returned to Canada initially to teach at Saskatchewan, and later at McGill in Montréal.

1.6.4   Other Notable Canadians Who Went Abroad

Like many others, French-language universities in Canada had a mixed experience sending students abroad. The Université de Montréal established its own PhD program and Alexis Zinger was its first statistics PhD student in 1957 (Rousseau, 2000). Zinger went on to become "Vice-recteur aux Communications" at UQAM. The mixed experience is seen in four notable undergraduate students. André Plante went to Toronto to obtain his PhD in 1962; Pierre Robillard and Robert Côté went to the University of North Carolina, graduating in 1968 and 1971, respectively; as for Robert Cléroux, he stayed in Québec and obtained his PhD from the Université de Montréal in 1965. André Plante was a Professor at UQAM for the better part of his career. Prior to his untimely death, Pierre Robillard was a central figure in trying to bring the two Canadian statistical societies together in the 1970s (Bellhouse and Genest, 1999); the SSC annual best thesis award is named after him. After his PhD, Robert Côté returned to Université Laval, where he had done his undergraduate studies; he helped to develop a very successful undergraduate program in statistics and was Chair of the Department of Mathematics and Statistics for several years. Cléroux was a Professor at the Université de Montréal and served as President of the SSC in 1988–89.

There were at least two students from Kingston who went on to the United States for graduate studies. Jack Graham (BSc 1955 Queen's) went to Iowa State University where he obtained his PhD in 1963. Upon returning to Canada he taught at Carleton for many years. Roger Davidson (BSc 1960 Queen's; MA 1961 Toronto) obtained his PhD from Florida State under Ralph Bradley. On his return to Canada, he taught at the University of Victoria.

Another two students bound for studies in the United States came out of McGill — one came back, the other did not. Donald Dawson (BSc 1958; MSc 1959 McGill) completed his PhD at MIT in 1963. He returned to Canada and taught at McGill before moving to Carleton in 1970. Edward Rothman (BSc 1965 McGill) went to Johns Hopkins for a PhD in 1969, after which he joined the University of Michigan.

Saskatchewan also produced two students. Ian MacNeill (BEd 1958; BA 1962 Saskatchewan; MSc 1964 Queen's) studied at Stanford, receiving his PhD in 1967. He founded the Department of Statistical and Actuarial Sciences at University of Western Ontario in 1980. James Tompkins (BA 1965 Saskatchewan) went on to do graduate studies at Purdue (MS 1967; PhD 1970). He taught at the University of Regina on his return to Canada and devoted much energy to promoting the SSC.

Other universities at this time produced a single but notable graduate who continued his studies, with one exception, in the US. The exception is Donald Watts (BASc 1956; MASc 1958 British Columbia) who went to Imperial College to obtain his PhD in 1962. He taught at the University of Wisconsin (1966–70) before taking a faculty position at Queen's University. James Zidek (BSc 1961; MSc 1962 Alberta) received his PhD from Stanford in 1967. He then joined the faculty at UBC where he remained for his entire career, and was a founding member of its Department of Statistics in 1983. Christopher Field (BSc 1964 Dalhousie) went to Northwestern where he obtained his PhD in 1968, before returning to the Dalhousie faculty where he became the leader of the statistics group within Dalhousie's Department of Mathematics. Both Zidek and Field were SSC presidents and Gold Medal winners. One person returned after a long hiatus. David Thomson (BSc 1965 Acadia) studied at the Polytechnic Institute of Brooklyn, where he received his PhD in 1971 in electrical engineering. Following a long career at Bell Laboratories, he returned to Canada to teach at Queen's in 2001. And there was another who got away. Barry Arnold (BSc 1961 McMaster) did his graduate work in statistics at Stanford (MS 1963; PhD 1965) and is a long-time faculty member at the University of California at Riverside.

As DeLury was the exception to out-of-country study in the 1930s and 1940s, there were also exceptions in the 1950s and 1960s, e.g., Andrews, Csörgő, and Guttman. David Sprott also studied at Toronto in mathematics (BA 1952; MA 1953) and completed a 1955 PhD there under DeLury, followed by postdoctoral study at the Galton Laboratory at University College London. After three years on the faculty at Toronto, he joined the fledgling program at Waterloo, where he remained for his entire career as a researcher and administrator. Ivan Fellegi completed three years of university in Hungary before emigrating to Canada in 1956 at the time of the Hungarian uprising. He joined the staff of DBS/Statistics Canada and completed his studies with night courses at Carleton, where he was the first MSc (1958) and PhD (1961) recipient. He spent his entire career at Statistics Canada, succeeding Martin Wilk as Chief Statistician in 1985, and serving until retirement in 2008.

1.7 A Paradigm Shift

From the 1910s to the 1960s, many Canadians went abroad to do their graduate work in statistics (or mathematics), the vast bulk of them to the United States. Beginning in the 1960s the whole dynamic began to change as departments in Canada developed and many began producing PhD graduates in statistics, initially in Montréal and Toronto. Both Canadians and students from other countries entered these programs. Many, both Canadian-born and not Canadian-born, remained in Canada enhancing the statistics profession in general and the faculty at Canadian universities in particular. Further, with the explosion in the size and number of Canadian universities in the 1960s and beyond, several statisticians were brought into Canada from other countries to take faculty positions in statistics, as well as many other disciplines. Many of these "imports" have contributed greatly to statistics in Canada. The general growth of the discipline can be seen in the creation of separate departments of statistics, with Manitoba and Waterloo leading the way in 1967. Later, several departments with the name "Department of Mathematics" were renamed to "Department of Mathematics and Statistics."

Although the development of statistics as a discipline in Canada lagged behind that in the US, the same could well be said for other academic disciplines. But what should be clear from this brief exposition is that the generations of statisticians who trained initially as undergraduates in Canada, and then went abroad for graduate study, mostly to the US, went on to shape the course of statistics as a field in Canada, and to some extent in the US as well.

Acknowledgments

We are grateful for the help provided by librarians and archivists at various universities in obtaining access to early university calendars. Unfortunately there was no systematic way to trace the paths of undergraduates from Canada to the US and elsewhere in the 1945–65 period, and we relied on others for assistance and information. Everything worth knowing turns out *not* to be easily available from a `Google` search. A number of old friends and colleagues graciously assisted us by identifying and helping track down information on Canadians who began their careers with an undergraduate education in Canada; but we are sure our coverage is incomplete.

About the Authors

David R. Bellhouse is a professor of statistics at Western University. His undergraduate work was done at University of Manitoba and his PhD at University of Waterloo. His earliest research was in the areas of survey sampling theory and methods, and he served on Statistics Canada's Advisory Committee on Statistical Methods from 1998 to 2005. Currently, his research is focused on the history of probability, statistics and actuarial science, with an emphasis on developments in the 18th century. He served as president of the Statistical Society of Canada in 1998–99.

Stephen E. Fienberg is the Maurice Falk University Professor of Statistics and Social Science at Carnegie Mellon University. He holds degrees in mathematics and statistics from the University of Toronto and Harvard University. His interests in statistical theory and methods include discrete data analysis, data confidentiality and privacy, the social sciences, statistics and law, and network data analysis. He has served as editor of the *Journal of the American Statistical Association* and *The Annals of Applied Statistics*. He is a member of the United States National Academy of Sciences, an elected fellow of the American Academy of Arts and Sciences and the Royal Society of Canada, and a fellow of the American Statistical Association and of the Institute of Mathematical Statistics, for which he served as president.

Bibliography

Agresti, A. and Meng, X.-L. (2013). *Strength in Numbers: The Rising of Academic Statistics Departments in the U.S.* Springer, New York.

Barnard, G. A. and Plackett, R. L. (1985). Statistics in the United Kingdom, 1939–45. In *A Celebration of Statistics*, pp. 31–55. Springer, New York.

Bellhouse, D. R. (1992). The statistical work of C. H. Goulden. *SSC Liaison*, 6(3):8–14.

Bellhouse, D. R. (2002). Isobel Loutit: Statistician of quality. *SSC Liaison*, 16(2):14–19.

Bellhouse, D. R. and Genest, C. (1999). A history of the Statistical Society of Canada: The formative year (with discusion). *Statistical Science*, 14:80–125.

Coats, R. H. (1939). Statistics and society. *Journal of the American Statistical Association*, 34:1–26.

DiCiccio, T. J. and Thompson, M. E. (2004). A conversation with Donald A. S. Fraser. *Statistical Science*, 19:370–386.

Fellegi, I. P. (1983). New honorary members of the Statistical Society of Canada. *The Canadian Journal of Statistics*, 11:1–8.

Fienberg, S. E. (1985). Statistical developments in World War II: An international perspective. In *A Celebration of Statistics*, pp. 25–30. Springer, New York.

Fienberg, S. E. (1992). A brief history of statistics in three and one-half chapters: A review essay. *Statistical Science*, 7:208–225.

Fienberg, S. E. (2006). When did Bayesian inference become "Bayesian"? *Bayesian Analysis*, 1:1–40 (electronic).

Fienberg, S. E. and Tanur, J. M. (2001). Sample surveys. In *International Encyclopedia of the Social and Behavioral Sciences*, pp. 13453–13458. Elsevier, Oxford.

Genest, C. and Brackstone, G. (2010). A conversation with Martin Bradbury Wilk. *Statistical Science*, 25:258–273.

Hald, A. (1998). *A History of Mathematical Statistics from 1750 to 1930*. Wiley, New York.

Huntington, E. (1919). Mathematics and statistics, with an elementary account of the correlation coefficient and the correlation ratio. *American Mathematical Monthly*, 26:421–435.

Maag, U. R. (1980). The first honorary members of the Statistical Society of Canada. *The Canadian Journal of Statistics*, 8:1–6.

Macdonald, P. D. M. (1988). A conversation with Charles W. Dunnett, SSC 1986 Gold Medallist. *SSC Liaison*, 3(1):25–33.

Magnello, M. E. (2009). Karl Pearson and the establishment of mathematical statistics. *International Statistical Review*, 77:3–29.

Mood, A. M. (1965). Samuel S. Wilks: His philosophy about his work. *Journal of the American Statistical Association*, 60:953–955.

Morton, N. W. (1956). A brief history of the development of Canadian military operations research. *Operations Research*, 4:187–192.

Porter, T. M. (1988). *The Rise of Statistical Thinking, 1820–1900*. Princeton University Press, Princeton, NJ.

Rankin, L. (2011). *Assessing the Risk: A History of Actuarial Science at the University of Manitoba*. Technical report, Warren Centre for Actuarial Studies and Research, Winnipeg, MB.

Rousseau, P. (2000). *Brève histoire de l'évolution de la statistique appliquée à l'Université de Montréal avant 1970*. Technical report, Université du Québec à Montréal, Montréal, QC.

Stigler, S. M. (1986). *The History of Statistics: The Measurement of Uncertainty Before 1900*. Harvard University Press, Cambridge, MA.

Walker, H. M. (1929). *Studies in the History of Statistical Method, with Special Reference to Certain Educational Problems*. Williams & Wilkins, Baltimore, MD.

Wallis, W. A. (1980). The Statistical Research Group, 1942–1945. *Journal of the American Statistical Association*, 75:320–335.

Watts, D. G. (1984). Teaching statistics in Canada: The early days. *The Canadian Journal of Statistics*, 12:237–239.

W.L.G.W. (1959). Wacław Kozakiewicz: In memoriam. *Canadian Mathematical Bulletin*, 2:148–149.

2

Some of Statistics Canada's Contributions to Survey Methodology

Jean-François Beaumont, Susie Fortier, Jack Gambino, Mike Hidiroglou, and Pierre Lavallée

Statistics Canada, Ottawa, ON

The conduct of large scale surveys in national statistical offices follows a well-established process that involves a large number of steps. These steps, which are described in Statistics Canada's *Quality Guidelines* (Statistics Canada, 2008), begin with the setting of objectives for the survey and end with dissemination and documentation. In this chapter, we describe some of the important developments that have taken place at Statistics Canada to improve the intermediate steps. Some of the steps that we will describe include record linkage, sample design, editing, imputation, weighting, estimation, variance estimation, data analysis, benchmarking, and the use of time series. Many of these developments were necessary to improve the quality of the resulting published estimates, to reduce costs, and to aid in the building of either specific or generalized software to automate the methodology.

2.1 Introduction

When most Canadians think of Statistics Canada, they probably think of one of three things: the monthly consumer price index, the monthly unemployment rate or the Census. It is not an exaggeration to say that they are seeing just the tip of an enormous information iceberg. Statistics Canada produces a large volume of information on a daily basis — so much so that since 1932 it has had a publication called *The Daily* (accessible on the web at http://www. statcan.gc.ca). *The Daily* was literally a daily (weekday) paper publication initially and is now published on the web. The web version of *The Daily* first appeared in 1995. On any given day, *The Daily* typically covers several major releases, and each release may include hundreds, sometimes thousands, of estimates. For example, when the Labour Force Survey results are published in the early part of each month, the media focus is on the national employment and unemployment numbers, but in fact, thousands of estimates are released, broken down by geography, age and sex, and covering not only employment and unemployment, but also type of occupation and industry of employment, wages and other variables.

The information published by Statistics Canada is of interest not only to the general public but also to policy makers at all levels of government, as well as to non-governmental organizations, business planners, social and economic researchers, and so on. The range of topics for which the agency produces information is very broad. It includes data related to the economy (labour, commerce, trade, transportation, energy, etc.) and society (health, education, travel, tourism, the justice system, etc.). Many data users may not realize how much lies behind the numbers that they use to make decisions and conduct their studies: where the data come from, how they are collected and aggregated, and what their strengths and weaknesses may be. One of our goals in this chapter is to bring to light the statistical methods underlying the information that is disseminated by the agency.

These methods have evolved over time to satisfy the changing needs of data users and to take advantage of innovations due to technological change. Statistics Canada's mathematical statisticians (often referred to internally as "survey methodologists" or simply "methodologists") have made major contributions to the branches of statistics known as survey methodology and sample survey theory and methods. A second goal of this chapter is to discuss these contributions and, in the process, introduce numerous domains that may be unfamiliar not only to the general public but to other statisticians as well. Some of these domains are familiar to all statisticians: estimation, variance estimation, data analysis, and time series. But some may not be: imputation, multiple frames, and record linkage.

As a result of its research and development program, Statistics Canada is viewed as a world leader in survey methodology and its methods and approaches to research and development have been used as models by numerous national statistical organizations throughout the world.

The chapter is organized as follows. Section 2.2 presents innovations related to sample design. Sample design is a very important part of a survey. It involves all the steps needed to define and specify the population frame, sample size, sample selection method, and sample estimation method. Once this is established, the data are collected and processed. At this stage, survey statisticians need to validate and sometimes replace incorrect or missing data. This is called edit and imputation and is discussed in Section 2.3. Section 2.4 highlights various contributions from Statistics Canada to the estimation process. This is where the data, possibly in combination with auxiliary information from other sources, are used to estimate the values of interest in the population. The quality of this estimation is often measured in terms of variance, and variance estimation is presented in Section 2.5. Section 2.6 focuses on how survey statisticians handle the challenging task of data analysis, especially in the context of a complex survey design. Finally, Statistics Canada's contributions to the field of time series — defined as sequences of estimates ordered in time — are featured in Section 2.7.

2.2 Sample Design

Sample surveys are carried out by selecting samples of persons, businesses or other entities such as farms (called units) that we survey in order to get information. Sample selection is often done by randomly selecting certain units from a list that we call a sampling frame. This sampling frame represents the set of units for which we want to produce information; this is what makes up the target population. Obtaining the sampling frame is often difficult and costly because we want it to be complete (all units of the target population are contained in the frame), unique (every unit is present only once in the frame), and exclusive (no units of the frame are outside the target population).

The sample design describes the way that the random sample is selected. For instance, we may decide to divide the sampling frame into subgroups (called strata) to deal with more homogeneous subpopulations. We may also choose to select the sample using selection probabilities proportional to the size of the units. All of this is contained in the sample design. We will describe more aspects of the sample design as we discuss some of the contributions of Statistics Canada in this domain.

2.2.1 Record Linkage

Data from different sources are increasingly being combined to augment the amount of information that we have. This is the case especially with the creation of sampling frames. Often, the sampling frame is built by combining different sources using record linkage. Another application of record linkage is to aid in data collection by using address information to match publicly available telephone numbers with sampled households.

The concepts of record linkage were introduced by Newcome et al. (1959), and formalized mathematically by Fellegi and Sunter (1969). As described by Bartlett et al. (1993), record linkage is the process of bringing together two or more separately recorded pieces of information pertaining to the same unit (individual or business). Record linkage is sometimes called exact matching, in contrast to statistical matching. In the latter case, linkages are based on similar characteristics rather than unique identifying information.

The purpose of record linkage is to link the records from two files. If the records contain unique identifiers, then the matching process is trivial. Unfortunately, often a unique identifier is not available and then the linkage process needs to use some probabilistic approach to decide whether two records, coming respectively from each file, are linked together or not. With this linkage process, the probability of having a real match between two records is calculated. Based on the magnitude of this probability, it is then decided whether the two records can be considered as really being linked together or not.

With the approach of Fellegi and Sunter (1969), we choose an upper threshold and a lower threshold to which each linkage probability is compared. These thresholds are chosen to reduce the linkage errors coming from accepting bad links and rejecting good links. During the linkage process, if the linkage probability between two records is greater than the upper threshold, the link between these records is considered a true link. If the linkage probability is lower than the lower threshold, we consider that we do not have a link. Last, if the linkage probability is between the lower and upper thresholds, then we have a possible link, and we will decide on this link by manual resolution. This is generally done by looking at the data, and also by using auxiliary information.

In addition to frame creation, Statistics Canada uses record linkage in various other applications. It is used, for example, to add information to files for sociological or economic analysis and in the coverage studies of the census.

2.2.2   Multiple Frames

To cover the target population well, one sampling frame may not be sufficient. For example, for agriculture surveys, we may have available a list of farms, which can be used as a list frame. But the list frame may be incomplete (for example, newer farms may not be on it). We can also select farms by first selecting small land areas and then farms within those areas; the set of land areas is an area frame. The survey may decide to use both the list and area frames to select its sample. When the units involved in each frame are the same, it may be possible to combine the frames into a single one using, for instance, record linkage. However, in the agriculture example, the different frames do not contain the same type of unit and, in this case, we may decide to use the frames separately. We then talk about the use of multiple frames.

When using multiple frames, we are typically selecting a sample from each frame independently. Different approaches are then possible to produce estimates from the resulting samples. In the approach by Hartley (1962), who was the first to formalize estimation for multiple frames, estimates from each sample portion (i.e., the sample in the intersection of both frames, and the samples from the non-overlapping portions of the frames) are combined to produce a global estimate.

Bankier (1986) developed at Statistics Canada an alternative method for producing estimates from a multiple frame survey. Even though he confined his method to sample designs for which a stratified simple random sample is selected independently from each frame, the estimation technique can be applied to more complex sample designs. Bankier (1986) viewed a multiple frame sample as a special case of selecting two or more samples independently from the same frame. The idea was to combine the different samples selected from the frames, and to compute the selection probability of each unit of the combined sample. As a result, standard techniques for estimating parameters from a single frame can be applied to multiple frame samples. The resulting estimators have been found to be simpler to compute and are easier to extend to three or more frames than those given in Hartley (1962).

2.2.3 Indirect Sampling

In practice, there may be no available sampling frame that directly corresponds to the desired target population. We then have to use a sampling frame that is indirectly related to this target population. An example of this is the following. Suppose that estimates are required for young children, but that the sampling frame is a list of their parents. In this case, the target population is young children; however, they are selected in two steps. First, a sample of the parents is selected using the list of the parents, and second, a sample of their children is selected. Note that the children of a particular family can be selected through the father or the mother. We then have two populations that are linked with one another: one is associated with the sampling frame and the other with the target population (parents and children respectively in our example). This type of sampling is referred to as indirect sampling (Lavallée, 2002, 2007).

Estimation of population parameters, such as totals and means using data selected by indirect sampling is not straightforward, particularly if the links between the units of the two populations (the sampling frame and the target population) are not one-to-one. If we consider the previous example of families, it can be very difficult to assign a selection probability to each child of a selected family. Indeed, we could have selected a family through one or more of the parents but the computation of the selection probability of the family, and consequently that associated with each child, requires that the selection probability of each parent is known, whether the latter is selected or not. Lavallée (1995) developed the generalized weight share method (GWSM) to solve this estimation problem. The GWSM is a generalization of the weight share method described by Ernst (1989). We can also consider indirect sampling and the GWSM as a generalization of network sampling as well as adaptive cluster sampling. These two sampling methods are described by Thompson (2002) and Thompson and Seber (1996).

2.2.4 Optimal Stratification

Populations are typically not homogenous, and sampling from them would result in estimates with high variability. They are therefore split into subpopulations (called strata) that are relatively homogeneous, and mutually exclusive. For example, a population of persons or businesses can be partitioned into low, medium and high income strata. The stratification of the population depends on the type of survey that is carried out. For example, in the case of household surveys, the stratification is mainly based on geography (e.g., provinces), whereas for business surveys it is based on a combination of geography, industrial classification and size. The stratification of a population may also be constrained by the level at which estimates are required. For example, if estimates for a household survey are required for the larger cities in a given province, the stratification of that province will account for that requirement. Once the strata are formed, sample selection takes place independently within each of them.

The use of a stratified sample plan involves five different design operations: (i) choosing a stratification variable(s); (ii) determining the number of strata; (iii) stratifying the population given the chosen stratification variable(s); (iv) allocating the total sample size to the strata; (v) selecting the units within each stratum according to the adopted sampling design. Note that although any sampling design can be used in the fifth step, simple random sampling (SRS) without replacement is often used as the method for selecting the units. This procedure is extensively used for sampling from list frames, especially for business surveys.

Some populations, such as populations of businesses, can be highly skewed, i.e., there will be many small and medium sized units and just a few very big units. Efficient sampling of highly skewed populations requires that they be stratified into a take-all stratum and a number of take-some strata. In our example, the take-all stratum would consist of the biggest units. All units in the take-all stratum are selected with certainty, whereas units in the take-some strata are selected via a probability mechanism, such as SRS. Given that a quantitative stratification variable is highly correlated with the variables of interest, we search for optimal stratification boundaries that split the population into subpopulations in such a way that large units will be selected with certainty (take-all stratum) and smaller units will be selected with some positive sampling fractions (take-some strata). For example, we may want to stratify the frame by small, medium and large businesses based on a revenue variable. The problem is to find the stratum boundaries that will categorize the small, medium and large businesses.

Approximate cut-off rules for stratifying a population into a take-all and a single take-some stratum were given by Hidiroglou (1986). Lavallée and Hidiroglou (1988) considered the more complex case where the population is to be stratified into a single take-all stratum and several take-some strata. The algorithm of Lavallée–Hidiroglou minimizes the overall sample size required for the survey, given the precision of the estimator and the allocation scheme of the sample to the take-some strata. The allocation scheme used by the algorithm is power allocation, as described in the next section.

The Lavallée–Hidiroglou algorithm has been used extensively in most business surveys at Statistics Canada, and elsewhere. The main program where it has been used is the Unified Enterprise Survey (Li et al., 2011), a business survey that unifies more than 60 surveys for different industries.

2.2.5 Power Allocation

In many surveys, reliable estimates are required both at the national level and for subnational areas. These subnational areas can be, for example, the provinces of Canada. The requirement for subnational estimates usually results in the sample being stratified geographically. When the subnational areas vary considerably in size or importance (e.g., in terms of population in a social survey or total revenue in a business survey), this variation must be taken into account when deciding how to allocate the sample to strata. For example, allocating sample to provinces proportional to their size may produce good precision at the national level. However, it may cause certain subnational estimators (e.g., for the smallest provinces) to have poor precision. Alternatively, a sample allocation that achieves nearly equal precision for the subnational estimators may result in the national estimator having much lower precision than under the above scheme. A compromise between these two types of allocations was developed by Bankier (1988), which has been called power allocation.

Power allocations were proposed in the past by Carroll (1970) and Fellegi (1981). The power allocation of Bankier (1988) creates a compromise between two different sample allocations by the use of a parameter appearing as a power (in the mathematical sense). By setting the power to different values, we get various allocations, including the two mentioned in the previous paragraph as special cases: they correspond to the power 0 and 1, respectively. By choosing a suitable power between 0 and 1, we create a compromise allocation that produces subnational estimators of approximately equal precision, without reducing too much the precision at the national level.

Power allocations have been used in the design of several surveys at Statistics Canada. For example, they have been applied to a sample survey of business tax returns where the stratification size measures were based on sales. In addition, they have been used for samples stratified by province that evaluate different aspects of the Canadian census.

2.3 Edit and Imputation

Once the sample is selected, there are numerous ways of collecting data. For instance, a paper questionnaire can be filled in by the respondent and returned by mail, responses can be recorded by an interviewer on a computer during a telephone or personal interview, or an Internet questionnaire can be filled in directly by the respondent. Regardless of the data collection mode, it is very likely that some returned questionnaires will contain errors and missing values. Computer-assisted data collection can reduce the occurrence of certain types of error by integrating edit rules in the computer application. Edit rules are used to verify that responses are valid and consistent. An example of an edit rule is that the age of a mother must be greater than the age of any of her children. However, there is a limit to the number of edit rules that can be integrated in the computer application without unduly lengthening the time needed to complete the questionnaire. As a result, errors in the collected data must be anticipated in most surveys.

After data collection, the goal of editing is to detect invalid responses and inconsistencies, and treat these errors either through a follow-up with respondents or through an automated algorithm. Statistical agencies long ago recognized that following up all the respondents with suspicious or erroneous values was a costly activity. In the context of business surveys, Latouche and Berthelot (1992) suggested coping with this issue by following up only a subset of these respondents, namely those that are expected to have a non-negligible impact on the survey estimates. They proposed three score functions to prioritize respondents to be followed up and concluded that their approach, often called selective editing, led to estimates that are similar to those obtained after following up all the respondents. Granquist and Kovar (1997) argued in favor of selective editing as a tool for reducing costs while maintaining the quality of estimates. They also suggested that editing should not only be viewed as a correction tool but also as a process that provides information about the quality of survey data. Such information could then serve as a basis for future improvements to the whole survey.

Although edit rules can be used to detect inconsistencies, they do not generally tell us how to resolve them. To do so, the first step is to determine which values need to be modified to resolve inconsistencies. This step is called error localization. Imputation, which is the process of replacing values that need modification or filling in missing values, can then be carried out. The goal of imputation is to produce a complete file that satisfies all the edit rules. Such a file simplifies the production of estimates for ultimate users who may not be familiar with more complex statistical techniques dealing with inconsistent and missing values.

Prior to the seminal paper of Fellegi and Holt (1976), there was no unified theory of edit and imputation. They developed such a theory and proposed the minimum-change principle for error localization. Their principle consists of changing the smallest number of values in a record so as to remove inconsistencies. They also proposed an algorithm to solve this minimization problem. The minimum-change principle is still used and is implemented in Statistics Canada's BANFF edit and imputation system. A description of BANFF can be found in Kozak (2005).

Inconsistencies represent one type of error. There are also other types of potential errors that need to be detected such as outliers, which are unusually large or small values. There are numerous univariate outlier detection techniques in the literature. In practice, survey analysts are often interested in detecting outliers in the ratio of collected values from one period to the next. Univariate outlier detection methods can be applied to ratios. However, ratios tend to be more volatile for smaller previous-period values than for larger previous-period values in business surveys. This implies that the straight application of univariate techniques would tend to detect too many outliers for smaller previous-period values and not enough for larger previous-period values. Hidiroglou and Berthelot (1986) proposed a simple method for tackling this problem that is used in many surveys in different statistical organizations and is implemented in BANFF. See Section 2.4.2 for a more general discussion on outliers.

We have already noted that, once errors have been identified and follow-ups have been completed, imputation can be carried out to complete data cleaning. Imputation replaces missing values or values that need modification by other values, such as predicted values from a linear regression model. A key feature of imputation methods is that they rely on auxiliary information available for all the sample records, including those with missing or erroneous values. In business surveys, different types of imputation methods are used. One popular imputation method is linear regression imputation. Another popular imputation method, especially in social surveys, is donor imputation. It replaces missing values or values that need modification by the value of a "donor" chosen among the set of records that are deemed to have correct values. Characteristics of different imputation methods in business surveys are discussed in Kovar and Whitridge (1995). Most of these methods are implemented in BANFF. Greater detail about imputation methods and the associated theory can be found in Haziza (2009).

In the context of donor imputation in the Canadian census, it was noted that some implementations of the minimum-change principle sometimes led to imputed records that seem implausible. A new imputation approach was developed to handle this issue; see, e.g., Bankier et al. (1999). Instead of localizing errors and then imputing, Bankier et al. proposed to reverse this order. In their approach, close donors are first determined for each record with missing or inconsistent values, which is called a recipient. Then, for each donor-recipient pair, imputation actions are found such that the chosen actions minimize a weighted average of two distances. An imputation action is chosen at random for a recipient among the set of best imputation actions for that recipient. This approach is an attempt to compromise between satisfying the minimum-change principle and plausibility. It has been implemented in the Canadian Census Edit and Imputation System (CANCEIS) developed at Statistics Canada.

2.4 Estimation

Once the data have been collected and validated, they are used to estimate parameters of interest, including simple statistics such as means, totals, and proportions as well as more complex parameters. The latter are usually linked to the analysis of the survey data. Examples include the estimation of regression parameters and chi-squared tests.

Formally, estimation is the survey process in which unknown population parameters are estimated using data from a sample, possibly in combination with auxiliary information from other sources. Estimation results are used to make inferences about these unknown parameters, that is, to draw conclusions about characteristics of the complete population.

Survey data are typically obtained via a sampling procedure, and estimators of the parameters of interest must reflect this. The first step is to assign a design weight to each of the responding sampled units. The design weight is the inverse of the probability of selection associated with this unit. For example, if one percent of a population is selected completely at random, then the probability of selection for each unit is 1/100 and the sampling weight is 100. If the sampling design is multi-stage or multi-phase, the design weight of a given unit will be the product of the design weights at each stage or phase. This ensures that the estimates are unbiased, or approximately unbiased, in the sense that the average value of the estimator over all possible samples is equal to the population parameter being estimated.

Once the design weights have been computed, they may need to be adjusted for nonresponse when all or almost all data for a sampled unit are missing. Nonresponse occurs when information for some units cannot be collected. This can happen for a variety of reasons, including refusals, inability to contact the unit (e.g., because of incorrect contact information or because the potential respondent is on vacation), natural disasters (e.g., floods, major winter storms) and so on. Because of nonresponse, design weights are adjusted upward based on the assumption that the responding units represent both responding and non-responding units. These modified weights are called survey weights. For a given sampled unit i, we denote its corresponding survey weight by w_i. Estimates are then obtained for the parameters for arbitrarily defined subsets of the population, i.e., for domains. Common domains of interest are the individual strata and the entire population. Domains may also cut across the design strata; for example, the domain may consist of all companies that have been in existence for less than five years. If we are interested in estimating totals for a given domain d, then the estimate for that domain is given by

$$\hat{Y}_{(d)} = \sum_{i \in s_d} w_i y_i,$$

where y_i is the variable of interest and s_d is the part of the sample s that belongs to that domain. In the example, s is the sample of companies and s_d is the set of companies in the sample that are less than five years old.

Auxiliary data, such as age-sex population counts and other information from external sources, are usually incorporated in the weighting process at Statistics Canada. There are two main reasons for using auxiliary data in estimation. The first reason is that it is often important for the survey estimates to match known population totals or estimates from another, more reliable, survey. This can be viewed as a desire for consistency. The second reason is that the resulting estimators will be more efficient (in the sense of smaller variance) than those only weighted with the survey weights, if the auxiliary data are well correlated with the variables of interest. Examples of estimators that incorporate auxiliary data range from the ratio estimator that uses a single auxiliary variable to more complex estimators that use several variables via regression or calibration. This range of estimators can be obtained via the calibration theory given in Deville and Särndal (1992). Calibration adjusts the survey weights by incorporating auxiliary variables. The resulting weights $\tilde{w}_i$ known as calibrated weights are obtained by minimizing a function of the calibrated weights and the survey weights subject to a benchmark constraint. Given that x_i are the auxiliary data, the constraint requires that the sum of the product between the calibrated weights $\tilde{w}_i$ and the auxiliary data ($\sum_{i \in s} \tilde{w}_i x_i$) equals the known population totals ($\sum_{i \in U} x_i$), where U represents the whole population.

An important application of this method is used by most household surveys. The weights of individuals in the sample are adjusted so they add up to population-level age-sex totals, which are the auxiliary variables in this case. For example, if there are P boys aged 5–9 in the population, then the weights of boys in the sample who are in this age group are adjusted so they add up to P (this P corresponds to the final summation in the previous paragraph, with x equal to 1 for boys aged 5–9 and 0 otherwise). This is done simultaneously for all age groups and both sexes. In practice, household surveys that produce both person-level and household-level estimates (such as the unemployment rate and average household income, respectively), may impose an additional constraint on the weights: the weights of all persons in a given household are forced to be equal. This is sometimes referred to as an integrated weight and is described by Lemaitre and Dufour (1987).

2.4.1 Generalized Estimation System

The development of the Generalized Estimation System (GES) in the early nineties allowed Statistics Canada to consolidate numerous estimation procedures for cross-sectional surveys. GES is currently applied to a multitude of Statistics Canada surveys. Specifications for a general estimation system were initially written in 1990 and 1991. The methodology that underpins GES is described in Estevao et al. (1995). GES is built around the following elements: the sampling plan, the population parameters to be estimated, the use of auxiliary information, and domains of interest. The sampling designs covered by GES include not only simple designs such as stratified simple random sampling, where all units in a stratum have the same probability of selection, but also a variety of complex designs, where units may be selected in several steps and need not have the same probability of selection (e.g., the probability may be related to the size of each unit). GES computes estimates of totals, means, and ratios with their associated measures of reliability given that auxiliary information has been incorporated in the estimation process. This auxiliary information can cut across design strata, or be included within them. This allows the computation of most of the commonly used estimators in survey sampling, including separate and combined ratio or regression estimators (or intermediate combinations), post-stratified estimators (separate, combined or a hybrid version of those two procedures), and others such as the raking ratio estimator. Estimates and their associated measures of reliability are computed by GES for user-specified domains of interest.

2.4.2   Outliers

Outliers are observations that are numerically distant from the rest of the data. Grubbs (1969) defined an outlier as follows: "An outlying observation, or outlier, is one that appears to deviate markedly from other members of the sample in which it occurs." Outliers can occur by chance in any survey, but they are often indicative of a population that has a heavy-tailed distribution. Such surveys run into the danger of including outliers in their sample if these populations have not been properly (or cannot be) stratified by size. In the case of business surveys, this will occur when there is tremendous growth in the size of some businesses that has not yet been recorded on the frame. For household surveys, this may occur when financial data are collected. The presence of outliers in the sample may result in unrealistically high or low estimates of population parameters, such as means and totals.

Even though an efficient sample design can minimize outlier problems, it cannot eliminate them. This means that outliers must be detected and treated. Outliers can be detected via univariate or multivariate methods depending on the dimensionality of the variables. Their impact on the estimates can be dampened by trimming their values (Winsorizing), reducing their weights, or using robust estimation techniques such as M-estimation. These three methods have been studied at Statistics Canada, and these procedures are discussed in Lee (1995) and in Beaumont and Rivest (2009).

In the context of simple random sampling without replacement, Rao (1971) suggested changing the weights of the outlier units to 1, and adjusting the weights of the remaining units in the sample. Hidiroglou and Srinath (1981) proposed alternative weight modifying procedures. Additional work on a Winsorization approach to outliers was done by Tambay (1988), who designed a procedure that performs well on data from skewed populations.

2.4.3   Composite Estimation

Monthly surveys such as the Canadian Labour Force Survey (LFS) deliberately retain a part of their sample in consecutive months. The LFS does this by changing (or rotating) only one sixth of its sample each month. The LFS sample is a set of dwellings (i.e., single family homes, apartment units, etc.). As a result, five sixths of the dwellings are the same in consecutive months, four sixths are the same two months apart, and so on. This is done primarily to reduce costs: interviews can be shorter in subsequent months, less travel is needed since subsequent interviews can usually be conducted by telephone, etc. A benefit of having some overlap in the sample is that it can be used to improve the quality of the estimates produced by the survey.

For example, to estimate the month-to-month change in employment, it is advantageous to focus on the five-sixths of the sample that is common because any observed differences are not due to sampling error. In fact, this estimate of change can be added to last month's overall estimate to update it, thereby giving us an alternative estimate of this month's employment. As a result, we then have two estimates for this month's total: the "usual" one based only on this month's sample, and a second estimate derived by updating last month's estimate using the common (or "matched" sample) to estimate change (Statistics Canada, 2008).

Whenever we have two estimates of the same quantity, it is natural to combine them in some way to obtain an estimate that we hope will be better than its parts. In the survey world this combined estimate is referred to as a composite estimate. In the case of the LFS, if we are interested in the number of employed people in the current month, then the composite estimate of employed is a linear combination of the above two estimates. It can be expressed as

$C \times$ ("usual" estimate)

$\qquad + (1 - C) \times$ (last month's estimate

$\qquad\qquad +$ estimate of change based on the common sample).

The constant C is chosen to make the variance of this combination as small as possible. The estimator used by the Canadian LFS is a more elaborate version of this and is described by Singh et al. (2001), Gambino et al. (2001), and Fuller and Rao (2001).

2.4.4 Small Area Estimation

Sample surveys are generally designed to provide reliable estimates for large areas or major domains of a population. Direct survey estimates such as those computed by GES are likely to yield unacceptably large sampling errors if the domain size is small. Examples of such domains include small communities and small subpopulations. For example, the Labour Force Survey is designed to produce good estimates of unemployment at the province level, but unemployment estimates for small cities and for occupations such as nursing are not expected to be of high quality because of the limited sample size in those domains. However, there is a need for such estimates, and the methodology that has been developed to meet this demand is known as small area estimation. Small area estimation, for a given small area, essentially combines in an optimal manner the associated direct estimates with model-based estimates. The model-based estimates involve known population totals (auxiliary data) and estimates of the regression between the variable of interest and the auxiliary data across the small areas. In general, these models are classified into two groups: unit level models and area level models. Unit level models are generally based on observation units (e.g., persons or companies) from the survey and auxiliary variables associated with each observation, whereas area level models are based on direct survey estimates aggregated from the unit level data and related area-level auxiliary variables; see Rao (2003) for an overview of small area models.

The basic area level model is based on the methodology proposed by Fay and Herriot (1979). This model takes into account the survey design through the use of direct survey estimates and related design-based variance estimates. A variation of the Fay–Herriot model was proposed by You and Rao (2002a) to deal with the estimation of proportions such as the unemployment rate. Their approach was used at Statistics Canada in two instances: (i) You and Rao (2002a) estimated the Canadian census undercount via a specific unmatched model; and (ii) You (2008) proposed a time series model to estimate unemployment rates using Labour Force Survey data.

You and Rao (2002b) also contributed to the unit level approach for small area estimation. Their proposed estimator has several desirable properties; in particular, it satisfies a benchmarking property in that its estimators for the domains add up to a regression estimator of the overall total. This method has been used in many official statistics programs for different surveys.

The methods in the above two paragraphs have been incorporated in a Generalized Small Area Estimation prototype system that Statistics Canada is currently building (Estevao et al., 2012, 2013).

2.5   Variance Estimation

Estimates produced from a survey are subject to errors that need to be quantified. It is important to quantify these errors to give users of the associated estimates an idea of how reliable these estimates are. The most widely used statistic to measure this reliability is the variance. For convenience, a related measure, the coefficient of variation (CV) is often presented. The CV is a relative measure that is usually expressed as a percentage, making it easier to compare different estimates.

The variance (and the CV) measure one particular type of error, namely, sampling error. In survey sampling, in addition to sampling error there are also non-sampling errors.

Sampling error is the error that results from estimating a population characteristic by measuring a portion of the population rather than the entire population. The most commonly used measure to quantify sampling error is sampling variance. Sampling variance measures the extent to which different possible samples of the same size and the same design yield differing estimates of the same characteristic. Factors affecting the magnitude of the sampling variance include:

a) the variability of the characteristic of interest in the population (the more variable the characteristic in the population, the larger the sampling variance);

b) the size of the population (the size of the population only has an impact on the sampling variance for small to moderate sized populations); and

c) the sample design and method of estimation.

For sample designs that use probability sampling, the magnitude of an estimate's sampling variance can be estimated on the basis of observed variation in the characteristic among sampled units (i.e., based on the values observed for the units in the one sample actually selected). The estimated sampling variance depends on which sample was selected and varies from sample to sample. In other words, the estimated variance itself has a variance.

A non-sampling error, on the other hand, is a catch-all term for all other possible errors associated with the sample. These include under or over-coverage of the sampling frame, nonresponse, measurement and recall errors, etc. Although the non-sampling errors can be measured individually, there are no methods that combine them all into a single measure. One exception is that the variation introduced into estimation by either nonresponse adjustments or imputation can be incorporated in the variance.

The estimation of the sampling variance can become difficult if the estimated parameters are statistics that are not simple linear functions of the observed data or if the estimators incorporate auxiliary data. Examples of complex statistics include medians and functions of totals or means. An example of an estimator that incorporates auxiliary data is the calibration estimator discussed in Section 2.4. In this section, we focus on a number of developments that took place at Statistics Canada regarding the estimation of the sampling variance, as well as the estimation of variance reflecting nonresponse adjustments or imputation. These developments can be split into two broad classes: (i) Procedures based on Taylor linearization; and (ii) Procedures based on replication.

2.5.1 Taylor Linearization

For linear statistics, the estimated variance incorporates the first order and second order inclusion probabilities of the sample, and the observed values of the sample. Here, first order refers to the probability that a unit will be included in the sample, and second order refers to the probability that a given pair of units both will be selected. For nonlinear statistics, however, variance estimation can be much more difficult. Fortunately, the variance estimation procedures for linear statistics can be used to approximate the variance of nonlinear ones by using Taylor's theorem to approximate (or "linearize") the nonlinear statistics and then obtaining the variance of this linear approximation.

Binder (1983) used linearization to obtain the estimated variance of complex statistics represented as weighted estimating equations. Special cases of his approach include the generalized linear models described by Nelder and Wedderburn (1972), logistic regression, and log-linear models for categorical data. The Binder procedure is highly quoted in the literature, and is used to obtain variances for parameters other than the ones just mentioned.

Demnati and Rao (2004) presented a unified approach to deriving Taylor linearization variance estimators and applied it to a variety of problems. Their methodology produces the correct variance estimates for many complex statistics used in survey sampling. A non-comprehensive list of those complex statistics includes proportions, ratio estimates, generalized linear regression models, and the Wilcoxon two-sample rank-sum test.

2.5.2 Replication Methods

Replication methods involve selecting several subsets of the sample that are representative of the population, i.e., each so-called replicate is a subset of the sample. The number of units included in each replicate is roughly the same. Each replicate is selected using the same design. Since each replicate is essentially a sample itself, it can be used to produce an estimate. Therefore, estimates of the parameter of interest can be computed for each replicate in exactly the same way that they were computed for the full sample. It turns out that the variation in these replicate estimates is related to the variance we are trying to estimate, and this fact is used to obtain a valid variance estimate. There are several ways of creating the replicates. The two most widely used replication procedures at Statistics Canada are the jackknife and the bootstrap.

The jackknife. A very common way of estimating the variance using the jackknife method is to form each replicate by removing a single sampled unit (cluster or element) from the original sample, and recomputing the estimate for the parameter of interest. This procedure, known as the delete-one jackknife, can become time consuming for a survey with many clusters or elements, whereas the corresponding Taylor procedure requires fewer computations. Given an estimator and its corresponding variance expression in terms of the jackknife procedure, Yung and Rao (1996) showed how to linearize the jackknife variance estimator to obtain a linearization-type variance estimator for post-stratified and regression estimators.

Two-phase sampling poses interesting variance estimation challenges for complex estimators. A two-phase sampling design works as follows. A large first-phase sample is selected and some minimal information is collected at a small cost. The information collected at the first-phase is then used as auxiliary data for the selection of a subsample (the second-phase sample) drawn from the units of the first-phase sample. For example, the information collected at the first phase can be used as stratification variables. In the context of agriculture surveys, for example, the first phase of sampling can identify farms with specific crops, and the second phase can measure pesticide use on those crops.

To estimate the variance of complex estimators under two-phase sampling, Kott and Stukel (1997) tackled the problem using the jackknife technique. They showed that the jackknife may be used to estimate the variance for one common type of estimator (the reweighted expansion estimator) under certain conditions; it is not generally effective as a variance estimator for another type of estimator (the double expansion estimator). They found that the jackknife variance estimator can be nearly unbiased for the reweighted variance estimator, while the jackknife can fail as a variance estimator for the double expansion estimator. These results are found to be very useful at Statistics Canada because several social surveys use multi-phase sample designs, and also because a two-phase sample design is planned for the Integrated Business Statistics Program (Turmelle et al., 2012).

The bootstrap. The bootstrap differs from the jackknife in that random subsamples are selected with replacement from the original sample a number of times. Each random subsample is viewed as a replicate. The weights associated with each replicate depend on the original sampling design. Rao and Wu (1988) provide details on how these weights can be generated for fairly complex sample designs. For each subsample, the same estimation procedure used for the entire sample is repeated. The resulting estimates for each replicate are then used to estimate the variance. The bootstrap method most commonly used at Statistics Canada was originally described by Rao et al. (1992).

In practice, instead of resampling sampled units, the bootstrap can be carried out by attaching a series of adjustment weights to each observation in the original sample. Beaumont and Patak (2012) generated these weights in a novel way that maintains key properties of the sampling error (specifically the first two moments). It can be applied to most sampling designs. Most previous bootstraps, such as the one by Rao and Wu (1988), are special cases of their procedure.

2.5.3 Variance Estimation in the Presence of Imputation

The above variance estimation methods can deal with the variability due to selecting a sample of the finite population. There are many other sources of error. One important source of error that can be accounted for at the variance estimation stage is the error due to missing values. Many developments were made at Statistics Canada on this topic over the last 25 years, especially when imputation is used to compensate for the missing values. Two excellent reviews are Lee et al. (2001) and Haziza (2009). The earliest work on this subject at Statistics Canada is described in handwritten notes by Mike Hidiroglou from

1989. In these unpublished notes, Hidiroglou provided the required variance estimator for some special cases of regression imputation. The methodology is based on modeling the variable to be imputed.

A dataset will most likely have been imputed using a mixture of imputation procedures. Beaumont and Bissonnette (2011) provided a simple methodology to estimate variances for datasets that were imputed using such a mixture of procedures. This methodology has been implemented in the System for the Estimation of Variance due to Nonresponse and Imputation (SEVANI) developed at Statistics Canada; see Beaumont et al. (2010).

Yung and Rao (2000) looked at jackknife variance estimation in the presence of missing values for both the post-stratified and the GREG estimator. They were able to develop consistent jackknife variance estimators under weighting adjustments for unit nonresponse and under weighted mean and hot deck stochastic imputation procedures for item nonresponse. Following Yung and Rao (1996), they also derived corresponding jackknife linearization variance estimators.

2.6   Data Analysis

Up to the 1970s, survey research was mainly focused on the estimation of simple descriptive parameters of a finite population such as population totals, population means and ratios of two population totals. In this context, the quality of estimators of descriptive parameters is often assessed by estimating their sampling variance, as discussed in Section 2.5.

After the seventies, survey analysts became more interested in studying relationships between variables by postulating models, such as regression models which relate a variable of interest y to one or more explanatory variables $x_1, \ldots, x_p$. In this case, the parameters to be estimated are no longer descriptive parameters, but model parameters, sometimes called analytical parameters. In the regression example, these parameters would be the regression coefficients and correlations between variables. Analysts are usually interested in drawing conclusions about the unknown model parameters using sample survey data; i.e., they are interested in making inferences.

Throughout this chapter we have mentioned the variability due to the selection of a sample from the finite population. When making inferences about model parameters, an additional source of variability, namely the variability due to the model, must be considered. Rubin-Bleuer and Schiopu-Kratina (2005) developed a formal theoretical framework for inferences about model parameters in the joint model-design space. If the overall sampling fraction is small, i.e., if the size of the sample is small compared to the size of the population, the model variability can often be neglected; see, e.g., Binder and Roberts (2003). This simplifies variance estimation. For instance, we can use the linearization technique of Binder (1983) or replication techniques that account for the sampling design. However, it is not always appropriate to ignore the model variability. Methods were developed at Statistics Canada to handle this issue. Demnati and Rao (2010) extended their linearization method to account for the model variability while Beaumont and Charest (2012) extended the generalized bootstrap.

Often, survey analysts are interested in testing different types of hypotheses about model parameters. For instance, it may be of interest to test whether education is related to the propensity to leave low-income status. There are numerous procedures that can be used to test hypotheses in classical statistics. For survey data, these procedures need some adaptation to account for the sampling design. An early example of this is Fellegi (1980), who adapted such a procedure. After this, Statistics Canada statisticians, often in collaboration with university researchers, made numerous contributions to the analysis of survey data; see Hidiroglou and Rao (1987), Roberts et al. (1987), and Roberts et al. (2009). Beaumont and Bocci (2009) developed bootstrap tests of hypotheses that can be obtained by replicating a standard test statistic several times.

2.7 Time Series

The information published by Statistics Canada is often in the form of time series, i.e., as a sequence of estimates available for a specific period and ordered in time. Over the years, Statistics Canada has contributed to the development of many methods related to the dissemination and analysis of time series data. Two areas are particularly worth emphasizing: seasonal adjustment and trend-cycle estimation as well as benchmarking and reconciliation methods.

2.7.1 Seasonal Adjustment and Trend-Cycle Estimation Method

Seasonal adjustment is the process used to identify, estimate and, when appropriate, remove seasonal and/or calendar effects from a time series. These are the repetitive patterns that normally occur at the same time and in about the same magnitude every year. For example, construction activity slows down every winter and students seek summer jobs at about the same time every year. These patterns yield little information on the underlying socio-economic situation and make the data more difficult to interpret. As such, most key economic indicators published by national statistical organizations are seasonally adjusted, i.e., the seasonal effects are removed. Some of the most popular statistical methods used to perform seasonal adjustment are based on the X–11 variant of the Census Method II Seasonal Adjustment program. This method was thoroughly reviewed and documented in Ladiray and Quenneville (2001), a book mostly written at Statistics Canada as part of an exchange with the "Institut national de la statistique et des études économiques" (INSEE). The book quickly became an important reference for government agencies, macroeconomists and other users of economic data.

While the core X–11 method itself was developed at the US Census Bureau in the 1950s and 1960s, it underwent many enhancements and extensions over the years. A major one was proposed by Dagum in the mid-1970s and concerns the use of so-called ARIMA time series models in combination with the X–11 method, as described in Dagum (1978). To estimate the seasonal component of a data series at any given point in time, statisticians use moving averages on previous, current and future observations. Because information on future observations is not available, seasonal adjustment can be conducted using only previous and current values, and asymmetric moving averages. Dagum showed that the asymmetric approach yields higher revisions than using a forecasted value with symmetric filters (Dagum, 1982). Under her leadership, the X–11–ARIMA method (Dagum, 1980) was developed at Statistics Canada. Various other enhancements such as extra diagnostics and treatment of calendar effects were included in subsequent years (Dagum, 1988, 2000). The idea of using ARIMA forecasts in combination with the X–11 method is still in use and available today with newer methods (X–12–ARIMA and X–13–ARIMA–Seats). Dagum (1996) also established a method to further extract the economic signal from a seasonally adjusted series. This method helps identify turning points in the economic signal by minimizing unwanted ripples (false turning points) and revisions.

2.7.2   Benchmarking and Reconciliation Methods

When they deal with time series, statistical agencies also face various coherence issues. For example, statistical programs often have two sources of data measuring the same target variable: (i) a more frequent measurement (e.g., monthly) with an emphasis on accurate estimation of the period-to-period movement; and (ii) a less frequent measurement (e.g., annually) with an emphasis on accurate estimation of the level. These two sources are not always perfectly aligned. Benchmarking refers to techniques used to ensure coherence between time series data of the same target variable measured at different frequencies, for example, monthly and annually. It entails imposing the level of the benchmark series while minimizing the revisions to the observed movement in the sub-annual series as much as possible. Statistics Canada has been developing and adapting methods to solve this problem for many years. An important contribution came from Cholette (1984) who modified a well-known procedure from Denton (1971) to better take into consideration assumptions about the first terms of the series. Techniques for related issues such as non-binding benchmarking, interpolation, temporal distribution, calendarization, linkage and reconciliation were developed at Statistics

Canada and documented through the years (Dagum and Cholette, 2006). The methods have also been implemented as SAS procedures in a generalized system initially called Forillon (now G-series) (Latendresse et al., 2006; Bérubé and Fortier, 2009). More recently, research and development has continued on topics such as calendarization — an advanced application of benchmarking techniques used to transform values from a time series observed over varying time intervals into values that cover calendar intervals such as day, week, month, quarter and year (Quenneville et al., 2013).

2.8 Conclusion

Survey statisticians have played an important role in fulfilling Statistics Canada's mandate to provide information to Canadians on the economy and society. In the process, they have made significant contributions to statistical methodology in a wide variety of areas, as outlined in this chapter. As society evolves, and in particular, as technological change continually affects the ways in which a statistical agency can interact with citizens, new challenges present themselves. We can expect statisticians to continue to play an important role in dealing with these challenges. For example, Statistics Canada is in the process of adopting the Internet response option for most of its surveys, following its successful use in the 2011 Census of Population. Giving Canadians this alternative to traditional response modes (telephone, personal visit, paper questionnaire) is both logical and inevitable, but it leads to methodological issues that need to be addressed.

Other examples where statisticians will continue to play an important role include dealing with decreasing response rates, meeting the increasing demand for estimates for small domains by developing methods for producing such estimates, and making more effective use of other sources of data such as tax data and administrative data, all while respecting privacy and confidentiality. It is also likely that massive administrative databases (often referred to as "big data") will start to play a bigger role in the statistical process. Ensuring that such data are used appropriately for official statistics will be another challenge for Statistics Canada's statisticians. With its well-established survey methodology program, Statistics Canada has the strengths needed to meet these challenges. As in the past, we can expect new developments that will benefit not only Statistics Canada, but the statistics community around the world as well.

About the Authors

Jean-François Beaumont is a chief in the Statistical Research and Innovation Division at Statistics Canada. He earned his MSc degree in statistics from Université Laval. His primary research and development activities are imputation of missing data, bootstrap variance estimation for complex sampling designs and estimation issues with survey data. He has been instrumental in the development of SEVANI, the System for the Estimation of Variance due to Nonresponse and Imputation.

Susie Fortier is chief of the Time Series Research and Analysis Centre at Statistics Canada. She received a BSc in statistics and an MSc in mathematics from the Université de Sherbrooke. She teaches various time series courses at Statistics Canada's Training Institute, and her research interests focus on time series reconciliation and benchmarking, seasonal adjustment and price index theory. She is production manager of the journal *Survey Methodology / Techniques d'enquête*.

Jack Gambino is director of the Household Survey Methods Division at Statistics Canada. He obtained an MSc from McGill University and a PhD in Statistics from the University of Toronto in 1982 and has worked at Statistics Canada since then. His primary research interests are in sample survey theory and methods and in Bayesian inference. He is an elected member of the International Statistical Institute.

Mike Hidiroglou is director of the recently created Statistical Research and Innovation Division (SRID) at Statistics Canada where he has worked since 1974, except for a period at the Office for National Statistics in the United Kingdom as Director of Methodology (2003–06). He obtained his PhD in 1974 from Iowa State University. His work at Statistics Canada has been dedicated to the development and implementation of methodology for business surveys, and his research covers a wide spectrum of topics. As director of SRID, he is responsible for the overall coordination of methodology research carried out at Statistics Canada, and the Editorship of Survey Methodology.

Pierre Lavallée is assistant director of the Business Survey Methods Division at Statistics Canada. He completed an MSc in 1988 at Carleton University, and a PhD at the Université libre de Bruxelles. He has worked with business surveys and agricultural surveys, as well as longitudinal social surveys such as the Survey of Labour and Income Dynamics. His published research includes the book *Le sondage indirect, ou la méthode généralisée du partage des poids* (Ellipses, 2002), which later appeared as *Indirect Sampling* (Springer, 2007). He gives courses and presentations regularly on subjects such as recent developments in indirect sampling and hard-to-reach populations.

Bibliography

Bankier, M. D. (1986). Estimators based on several stratified samples with applications to multiple frame surveys. *Journal of the American Statistical Association*, 81:1074–1079.

Bankier, M. D. (1988). Power allocations: Determining sample sizes for subnational areas. *The American Statistician*, 42:174–177.

Bankier, M. D., Lachance, M., and Poirier, P. (1999). A generic implementation of the new imputation methodology. In *Proceedings of the Survey Research Methods Section*, pp. 548–553. American Statistical Association, Alexandria, VA.

Bartlett, S., Krewski, D., Wang, Y., and Zielinski, J. M. (1993). Evaluation of error rates in large scale computerized record linkage studies. *Survey Methodology*, 19:3–12.

Beaumont, J.-F. and Bissonnette, J. (2011). Variance estimation under composite imputation: The methodology behind SEVANI. *Survey Methodology*, 37:171–179.

Beaumont, J.-F., Bissonnette, J., and Bocci, C. (2010). *SEVANI, Version 2.3, Methodology Guide*. Technical Report, Statistics Canada, Ottawa, ON.

Beaumont, J.-F. and Bocci, C. (2009). A practical bootstrap method for testing hypotheses from survey data. *Survey Methodology*, 35:25–35.

Beaumont, J.-F. and Charest, A.-S. (2012). Bootstrap variance estimation with survey data when estimating model parameters. *Computational Statistics & Data Analysis*, 56:4450–4461.

Beaumont, J.-F. and Patak, Z. (2012). On the generalized bootstrap for sample surveys with special attention to Poisson sampling. *International Statistical Review*, 80:127–148.

Beaumont, J.-F. and Rivest, L.-P. (2009). Dealing with outliers in survey data. In *Handbook of Statistics, Vol. 29A, Sample Surveys, Design Methods and Applications*, pp. 247–280. Elsevier, Amsterdam.

Bérubé, J. and Fortier, S. (2009). PROC TSRAKING: A user-defined SAS procedure for balancing time series. In *Proceedings of the Business and Economic Statistics Section of the American Statistical Association, Washington, DC*.

Binder, D. A. (1983). On the variances of asymptotically normal estimates from complex surveys. *International Statistical Review*, 51:279–292.

Binder, D. A. and Roberts, G. R. (2003). Design-based methods for estimating model parameters. In *Analysis of Survey Data*, pp. 29–48. Wiley, Chichester.

Carroll, J. (1970). *Allocation of a Sample between States*. Technical Report, Australian Bureau of Census and Statistics, Canberra.

Cholette, P. A. (1984). Adjusting sub-annual series to yearly benchmarks. *Survey Methodology*, 10:35–49.

Dagum, E. B. (1978). Modelling, forecasting and seasonally adjusting economic time series with the X–11–ARIMA. *Journal of the Royal Statistical Society, Series D*, 27:203–216.

Dagum, E. B. (1980). *The X–11–ARIMA Seasonal Adjustment Method.* Technical Report, Statistics Canada, Ottawa, ON.

Dagum, E. B. (1982). The effects of asymmetric filters on seasonal factor revision. *Journal of the American Statistical Association*, 77:732–738.

Dagum, E. B. (1988). *The X–11–ARIMA/88 Seasonal Adjustment Method: Foundations and User's Manual.* Technical Report, Statistics Canada, Ottawa, ON.

Dagum, E. B. (1996). A new method to reduce unwanted ripples and revisions in trend-cycle estimates from X–11–ARIMA. *Survey Methodology*, 22:77–83.

Dagum, E. B. (2000). *The X–11–ARIMA/2000 Seasonal Adjustment Method: Foundations and User's Manual.* Technical Report, Statistics Canada, Ottawa, ON.

Dagum, E. B. and Cholette, P. A. (2006). *Benchmarking, Temporal Distribution and Reconciliation Methods for Time Series.* Springer, New York.

Demnati, A. and Rao, J. N. K. (2004). Linearization variance estimators for survey data. *Survey Methodology*, 30:4–13.

Demnati, A. and Rao, J. N. K. (2010). Linearization variance estimators for model parameters from complex survey data. *Survey Methodology*, 36:193–201.

Denton, F. T. (1971). Adjustment of monthly or quarterly series to annual totals: An approach based on quadratic minimization. *Journal of the American Statistical Association*, 66:99–102.

Deville, J.-C. and Särndal, C.-E. (1992). Calibration estimators in survey sampling. *Journal of the American Statistical Association*, 87:376–382.

Ernst, L. (1989). Weighting issues for longitudinal household and family estimates. In *Panel Surveys*, pp. 135–159. Wiley, New York.

Estevao, V., Hidiroglou, M. A., and Särndal, C.-E. (1995). Methodological principles for a generalized estimation system at Statistics Canada. *Journal of Official Statistics*, 11:181–204.

Estevao, V., Hidiroglou, M. A., and You, Y. (2012). *Small-Area Estimation Unit-Level Model with EBLUP and Pseudo-EBLUP Estimation Methodology Specifications.* Technical Report, Statistics Canada, Ottawa, ON.

Estevao, V., Hidiroglou, M. A., and You, Y. (2013). *Fay–Herriot Area-Level Model with EBLUP Estimation Methodology Specifications.* Technical Report, Statistics Canada, Ottawa, ON.

Fay, R. E. and Herriot, R. A. (1979). Estimates of income for small places: An application of James–Stein procedures to census data. *Journal of the American Statistical Association*, 74:269–277.

Fellegi, I. P. (1980). Approximate tests of independence and goodness of fit based on stratified multistage samples. *Journal of the American Statistical Association*, 75:261–268.

Fellegi, I. P. (1981). Should the census counts be adjusted for allocation purposes? Equity considerations. In *Current Topics in Survey Sampling*, pp. 47–76. Academic Press, New York.

Fellegi, I. P. and Holt, D. (1976). A systematic approach to automatic edit and imputation. *Journal of the American Statistical Association*, 71:17–35.

Fellegi, I. P. and Sunter, A. (1969). A theory for record linkage. *Journal of the American Statistical Association*, 64:1183–1210.

Fuller, W. A. and Rao, J. N. K. (2001). A regression composite estimator with application to the Canadian Labour Force Survey. *Survey Methodology*, 27:45–51.

Gambino, J., Kennedy, B., and Singh, M. P. (2001). Regression composite estimation for the Canadian Labour Force Survey: Evaluation and implementation. *Survey Methodology*, 27:65–74.

Granquist, L. and Kovar, J. G. (1997). Editing of survey data: How much is enough? In *Survey Measurement and Process Quality*, pp. 415–435. Wiley, New York.

Grubbs, F. E. (1969). Procedures for detecting outlying observations in samples. *Technometrics*, 11:1–21.

Hartley, H. O. (1962). Multiple frame surveys. In *Proceedings of the Survey Research Methods Section*, pp. 203–206. American Statistical Association, Alexandria, VA.

Haziza, D. (2009). Imputation and inference in the presence of missing data. In *Handbook of Statistics, Sample Surveys: Theory, Methods and Inference*, pp. 215–246. Elsevier, Amsterdam.

Hidiroglou, M. A. (1986). The construction of a self-representing stratum of large units in survey design. *The American Statistician*, 40:27–31.

Hidiroglou, M. A. and Berthelot, J.-M. (1986). Statistical editing and imputation for periodic business surveys. *Survey Methodology*, 12:73–83.

Hidiroglou, M. A. and Rao, J. N. K. (1987). Chi-squared tests with categorical data from complex surveys. *Journal of Official Statistics*, 3:117–140.

Hidiroglou, M. A. and Srinath, K. P. (1981). Some estimators of the population total from simple random samples containing large units. *Journal of American Statistical Association*, 76:690–695.

Kott, P. S. and Stukel, D. M. (1997). Can the jackknife be used with a two-phase sample? *Survey Methodology*, 23:81–89.

Kovar, J. G. and Whitridge, P. (1995). Imputation of business survey data. In *Business Survey Methods*, pp. 403–423. Wiley, New York.

Kozak, R. (2005). The BANFF system for automated editing and imputation. In *Proceedings of the Survey Methods Section*. Statistical Society of Canada, Ottawa, ON.

Ladiray, D. and Quenneville, B. (2001). *Seasonal Adjustment with the X–11 Method*. Springer, New York.

Latendresse, E., Djona, M., and Fortier, S. (2006). *Benchmarking Sub-Annual Series to Annual Totals: From Concepts to SAS Procedure and SAS Enterprise Guider Custom Task*. Technical Report, Paper presented at the 2007 SAS Global Forum, Orlando, FL, April 2007.

Latouche, M. and Berthelot, J.-M. (1992). Use of a score function to prioritize and limit recontacts in editing business surveys. *Journal of Official Statistics*, 8:389–400.

Lavallée, P. (1995). Cross-sectional weighting of longitudinal surveys of individuals and households using the weight share method. *Survey Methodology*, 21:25–32.

Lavallée, P. (2002). *Le sondage indirect, ou la méthode généralisée du partage des poids*. Éditions de l'Université de Bruxelles (Belgique) and Éditions Ellipses (France).

Lavallée, P. (2007). *Indirect Sampling*. Springer, New York.

Lavallée, P. and Hidiroglou, M. A. (1988). On the stratification of skewed populations. *Survey Methodology*, 14:33–43.

Lee, H. (1995). Outliers in business surveys. In *Business Survey Methods*, pp. 503–521. Wiley, New York.

Lee, H., Rancourt, É., and Särndal, C.-E. (2001). Variance estimation from survey data under single imputation. In *Survey Nonresponse*, pp. 315–328. Wiley, New York.

Lemaitre, G. and Dufour, J. (1987). An integrated method for weighting persons and families. *Survey Methodology*, 13:199–207.

Li, Y., Liu, W., Mathieu, P., Stardom, J., and Yeung, C. (2011). *Sampling for the Unified Enterprise Survey*. Technical Report, Statistics Canada, Ottawa, ON.

Nelder, J. A. and Wedderburn, R. W. M. (1972). Generalized linear models. *Journal of the Royal Statistical Society, Series A*, 135:370–384.

Newcome, H. B., Kennedy, J. M., Axford, S. J., and James, A. P. (1959). Automatic linkage of vital records. *Science*, 130:954–959.

Quenneville, B., Picard, F., and Fortier, S. (2013). Calendarization with interpolating splines and state space models. *Journal of the Royal Statistical Society, Series C*, 62:371–399.

Rao, C. R. (1971). Some aspects of statistical inference in problems of sampling from finite populations. In *Foundations of Statistical Inference*, pp. 177–202. Holt, Rinehart and Winston, New York.

Rao, J. N. K. (2003). *Small Area Estimation*. Wiley, New York.

Rao, J. N. K. and Wu, C. F. J. (1988). Resampling inference with complex survey data. *Journal of the American Statistical Association*, 83:231–241.

Rao, J. N. K., Wu, C. F. J., and Yue, K. (1992). Some recent work on resampling methods for complex surveys. *Survey Methodology*, 18:209–217.

Roberts, G., Rao, J. N. K., and Kumar, S. (1987). Logistic regression analysis of sample survey data. *Biometrika*, 74:1–12.

Roberts, G., Ren, Q., and Rao, J. N. K. (2009). Using marginal mean models for data from longitudinal surveys with a complex design: Some advances in methods. In *Methodology of Longitudinal Surveys*, pp. 351–366. Wiley, Chichester.

Rubin-Bleuer, S. and Schiopu-Kratina, I. (2005). On the two-phase framework for joint model and design-based inference. *The Annals of Statistics*, 33:2789–2810.

Singh, A. C., Kennedy, B., and Wu, S. (2001). Regression composite estimation for the Canadian Labour Force Survey with a rotating panel design. *Survey Methodology*, 27:33–44.

Statistics Canada (2008). *Methodology of the Canadian Labour Force Survey*. Technical Report, Catalogue Number 71–526, Statistics Canada, Ottawa, ON.

Tambay, J.-L. (1988). Integrated approach for the treatment of outliers in sub-annual economic surveys. In *Proceedings of the Survey Research Methods Section*, pp. 229–234. American Statistical Association, Alexandria, VA.

Thompson, S. K. (2002). *Sampling,* Second Edition. Wiley, New York.

Thompson, S. K. and Seber, G. A. (1996). *Adaptive Sampling*. Wiley, New York.

Turmelle, C., Godbout, S., and Bosa, K. (2012). Methodological challenges in the development of Statistics Canada's new integrated business statistics program. In *Paper presented at the International Conference on Establishment Surveys IV*. Montréal, QC, June 2012.

You, Y. (2008). An integrated modeling approach to unemployment rate estimation for sub-provincial areas of Canada. *Survey Methodology*, 34:19–27.

You, Y. and Rao, J. N. K. (2002a). Small area estimation using unmatched sampling and linking models. *The Canadian Journal of Statistics*, 30:3–15.

You, Y. and Rao, J. N. K. (2002b). A pseudo empirical best linear unbiased prediction approach to small area estimation using survey weights. *The Canadian Journal of Statistics*, 30:431–439.

Yung, W. and Rao, J. N. K. (1996). Jackknife linearization variance estimators under stratified multi-stage sampling. *Survey Methodology*, 22:23–31.

Yung, W. and Rao, J. N. K. (2000). Jackknife variance estimation under imputation for estimators using poststratification information. *Journal of the American Statistical Association*, 95:903–915.

3

Watching Children Grow Taught Us All We Know

James O. Ramsay

McGill University, Montréal, QC

Michael Hermanussen

Christian-Albrechts-Universität, Kiel, Germany

A statistician, like Alfred Hitchcock, aims to reveal the shocking in seemingly ordinary data. His or her technology is a large variety of mathematical models, powerful computational devices and a wide selection of data display methods. But in the end, great statistics require great data; meaning information that many people care deeply about, that has a structure that is genuinely informative, and in a quantity that provides answers with useful precision. Data on growth of children works on all three counts; and in this chapter a Canadian statistician and a German pediatrician/auxologist team up to surprise the reader and record our own discoveries along the way.

3.1 Introduction

It is often not the reader of statistical analyses who gets the first shocks. The statistician arrives at the construction site, opens his toolbox, and discovers that there is nothing there that precisely meets the requirements of the job. This collision between science and data is where much of the evolution in the discipline of statistics happens, and we highlight a number of challenges that we encountered in our work together. It was unexpected that the study of human growth would require so many new ideas and approaches. These are now bundled into the larger enterprise of functional data analysis, which is methodology for the analysis of data distributed over continua such as time and space.

What, then, are functional data? The left panel of Figure 3.1 illustrates much of the answer. The data, from Hermanussen et al. (1998), are measurements, accurate to a tenth of a millimeter, of the length of a baby's lower leg

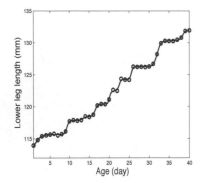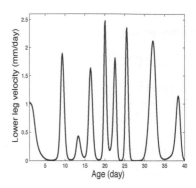

FIGURE 3.1: The left panel contains the measurements of lower leg length for a baby on its first forty days, along with a smooth monotonic curve estimated from these data. The right panel displays the first derivative or height velocity for the fitted curve.

over its first 40 days, the lower leg length being around eleven centimeters at birth. The plot also contains a smooth curve that approximates each point. Of course we have many babies available for measurement, and so we can think of these data as representative of a reasonably large sample of growth curves. We were amazed by the pulse-like nature of growth, with changes over a single day of two or more millimeters, or about three percent. These pulses are separated by three or four days of negligible growth, which seems essential since a sustained 3% per day growth would produce in a year a lower leg length of over three kilometers! If this were a heating system, we'd look for the thermostat that shuts the furnace down to prevent overheating.

The right panel of Figure 3.1 presents the first derivative of the curve in the left panel, which we will call velocity. There we see eight complete growth peaks, suggesting a growth spurt roughly every five days. There is some suggestion that later peaks tend to be smaller, broader, and spaced farther apart.

Functional data analysis is the exploration of variation in samples of curves, most often defined over time, but also over space, wavelength, and other continua. The data themselves are seldom continuous, but we assume that they are sufficiently accurate and sufficiently densely sampled that a curve fit to them can be considered to have an error that is small relative to the inter-curve variation that we wish to study. Functional data analysis has a short but vigorous history. The first paper to use the term was by Ramsay and Dalzell (1991). The field was defined more completely in the monographs by Ramsay and Silverman (1997), Ramsay and Silverman (1997, 2005) and Ramsay et al. (2009).

3.2 Modeling Growth

Since data, like our lower leg lengths, do not conform to any obvious mathematical model, we need a flexible strategy for expressing an arbitrary curve in mathematical notation to any desired accuracy. The oldest and still the most useful approach is to express a function $x(t)$ as a blend of known functions $\phi_k(t)$, called basis functions. That is, $x(t)$ is expressed as the linear combination

$$x(t) = \sum_{k=1}^{K} c_k \phi_k(t). \tag{3.1}$$

By manipulating and estimating the coefficients c_k, and making K large enough, we hope to fit data as well as we wish. A function family very much in vogue for this purpose are splines, constructed by joining polynomials together at fixed locations called knots, in such a way as to ensure a smooth transition over a knot from one polynomial to another. The more knots there are over the interval of approximation (the first 40 days in this example), the more flexible the fit to the data.

But here's the first nasty surprise. We expect growth to be positive, so we need a curve that is monotone, or everywhere at least non-decreasing. Series expansions like (3.1) don't do this well; if they fit the data closely, they will almost surely violate monotonicity. Or, to put the problem another way, monotonicity is a powerful but demanding constraint on a function's shape, and the science of human growth, auxology, requires that growth models have this feature.

Growth, as we have said, is all about change in a data-fitting curve x, and we keep our notation clean by using $v(t)$ (for velocity) rather than dx/dt. Now since v is required for our purposes to be everywhere positive, it can be expressed as the exponential of an unconstrained function W, i.e.,

$$v(t) = e^{W(t)} \quad \text{or} \quad W(t) = \log v(t). \tag{3.2}$$

This is a differential equation in x, and the solutions to the equation can be expressed as

$$x(t) = C_0 + \int_0^t e^{W(v)} dv, \tag{3.3}$$

where C_0 is a constant indicating the lower limb length of the baby at birth. We have now expressed two constrained functions, x and v, in terms of an unconstrained function W, and this function in turn can be expressed in terms of a basis function series. Moreover, this is a universal expression for a strictly monotonic differentiable function. Ramsay (1998) first articulated this approach, and provided fast and reasonably accurate numerical methods for computing values of x.

3.3 Phase/Plane Plots of Derivative Interactions

Conventional plots like those in Figure 3.1 display what we see at the level of the data or a single derivative, but don't convey much information about how the function and one or more of its derivatives interact. In part, this limitation arises because one of the axes is used to display variation over something that we already understand, namely clock time. A good graph highlights the informative but downplays the obvious. Can we rise to this second challenge by replacing time by something more revealing?

The capacity to accurately estimate, inspect and model features of functions like their slope and curvature is the aspect of functional data analysis that most separates it from multivariate data analyses of the raw discrete data. The concept of energy, defined in physics as the capacity to do work, leaps to our attention in Figure 3.1 and seems natural in a context where growth represents work done in terms of adding to a baby's mass and size. A baby adds to its energy pool by ingesting nutrients in order to compensate for energy lost through growth.

We now want to consider acceleration in lower leg length, denoted here as $a(t) = d^2x/dt^2$, as offering further information on growth. The following equation offers a useful breakdown of what produces a change $v(t_2) - v(t_1)$ in velocity over two times $t_2 > t_1$:

$$v(t_2) - v(t_1) = w(t_1)v(t_1)(t_2 - t_1).$$

That is, a change in velocity is proportional to (i) the current velocity (usually the bigger the velocity, the bigger the change) and (ii) the time difference (the bigger this difference, the bigger the potential change). The instantaneous constant of proportionality is $w(t)$, which may be positive, negative or zero, but will most likely vary with t. Dividing both sides by the time difference and allowing this difference to go to zero leads to

$$a(t) = w(t)v(t). \tag{3.4}$$

This is a first order differential equation in velocity $v(t)$, the solutions to which are of the form (3.2), where $W(t)$ in (3.3) is given, for some constant C_1, by

$$W(t) = C_1 + \int_0^t w(z)\mathrm{d}z.$$

Figure 3.2 shows how velocity and acceleration are coupled together over these forty days by plotting acceleration as a function of velocity. Each peak appears as a balloon shape with the mouthpiece at point $(0,0)$ where both velocity and acceleration vanish. On the up-side of each peak, both the acceleration and velocity are positive, whereas the down-side has negative acceleration. The symmetry of each peak shows up as symmetry of these balloons

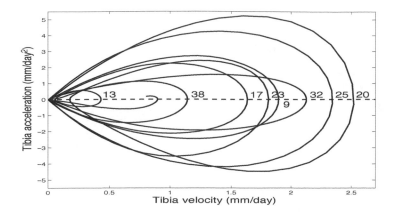

FIGURE 3.2: The second derivative or height acceleration for the baby plotted against height velocity. This is called a phase/plane plot. The numbers to the right of each loop are the day of the corresponding peak velocity in the right panel of Figure 3.1.

about a horizontal zero axis. The "power" of a peak, or work done per unit time, corresponds roughly to the area of the corresponding balloon. The two thinnest loops correspond to the last two peaks, which seem here as well as in Figure 3.1 to have less power. Time now plays the much humbler role of separating one balloon from another. We use the term phase/plane plot, borrowed from dynamic systems jargon, for a plot of one derivative against another.

How long does the bursty nature of growth that we see in Figures 3.1 and 3.2 persist? Thalange et al. (1996) report daily data on the growth of a ten-year old boy, and we estimate from their data the same phase/plane diagram shape as Figure 3.2, except that the velocity peaks reach only .5 mm/day, the acceleration swings through $\pm.05\,\text{mm/day}^2$, and the peaks are about 100 days apart. It is remarkable that velocities between the peaks are no more than a tenth of the peak heights. Applications like these of functional statistical methods have led auxologists to some profound rethinking about the nature of human growth.

Let's see what the phase/plane plot brings to the study of growth over the entire first 18 years of a child's life. The left panel of Figure 3.3 shows the heights of ten girls in the Berkeley growth study (Tuddenham and Snyder, 1954) along with strictly monotone fitting curves, and the right panel displays the corresponding height velocities. Notice the pubertal growth spurt, which is a peak in velocity for each girl at times varying around about 12 years. Figure 3.4 presents the acceleration curves. Data like these in Figure 3.3 are typical of older growth studies, including the Fels (Roche, 1992) and Zürich (Falkner, 1960) studies, where the focus was on long-term skeletal growth, and

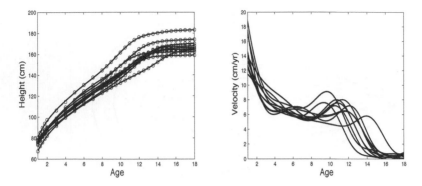

FIGURE 3.3: The left panel shows the height measurements and fitted curves for the first ten girls in the Berkeley growth data. The right panel displays the corresponding height velocity functions.

it was not considered necessary to measure heights any more frequently than twice a year. We now know better.

Figure 3.5 plots acceleration $a(t)$ against velocity $v(t)$ for each girl. Now we see that there are three distinct phases to long-term growth. The first, starting from the lower right of the figure, shows a nearly linear relation between acceleration and velocity such as would be exhibited by an exponential decay process, and that is precisely what we see in the left panel of Figure 3.3 where velocity decreases exponentially up to the age at which the pubertal growth spurt (PGS) begins. The second phase, which begins quite abruptly for some girls, consists of a loop spanning the pubertal growth spurt, quite similar in shape to the balloon in the baby's phase/plane plot. In this case, however, there is no $(0,0)$ point in the loop, but rather the loops center on

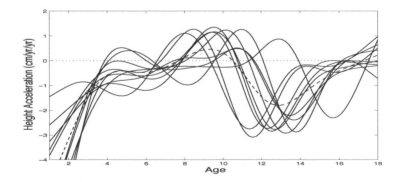

FIGURE 3.4: The height acceleration functions for the first ten girls in the Berkeley growth data. The dashed line indicates the mean of these curves.

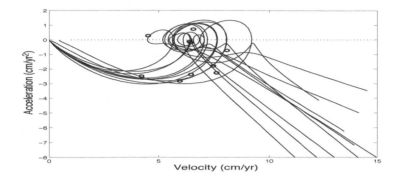

FIGURE 3.5: A phase/plane plot of the fitted curves for the first ten girls in the Berkeley growth data. The circle on each curve indicates the mean center of the pubertal growth spurt, 11.7 years.

a velocity of about seven cm/year and zero acceleration. Finally, there is the half-balloon phase in which both velocity and acceleration decay to zero as the girl approaches her adult height. Of course, we now know that all this is too simple, since right up to puberty there are distinct cycles between high and low intensity growth. The PGS is only the final growth blowout, after which energy is diverted to, among other things, reproductive behavior.

We will say more about this interesting plot in the next section, but already we can conclude that plots of one derivative against another can highlight a lot of information that might escape our attention in ordinary data plots. In effect, a phase/plane plot is a graphical analog of a differential equation like (3.4), where a higher order derivative is expressed as a function of lower order derivatives and the function itself.

3.4 Identification and Analysis of Phase or Tempo Variation

The PGS is an example of what we call a feature in a curve, typically seen as a peak, valley or crossing of a fixed threshold. In Figure 3.5 we have put a small circle on each curve at the average PGS age, 11.7 years; girls having an early PGS will, at 11.7 clock years, be already in the final wind-up phase of growth in the left portion of their curves, while girls with late PGS's will have the small circle in the right lead-up portion prior to the loop. We see two types of variation across curves: in intensity or size of features, called amplitude variation, and in timing of the feature, called phase variation in classical mathematics. Because "phase" carries such a heavy semantic burden in the mathematical sciences, we rather like borrowing the term "tempo" from music for the idea of time as an elastic medium.

The drastic consequences for statistical methodology of mixing tempo and amplitude variation were not obvious to Ramsay and Silverman (1997, 2005) when they began their work. These consequences are nicely illustrated by the dashed line in Figure 3.4, which displays the cross-sectional mean of the ten acceleration curves. The mean, which is supposed to capture typical behavior in data, fails miserably here. The amplitude of the mean PGS acceleration is less than that of any observed curve, and the duration of the PGS shown in the mean as peak-to-valley time substantially exceeds the four years that most girls actually take.

One can presume that, if the mean doesn't work, not much else in classical statistics will either. Such is the case; standard deviation functions, correlation surfaces and other summaries are distorted by tempo variation, and data exploration tools such as principal components analysis show that even a small amount of tempo variation in the data inflates the number of modes of important variation detected. An approach to quantifying the amount of tempo variation developed by Kneip and Ramsay (2008) estimates that over 50% of the variation in the Berkeley growth data is in tempo, and similar percentages have been reported for the Fels and Zürich data. Moreover, one sees tempo variation in almost all sets of functional data; for example, about 35% of seasonal variation in temperature is due to tempo, which is mostly attributable to the early or late arrival of winter.

Our third challenge, then, was to discover either how to simultaneously estimate tempo and amplitude variation, or to first estimate tempo variation and then remove it from the data so as to leave pure amplitude variation. We called the process curve registration. Conceptually, at least in the growth context, tempo variation is due to growth evolving at different rates for different girls. This is to say that there is a physiological or growth time associated with each girl's growth relative to which the PGS is always centered on 11.7 growth years; and growth time and clock time are nonlinearly related. We denote this relationship as "clock time = h_i(growth time)" for the ith girl, and we call function h_i that girl's time-warping function. Each function h is, of course, strictly monotonic, so that we can consider time, too, as a growth process, and we can, at least for convenience, add the additional restriction $h(0) = 0$ and $h(18) = 18$.

An elementary tempo estimation method called landmark registration calculates for each girl a smooth transformation of clock time such that transformed feature times all occur at corresponding fixed values. For the growth curves, the single landmark time is the time at which the acceleration curve crosses zero with negative slope, and this is visible in all curves, so that landmark registration works well and is easy to apply. The left panel of Figure 3.6 shows the warping functions h for each of the ten girls; values above the diagonal indicate late growth, and values below this diagonal indicate early growth. The right panel of Figure 3.6 shows the registered velocities, along with the

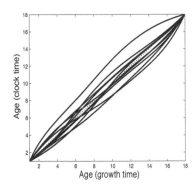 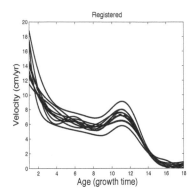

FIGURE 3.6: The left panel contains the warping functions $h(t)$ transforming clock time into growth time for the first ten girls in the Berkeley growth data. Warping functions above the dashed diagonal line correspond to late growth, and those below correspond to early growth. The right panel shows the registered velocity functions for the first ten girls in the Berkeley growth data. The dashed line indicates the mean of these velocity curves.

mean of these velocities. These curves represent pure amplitude variation, and now we see that the mean does a nice job of summarizing the data.

However, because landmarks are not always distinct, and registering all features to a fixed time can seem artificial, more powerful continuous registration methods have been developed by an ever-increasing number of statisticians beyond what is detailed in Ramsay and Li (1998) and Ramsay and Silverman (2005). Indeed, the mathematical analysis of tempo and amplitude variation is now taking us deeply into differential geometry, and it appears that there is much yet to learn about this fascinating problem.

Once amplitude and tempo variation are separated, it is natural ask if there are any patterns in their joint variation. In both Figure 3.4 and Figure 3.5 we notice that early PGS's tend to be intense, as reflected in the large swing between positive and negative acceleration and the larger loops in the phase/plane plot. Conversely, late PGS's are usually mild. This makes sense from the standpoint of growth energy: girls with early PGS's need large growth power during puberty in order to compensate for the extra years of growth that late puberty girls enjoy, and as a consequence adult height does not depend very much on the timing of the PGS.

3.5 Conclusion

The fundamental theme in this chapter is the power of differential equations for modeling functional data. The importance of differential equations as models for data has led to new approaches to the estimation of the parameters that define these equations Ramsay et al. (2007). Growth data have inspired much of the functional data analysis enterprise, but many other data contexts could have served the purpose of motivating dynamic systems models in statistics just as well. Ramsay and Silverman (2005), Ferraty and Romain (2011) and many more current publications promise a new era of partnerships between statisticians and the rapidly increasing number of researchers working with functional data.

About the Authors

James O. Ramsay is a professor emeritus of psychology and an associate member in the Department of Mathematics and Statistics at McGill University in Montréal. He received a PhD from Princeton University in 1966 in quantitative psychology. His research interests include pyschometrics, dynamic systems and functional data analysis, of which he was a founder. He has been president of the Psychometric Society and the Statistical Society of Canada, and is a fellow of the Canadian Psychological Association. He was awarded the SSC Gold Medal for research in 1998, and honorary membership by the Statistical Society of Canada in 2012.

Michael Hermanussen is a German pediatrician and professor at Christian-Albrechts-Universität zu Kiel. He studied medicine and worked as a pediatrician at the University of Kiel from 1982 until 1989. He investigated growth and child development (auxology) and first described mini growth spurts. Since 1990 he has cooperated in international projects, and he organizes national and international meetings on growth and nutrition. From 2003 to 2011 he was a member of the scientific board of the Deutsche Gesellschaft für Anthropologie and chief editor of *Anthropologischer Anzeiger*. He is the founder and head of the Deutsche Gesellschaft für Auxologie.

Bibliography

Falkner, F. (1960). *Child Development: An International Method of Study.* Karger, Basel.

Ferraty, F. and Romain, R. (2011). *The Oxford Handbook of Functional Data Analysis.* Oxford University Press, Oxford.

Hermanussen, M., Thiel, C., von Büren, E., Rol de Lama, M., Pérez Romero, A., Ariznaverreta Ruiz, C., Burmeister, J., and Tresguerres, J. A. F. (1998). Micro and macro perspectives in auxology: Findings and considerations upon the variability of short term and individual growth and the stability of population derived parameters. *Annals of Human Biology*, 25:359–385.

Kneip, A. and Ramsay, J. O. (2008). Combining registration and fitting for functional linear models. *Journal of the American Statistical Association*, 103:1155–1165.

Ramsay, J. O. (1998). Estimating smooth monotone functions. *Journal of the Royal Statistical Society, Series B*, 60:365–375.

Ramsay, J. O. and Dalzell, C. (1991). Some tools for functional data analysis (with discussion). *Journal of the Royal Statistical Society, Series B*, 53:539–572.

Ramsay, J. O., Hooker, G., Campbell, D., and Cao, J. (2007). Parameter estimation for differential equations: A generalized smoothing approach (with discussion). *Journal of the Royal Statistical Society, Series B*, 69:741–796.

Ramsay, J. O., Hooker, G., and Graves, S. (2009). *Functional Data Analysis with R and MATLAB.* Springer, New York.

Ramsay, J. O. and Li, X. (1998). Curve registration. *Journal of the Royal Statistical Society, Series B*, 60:351–363.

Ramsay, J. O. and Silverman, B. W. (1997). *Functional Data Analysis.* Springer, New York.

Ramsay, J. O. and Silverman, B. W. (2005). *Functional Data Analysis*, Second Edition. Springer, New York.

Roche, A. (1992). *Growth, Maturation and Body Composition: The Fels Longitudinal Study 1929–1991.* Cambridge University Press, Cambridge.

Thalange, N. K., Foster, P. J., Gill, M. S., Price, D. A., and Clayton, P. E. (1996). Model of normal prepubertal growth. *Archives of Disease in Childhood*, 75:427–431.

Tuddenham, R. D. and Snyder, M. M. (1954). Physical growth of California boys and girls from birth to eighteen years. *University of California Publications in Child Development*, 1:183–364.

4

Modeling Dependence beyond Correlation

Christian Genest and Johanna G. Nešlehová
McGill University, Montréal, QC

As scientists, looking for relationships between variables is an essential part of our efforts to understand the world, identify the causes of illnesses, assess climate change, predict economic cycles or guard against catastrophic events such as tsunamis or financial crises. Statistical models are often used for this purpose. To many scientists and engineers, this is synonymous with correlation and regression because these techniques are typically the first, if not the only ones, to which they were exposed. However, thanks in part to the work of Canadian statisticians, we now know that there is much more to dependence than correlation, regression, and the omnipresent "bell-shaped curve" of basic statistics textbooks. As we will demonstrate in this chapter, those who are oblivious to new tools may miss important messages hidden in their data.

4.1 Beyond the Normal Brave Old World

It is often said that a little knowledge is a dangerous thing, and so it is with those who believe that the Normal law, or "bell-shaped curve," is the appropriate model for just about every phenomenon measured on a continuous scale. In reality, many variables of interest just aren't Normally distributed, whatever scale you use. This is typically the case for financial losses, insurance claims, precipitation amounts and the height of storm surges, among others. The problem is even more pervasive when several variables are involved.

For example, McNeil (1997) analyzed losses above one million Danish kroner arising from fire claims made to the Copenhagen Re insurance company between 1980 and 1990. Each of them, adjusted for inflation, can be divided into a claim amount for damage to buildings (X), loss of contents (Y), and loss of profits (Z). For simplicity, let's focus on the 517 cases in which these three amounts were non-zero. Because some losses are very large, these data are best displayed on the logarithmic scale. This is done in Figure 4.1, where the best fitting Normal bell-shaped curve is superposed on the histogram of losses of each type. A data analyst who is satisfied with the fit would conclude that the distributions of X, Y, Z are well approximated by the "Log-Normal distribution," i.e., a bell-shaped curve on the logarithmic scale. The use of the logarithmic scale is not just ad hoc in this context; it has been found that the Log-Normal distribution often provides a good first approximation for many monetary insurance losses.

However, the amounts claimed for damage to buildings, loss of contents and loss of profits are naturally related because they are generated by the same fire. More importantly, a severe conflagration is likely to cause substantial damage on all accounts. When this happens, the insurer may incur large losses. The dependence between the claim amounts X, Y, Z is thus of paramount interest to the company. A statistical model that captures the simultaneous variation of these amounts can be used, among others, to determine sufficient capital reserves in order to avoid insolvency.

To assess this dependence, people typically start by looking at the plots of all pairs of variables. To visualize the relationship between damage to buildings and loss of contents, for example, we can plot the pairs $(\log X_i, \log Y_i)$, where the subscript refers to fire event #i, so that $i = 1, \ldots, 517$. The scatterplots

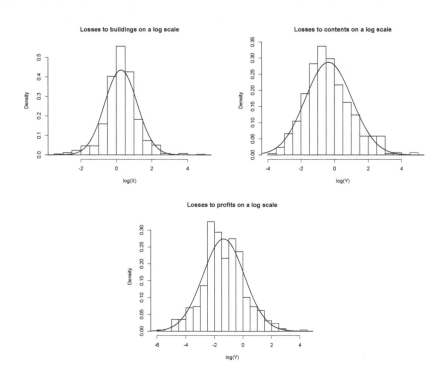

FIGURE 4.1: Histograms for X, Y, Z after transformation to the logarithmic scale.

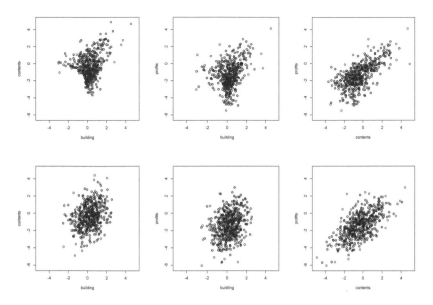

FIGURE 4.2: Top: pairwise scatterplots for X, Y, Z on the logarithmic scale; bottom: random sample of the same size from a fitted trivariate Normal law.

corresponding to the three possible pairs are displayed in Figure 4.2. The point clouds clearly show that the losses are positively associated. This is plausible because small fires generate smaller losses than big ones do.

Traditionally, this association is summarized by Pearson's correlation coefficient, which measures how close a point cloud is to a straight line. This coefficient always lies between -1 and 1; it equals either of these two values if the scatterplot is a perfect line. Here, the observed correlations are

$$\text{corr}(\log X, \log Y) \approx .282,$$
$$\text{corr}(\log X, \log Z) \approx .307,$$
$$\text{corr}(\log Y, \log Z) \approx .667.$$

Thus, on the logarithmic scale, losses of contents (Y) and profits (Z) are much more highly correlated with each other than they are with damage to buildings (X). Because the distributions of $\log(X)$, $\log(Y)$ and $\log(Z)$ appear to be Normal, the simplest joint statistical model for these three variables then consists of assuming that they are jointly Normal. Among others, this implies that the 3D histograms of all pairs of losses are bell-shaped on the logarithmic scale. What makes this model especially seductive is that the relationship among the three losses is then fully explained by the correlation coefficients computed above.

But is such a joint model plausible? To address this question, we simulated an artificial sample of 517 data points from the best fitting trivariate Normal distribution. The resulting pairwise scatterplots are displayed in the bottom row of Figure 4.2. The shapes of the point clouds are strikingly different from those of the original data that appear in the top row of the same figure. This suggests that the data are not jointly Normal, a fact that is confirmed by formal statistical tests (Cox and Small, 1978). Of course, more sophisticated transformations might improve the fit, at the cost of interpretability. However, we shall see that try as we may, no transformation of these data can make them jointly Normal.

4.2   Dangers of Ignoring Non-Normality

For many years now, statisticians have been pointing out that modelers and data analysts may miss something important, or worse, reach disastrous conclusions, when they assume that the Normal distribution applies universally. Inadequate protection against insolvency is a case in point. Other examples include the underestimation of risks associated with floods and catastrophic storms. The paper by Embrechts et al. (2002) and the subsequent book by McNeil et al. (2005) have been particularly effective at delivering this message. In spite of this admonition, however, the notions of linear correlation and Normality are so deeply entrenched that many people remain blind to their limitations.

First, Pearson's correlation is merely a number, and as such, it cannot possibly tell the whole story. In Figure 4.2, each of the three datasets displayed in the top row has the same correlation as the Normal dataset appearing just below it, yet their shapes are strikingly different. In particular, there are many more points in the upper right corner of the real data plots than in the plots generated from the trivariate Normal model. In other words, the chances of large losses occurring simultaneously is much higher in reality than the Normal model would predict. An insurance company that fixes its premiums based on the Normal paradigm would thus be underestimating the risk of large claims.

Second, linear correlation is not easy to interpret outside of a specific model. People often view it as a measure of association between two variables, but they do not always realize that it is affected by the choice of scale. In the Danish fire loss data, for instance, the correlations between the original variables are

$$\mathrm{corr}(X, Y) \approx .627, \quad \mathrm{corr}(X, Z) \approx .791, \quad \mathrm{corr}(Y, Z) \approx .617.$$

These figures are very different from those computed from the same data once they have been transformed to the logarithmic scale. While the losses of contents and profits had the largest correlation on the latter scale, it is the

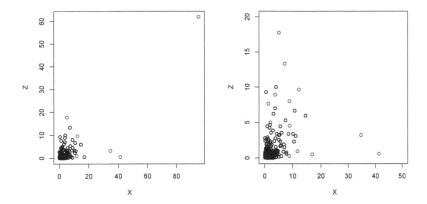

FIGURE 4.3: Scatterplot of X and Z on the original scale (left) and after removing the largest observation (right).

damage to buildings and losses of profits that are the most correlated on the original scale. Given that $\text{corr}(X, Z)$ is large, we might be tempted to think that the relationship between X and Z is nearly linear. However, the left panel of Figure 4.3 reveals that this is not at all so. The high correlation between the two variables is actually driven by a few influential points in the upper right-hand corner of the graph. Yet these points should not be ignored. In fact, they are exactly the type of event that the insurer wants to anticipate and guard against! But even if these points are removed, as in the right panel of Figure 4.3, a linear pattern remains hard to discern.

In effect, linear correlation does not make all that much sense beyond the Normal paradigm. As additional evidence, consider the widespread misconception that any degree of correlation between -1 and 1 can be achieved. For instance when two losses are Normally distributed on the logarithmic scale, it can be shown that the theoretical value of the correlation coefficient actually lies between $\varrho_{\min}$ and $\varrho_{\max}$, where

$$\varrho_{\min} = \frac{e^{-\sigma_1 \sigma_2} - 1}{\sqrt{(e^{\sigma_1^2} - 1)(e^{\sigma_2^2} - 1)}}, \quad \varrho_{\max} = \frac{e^{\sigma_1 \sigma_2} - 1}{\sqrt{(e^{\sigma_1^2} - 1)(e^{\sigma_2^2} - 1)}}.$$

Here, $\sigma_1 > 0$ and $\sigma_2 > 0$ quantify the variability of the Normal distributions about their respective means. These so-called standard deviations are also related to the height of the bell-shaped curves. For the Danish fire loss data, with σ_1 and σ_2 estimated from the sample, the correlations should satisfy

$$-.12 \leq \text{corr}(X, Y) \leq .63,$$
$$-.09 \leq \text{corr}(X, Z) \leq .52,$$
$$-.02 \leq \text{corr}(Y, Z) \leq .98.$$

The fact that the observed correlations did not all fall within these bounds could simply be due to the "luck of the draw" or sampling variation but in view of the large sample size, it is more likely a sign that the losses are not Log-Normal after all. Indeed, the presence of very large claims, visible in the left panel of Figure 4.3, suggests that the individual loss distributions are heavy-tailed. This suspicion can be confirmed with a Q–Q plot or the test of Normality due to American statistician Samuel Shapiro and Canadian Martin Wilk (Shapiro and Wilk, 1965). If the loss distributions are really heavy-tailed, the sample linear correlation coefficient is trying to estimate a quantity that may not even be defined theoretically. The observed correlations would then be meaningless!

4.3   Copulas to the Rescue

When dealing with non-Normal data, it is crucial to think beyond correlation, linear regression, and joint Normality. An interesting concept that makes it possible to study dependence in broader terms was proposed by the American mathematician Abe Sklar in response to a question posed by his French colleague Maurice Fréchet.

4.3.1   Fréchet's Problem

Cast in the simplest possible terms, Fréchet's problem is the following. Suppose that X and Y are two fire claim amounts, say, for which we know how to compute the probabilities $\Pr(X \leq x)$ and $\Pr(Y \leq y)$ for any values x and y. Viewed as functions of x and y, the probabilities $F(x) = \Pr(X \leq x)$ and $G(y) = \Pr(Y \leq y)$ are called the marginal distribution functions, or margins, of X and Y. The question is then how to construct a model for the probability of the events $X \leq x$ and $Y \leq y$ occurring simultaneously, denoted $\Pr(X \leq x \text{ and } Y \leq y)$, while ensuring that X has distribution F and Y has distribution G.

Sklar's solution can in fact be traced back to work done in the 1940s by Wassily Hoeffding, a statistician of Danish origin. The idea is to write

$$\Pr(X \leq x \text{ and } Y \leq y) = C\{\Pr(X \leq x), \Pr(Y \leq y)\}, \qquad (4.1)$$

where C is a specific function of two variables called a copula. The purpose of a copula is to "glue together the margins" or "couple the individual probabilities" (hence the Latin term "copula") in order to generate dependence between the variables. Any copula can be expressed as

$$C(u, v) = \Pr(U \leq u \text{ and } V \leq v) \tag{4.2}$$

in terms of variables U and V that are uniformly distributed on the interval $[0, 1]$, meaning that they are equally likely to lie anywhere between 0 and 1.

While regression imposes a linear relationship between the variables (after transformations), Sklar's formula provides complete flexibility in capturing the relation between X and Y because any copula C can be combined with any margins F and G to construct a valid joint model for the pair (X, Y). Conversely, any joint model for a pair (X, Y) can be expressed in terms of its margins and some copula C. This is "the joy of copulas" (Genest and MacKay, 1986a,b).

4.3.2   Measuring Association

Sklar (1959) showed that the copula C of a pair (X, Y) is unique when the variables are measured on a continuous scale. It then contains *all* the relevant information about the dependence between X and Y. Thus it makes sense to measure their association using C or, equivalently, in terms of the uniform variables U and V appearing in formula (4.2). These variables, which satisfy the relations $U = F(X)$ and $V = G(Y)$, govern the dependence between X and Y.

Surprisingly, the fact that association should be measured in terms of U and V, rather than X and Y, was recognized as early as 1904 by the British psychologist Charles Spearman. His measure of association, now termed Spearman's rank correlation coefficient (Spearman, 1904), turns out to be the sample analog of Pearson's linear correlation between U and V. It can be computed from C alone from a formula involving a double integral, viz.

$$\varrho(X, Y) = \text{corr}(U, V) = -3 + 12 \int_0^1 \int_0^1 C(u, v) dv du.$$

Similarly, the difference between the proportions of "concordant" and "discordant" pairs of observations popularized by the British statistician Maurice Kendall is a sample analog of another copula-based quantity called Kendall's tau (Kendall, 1938). [Two observations (X_1, Y_1) and (X_2, Y_2) are called concordant if the ranking of X_1 and X_2 is the same as that for Y_1 and Y_2; otherwise, the observations are said to be discordant.] As mentioned earlier, working on the uniform scale was also favored by Hoeffding in his seminal dissertation (Hoeffding, 1940).

The fundamental role of copulas in measuring association was crystallized in the work of Schweizer and Wolff (1981), who highlighted the fact that copulas are invariant with respect to increasing transformations of the variables, and hence scale-free. In the context of the Danish fire loss data, for instance, this means that the pairs (X, Y) and $(\log(X), \log(Y))$ share the same copula. As a consequence of this fact, Spearman's rho, Kendall's tau and other measures of dependence based on copulas yield the same values whether they are computed from the original data or after transforming them while preserving their order. In all cases, the sample values of Spearman's rho and Kendall's tau are

$$\varrho(X, Y) \approx .19, \quad \varrho(X, Z) \approx .29, \quad \varrho(Y, Z) \approx .64,$$
$$\tau(X, Y) \approx .12, \quad \tau(X, Z) \approx .2, \quad \tau(Y, Z) \approx .46.$$

By focusing on the copula, what we have achieved is a complete separation of the dependence pattern of the variables from their individual behavior. This has several advantages. For example, both Spearman's rho and Kendall's tau can now reach all possible values between -1 and 1; the bounds are attained if one variable is a monotone (but not necessarily linear) function of the other.

Moreover, there are many other aspects of dependence that can be studied by looking at the copula. For example, UBC professor Harry Joe (Joe, 1997) used copulas to construct a coefficient $\lambda(X, Y)$ which quantifies the tendency of variables X and Y to take large values simultaneously. His upper-tail coefficient, which varies between 0 (no tail dependence) and 1 (comonotonic dependence), is especially relevant for measuring the riskiness of a financial position. In the fire loss data, for example, tail dependence turns out be quite strong, viz.

$$\lambda(X, Y) \approx .56, \quad \lambda(X, Z) \approx .5, \quad \lambda(Y, Z) \approx .68.$$

This nicely echoes the simultaneous presence of large values of X, Y, Z which is clearly visible in the plots in the top row of Figure 4.2, and in Figure 4.3.

Given the complexity of relationships between variables, dependence can be quantified in more sophisticated ways than merely through one-number summaries. Originating in the work of Erich Lehmann, a professor of statistics at the University of California at Berkeley, many dependence concepts and orderings have been proposed (Lehmann, 1966; Müller and Stoyan, 2002) and much of this methodology is still under development. Many Canadian statisticians have contributed to this line of research, including Philippe Capéraà and Louis-Paul Rivest at Université Laval and Mhamed Mesfioui in Trois-Rivières. Many actuarial implications of these results have also been considered, notably by Hélène Cossette and Étienne Marceau, also at Laval (Denuit et al., 2005).

4.3.3 Stress Testing

When the individual behavior of risks is well understood but the way in which they interact is uncertain, Sklar's formula is the ideal tool to explore the impact of various dependence scenarios on quantities of interest.

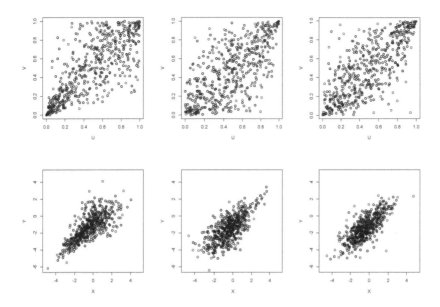

FIGURE 4.4: Top: scatterplots of random samples of size 517 from the Clayton (left), Gumbel (middle) and Student t_4 copulas (right); bottom: corresponding samples with fitted Normal margins.

For example, suppose that the losses of contents (Y) and profits (Z) are indeed Log-Normal. By calibrating these distributions to the data at hand, we can compute $\Pr(Y \leq y)$ and $\Pr(Z \leq z)$ for any real numbers y and z. By varying C in expression (4.1), we can then generate a host of models that can be used to investigate, say, the distribution of the total loss of contents *and* profits, $Y + Z$.

To illustrate, consider the Clayton, Gumbel, and Student t_4 copulas. These are merely three common choices among the wealth of copula families listed in the popular 1999 book *An Introduction to Copulas* by the American mathematician Roger Nelsen. The word "family" is used here because copulas typically involve a constant θ, called a parameter, that can be used to adjust the degree of dependence. For example, the copula defined, for all $u, v \in (0, 1)$, by

$$C(u, v) = (u^{-\theta} + v^{-\theta} - 1)^{-1/\theta} \qquad (4.3)$$

is the Clayton copula with parameter $\theta > 0$. The theoretical value of Kendall's tau for this model is $\tau = \theta/(\theta + 2)$, which increases from 0 to 1 as $\theta \to \infty$.

Displayed in Figure 4.4 are simulated samples of size 517 from the models corresponding to a Clayton (left), Gumbel (middle) or t_4 (right) copula. The top row displays samples from the copulas themselves; their parameters were calibrated so that Kendall's tau is the same as in the data. For example, to adjust the Clayton copula to the pair (Y, Z), we solved the equation $\theta/(\theta+2) = .46$, leading to $\theta = 1.7$. The very same samples were then transformed so as to reproduce the marginal behavior of Y and Z. To generate a pair from (X, Y), one generates a pair $(U, V) = (u, v)$ from the chosen copula and then one finds the values of x and y for which $F(x) = u$ and $G(y) = v$; this is done by numerical inversion. One then sets $X = x$ and $Y = y$.

The result of this transformation is shown on the logarithmic scale in the bottom row of Figure 4.4. The shape of the point clouds makes it abundantly clear that different copulas can induce very different dependence patterns. The same conclusion applies when the models are used to compute probabilities or make predictions. This is illustrated in two different ways in Tables 4.1–4.2.

Table 4.1 reports estimates of the probability that the total loss of contents and profits, $Y + Z$, exceeds either 50 or 100 million (M) Danish kroner. All the estimates are based on the same marginal distributions; only the copula varies. In addition to the Clayton, Gumbel and Student t_4 copulas, we included the Normal copula which corresponds to the joint Normal model. The estimates were obtained by simulating one million pairs from each of these models. As one can see, the probability that $Y + Z > 50M$ varies by a factor of about 1.75 between the most (Clayton) and the least (Gumbel) optimistic scenario. This factor grows to 2.18 for the probability that the total loss exceeds 100M.

At a first glance, the probabilities reported in Table 4.1 may seem very small and the differences between them appear to be inconsequential. Nevertheless, these differences can have a substantial impact on the financial reserves that regulators require the insurance companies to keep in order to cover potentially large claims. A quantity that often enters the calculations of such reserves is the Value-at-Risk, commonly abbreviated VaR. In statistical terms, the VaR at level α is the upper αth percentile of the loss distribution or, in other words, the "maximum loss which is not exceeded with probability α"; see Section 2.2.2 of McNeil et al. (2005) for further details and discussion.

Typically, the VaR is computed at $\alpha = .99$ or even .999. Table 4.2 reports the VaR (in millions of Danish kroner) at both levels when Log-Normal marginal loss distributions are linked by different copulas. While the VaR estimates are quite close at the level .99, they vary much more when $\alpha = .999$. These differences show how important is the choice of dependence structure.

TABLE 4.1: Estimates of the probability of a total loss of contents and profits $(Y + Z)$ based on four different copulas and Log-Normal margins.

	Copula			
	Clayton	Normal	Student t_4	Gumbel
$10^3 \times \Pr(Y + Z > 50M)$	1.3	1.8	2.0	2.3
$10^3 \times \Pr(Y + Z > 100M)$	.22	.3	.37	.48

TABLE 4.2: Estimates of the Value-at-Risk for the total loss of contents and profits $(Y + Z)$ based on four different copulas and Log-Normal margins.

	Copula			
	Clayton	Normal	Student t_4	Gumbel
VaR at $\alpha = .99$	21.44	23.07	23.61	24.21
VaR at $\alpha = .999$	55.05	63.42	68.75	71.69

Imagine, for instance, that a financial reserve is determined from the VaR at level .999 using a Clayton copula. If the true dependence structure happens to be Gumbel, the reserve would then be underestimated by over 16M Danish kroner, or nearly 3M Canadian dollars!

Needless to say, errors of judgment in the choice of margins can also lead to serious underestimation of the VaR or other measures of risk at such high levels. In the Danish fire loss example, there is evidence that the upper tails of the marginal distributions are much heavier than those of the Log-Normal distribution. If this were taken into account, the resulting VaR estimates would be quite a bit higher. More details and additional illustrations can be found in McNeil et al. (2005); see also Chapter 15 by Bruno Rémillard.

4.3.4   Validate, Validate, Validate!

The financial crisis of 2008 is arguably the most famous case of risk underestimation due to an inappropriate choice of copula and marginal distributions. It was described by the English journalist Felix Salmon in an award winning article entitled "Recipe for disaster: The formula that killed Wall Street." Originally published online (*Wired Magazine*: 17.03) and later reproduced in print (Salmon, 2012), the article recounts how quants (i.e., quantitative analysts from the financial sector) underestimated a joint probability of default by assuming that loans in different pools, say A and B, were linked by a Normal copula. In his article, Salmon cites

$$\Pr(T_A < 1, T_B < 1) = \Phi_2[\Phi^{-1}\{F_A(1)\}, \Phi^{-1}\{F_B(1)\}, \gamma].$$

as the formula that was "instrumental in causing the unfathomable losses that brought the world financial system to its knees." This formula simply states that the probability of observing loan defaults in pools A and B within one year is given by a distribution of the form (4.1) in which the margins are denoted F_A and F_B rather than F and G, and

$$C(u, v) = \Phi_2\{\Phi^{-1}(u), \Phi^{-1}(v), \gamma\}$$

is the Normal (or Gaussian) copula of the bivariate Normal distribution Φ_2 with standard Normal margins Φ and Pearson's correlation γ.

David X. Li, a quant of Chinese origin, proposed this model (Li, 2000). While he and his formula were repeatedly blamed for bringing on the financial crisis, this is clearly an oversimplification, though it makes a good story with a Canadian twist, given that Li did his graduate work at Laval and Waterloo. In effect, Li himself was aware of the limitations of his model (Meissner, 2008, p. 71), but this did not prevent the bankers from using it extensively. They failed to realize that it led to an underestimation of the risk of joint defaults because it has no tail dependence (i.e., Joe's coefficient is 0).

Compounded with the fact that in financial markets, dependence often changes over time, the static Normal copula model led to prices and capital reserves that were too low, and therefore unduly attractive for traders and bankers in quest of financial gain. Does this mean that the Normal copula should be locked up, never to be used again? Hardly. It may not be appropriate for modeling default times or fire insurance claims, but there are situations in which it makes perfect sense. Copula models are versatile tools that must be used wisely. What we need is to understand their limitations and, even more important, to make sure that they are fit for use in the situation at hand. This requires good data and proper model validation.

4.4   Proof of the Pudding Is in the Ranking

The strategy for building a statistical model is always more or less the same. First look at the data and choose a class of models that captures its main characteristics. Next, use the data to estimate the model parameters efficiently and check whether the fit is satisfactory. If it isn't, try over again and make sure to validate your choice before you use the model to compute probabilities or make predictions that inform a decision process. What is meant by "efficient" and "satisfactory" depends on the context and involves a fair dose of pragmatism.

When it comes to copula modeling, the main stumbling block is the fact that except in the rare instance where the individual behavior of the variables is known with absolute certainty, we never actually observe data from the copula. Remember from (4.2) that C is the distribution function of the uniform variables $U = F(X)$ and $V = G(Y)$. What we collect and observe, however, are data from X and Y, say $(X_1, Y_1), \ldots, (X_n, Y_n)$. If the formulas for F and G were available (that is, if we knew the exact distributions of X and Y), we could simply transform these data into a sample from (U, V) by setting $U_1 = F(X_1)$ and $V_1 = G(Y_1)$, etc. We could then plot the pairs $(U_1, V_1), \ldots, (U_n, V_n)$ to start the model building process. However, we simply don't have this luxury.

It was only in the late 1970s and 1980s that statisticians gradually realized that one could uncover the concealed copula by ranking the observations in ascending order and plotting the resulting pairs of ranks divided by the sample size n. As a simple illustration, suppose that only $n = 3$ fire claims are observed and that fictitious amounts for losses of contents and profits are as follows:

$$(Y_1, Z_1) = (2900, 50), \quad (Y_2, Z_2) = (80, 45), \quad (Y_3, Z_3) = (150, 120).$$

The ranks of the Y values are then $R_1 = 3$, $R_2 = 1$, $R_3 = 2$ while those of the Z values are $S_1 = 2$, $S_2 = 1$, $S_3 = 3$. The rank plot would then show the points

$$\left(\frac{R_1}{n}, \frac{S_1}{n}\right) = \left(\frac{3}{3}, \frac{2}{3}\right), \quad \left(\frac{R_2}{n}, \frac{S_2}{n}\right) = \left(\frac{1}{3}, \frac{1}{3}\right), \quad \left(\frac{R_3}{n}, \frac{S_3}{n}\right) = \left(\frac{2}{3}, \frac{3}{3}\right).$$

This does not amount to much of a plot because the sample is impractically small. However, look at Figure 4.5 and see what happens when this procedure is used on the Danish fire loss data. It transpires from the graphs that the dependence structure differs from one pair to the next.

For simplicity, focus on the dependence between losses of contents and profits. Comparing the point cloud in the lower panel of Figure 4.5 to the three scatterplots in the top row of Figure 4.4, would you rather pick the Clayton, Gumbel or Student t_4 copula as a plausible dependence structure between Y and Z? Though this strategy may sound simplistic, this is in essence what model builders need to do; and it isn't just black magic. Expressed in mathematical terms, the points on the rank plot constitute the support of what statisticians call the empirical copula, a "random function" defined by

$$
\begin{aligned}
C_n(u, v) &= \text{proportion of fire events for which } R_i \le nu \text{ and } S_i \le nv, \\
&= \frac{1}{n} \sum_{i=1}^{n} 1\left(\frac{R_i}{n} \le u, \frac{S_i}{n} \le v\right).
\end{aligned}
$$

As it turns out, C_n gets closer and closer to the true C as the sample size n becomes infinitely large. In the limit, $\sqrt{n}\,(C_n - C)$ is related to a multi-dimensional version of the well-known Brownian motion. This interesting phenomenon was first studied by the German probabilist Ludger Rüschendorf (Rüschendorf, 1976) and the French academician Paul Deheuvels (Deheuvels, 1979). Many refinements and extensions have since been developed, notably by Canadian statisticians Bruno Rémillard and Kilani Ghoudi (Ghoudi and Rémillard, 1998).

At first sight, it seems that in the Danish fire loss data, the Gumbel copula is preferable to the Clayton or the Student t_4 as a descriptor of the dependence between losses of contents (Y) and profits (Z). The formula for the Gumbel copula involves a parameter θ, viz.

$$C(u, v) = e^{-\{|\log(u)|^\theta + |\log(v)|^\theta\}^{1/\theta}}$$

and the value of $\theta \ge 1$ is linked to Kendall's tau via the relation $\tau = 1 - 1/\theta$.

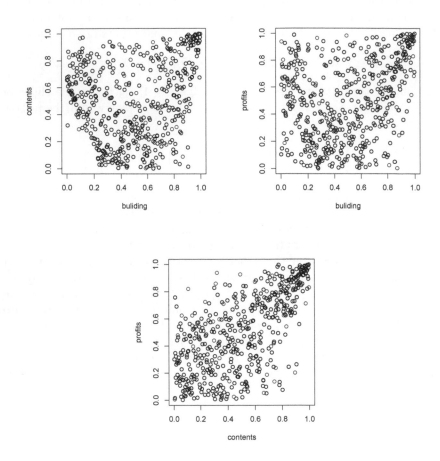

FIGURE 4.5: Rank plots for (X, Y) (top left), (X, Z) (top right) and (Y, Z) (bottom).

Proceeding as we did earlier with the Clayton copula, we could calibrate θ from the data by solving the equation $1 - 1/\theta = .46$, which yields $\theta = 1/.54 \approx 1.85$. This estimate, denoted by $\hat{\theta}_n$ to remind us that it is based on a data sample of size n, gives us a good idea about the value of the unknown parameter θ. To make it even more precise, we can exploit the so-called rank-based likelihood. If

$$c_\theta(u, v) = \frac{\partial^2}{\partial u \partial v} C_\theta(u, v)$$

is the density of the copula C_θ, obtained by differentiating C_θ once with respect to u and once with respect to v, the rank-based likelihood is a function ℓ of the parameter θ given by

$$\ell(\theta) = \sum_{i=1}^{n} \log \left\{ c_\theta \left(\frac{R_i}{n}, \frac{S_i}{n} \right) \right\}.$$

A much more efficient estimate is then given by the value of θ that maximizes $\ell(\theta)$. This estimation technique, originally studied by Genest et al. (1995) and Shih and Louis (1995), is now deemed the "industry standard." We could also maximize the full likelihood involving the margins — either globally or by steps as in Joe (2005) — but proceeding from ranks protects us against errors in the choice of the marginal distributions (Kim et al., 2007).

Having settled on an estimate $\hat{\theta}_n$ of the parameter, we need to check whether the selected copula C_θ with $\theta = \hat{\theta}_n$ does indeed reproduce the pattern observed in the data, represented by the empirical copula C_n. The closer $C_{\hat{\theta}_n}$ is to C_n, the better the model. This proximity can be conveniently measured by the statistic

$$S_n = \sum_{i=1}^{n} \left\{ C_n \left(\frac{R_i}{n}, \frac{S_i}{n} \right) - C_{\hat{\theta}_n} \left(\frac{R_i}{n}, \frac{S_i}{n} \right) \right\}^2$$

advocated and investigated by Genest et al. (2009b).

For the losses of contents (Y) and profits (Z), one gets $S_n = .0799$, which seems small, but beware of appearances! What do we know of the sampling variation of S_n in datasets of size 517 when the copula is Gumbel with parameter $\theta = 1.85$? One way to find out is to generate thousands of datasets from that copula and compute the value of the statistic for each one of them. This technique, called the "parametric bootstrap," was shown to be valid for copula models by Genest and Rémillard (2008). Another popular computer-intensive method is based on the Multiplier Central Limit Theorem first applied in this context by Rémillard and Scaillet (2009); see also Kojadinovic et al. (2011).

At the end of the day, it turns out that a value of .0799 or bigger is a pretty rare occurrence. The probability that S_n is at least .0799 when the true copula is Gumbel is then about .0005 or about 5 in 10,000. Because this value is so small, the Gumbel copula no longer seems satisfactory and the search for a more adequate dependence structure must go on. One viable option is the "survival Clayton," defined, for all $u, v \in (0, 1)$, by

$$\bar{C}_\theta(u, v) = 1 - u - v + C_\theta(1 - u, 1 - v)$$

with C_θ as in Equation (4.3). In that case, $S_n = .0345$ and $\Pr(S_n \geq .0345) \approx$ .0892 or 8.92%. This probability is larger than the conventional 5% significance level, meaning that there is no strong statistical evidence against the survival Clayton model.

But just as with the choice of marginal distributions, the selection of an appropriate copula requires much care and all sorts of checks and balances. Many other questions can be asked. For instance, is the copula exchangeable, that is, does $C(u,v) = C(v,u)$ always hold? Does the copula exhibit tail dependence? Does it belong to one of the general classes of copulas proposed in the literature, such as Archimedean, elliptical or extreme-value? Procedures are actively being developed to answer these and other questions, and here again many Canadians are contributing, including Jean-François Quessy in Trois-Rivières; see, e.g., Genest and Nešlehová (2013).

4.5   Applications and Future Challenges

Since the year 2000, the popularity of copula-based modeling has expanded considerably. For a brief history of this development, see Genest et al. (2009a). Today, copulas are commonly used in actuarial science, biostatistics and health sciences, environmental sciences, hydrology, finance, risk management, and financial economics, among others. Major books and review articles that discuss copula modeling at length include Joe (1997), Frees and Valdez (1998), Nelsen (2006), Cherubini et al. (2004), Denuit et al. (2005), McNeil et al. (2005), Genest and Favre (2007), Salvadori et al. (2007), and Patton (2012).

In many fields, researchers encounter data that pose challenges far beyond what we sketched here. For example, financial and environmental data typically exhibit time trends and the dependence patterns may also change over time. In medical studies, variables are often measured on a discrete scale and may be only partially observed. Furthermore, it is often necessary to include explanatory variables in the construction of the marginal distributions. Some of these issues are considered in Genest and Nešlehová (2007), Kolev and Paiva (2009) and researchers at the University of Waterloo (Cook et al., 2010), among others. New copula-based tools are constantly being developed to meet these challenges and many researchers across Canada are at the forefront of this work.

With the proliferation of large datasets from a variety of sources, perhaps the most pressing and ubiquitous challenge is posed by the need to "leverage big data." This calls for ways to build dependence models involving hundreds, and even thousands of variables. Most of what we described here can be extended to any dimension; we limited ourselves to two or three variables for simplicity only. However, the use of these techniques in high-dimensional contexts poses serious statistical and numerical challenges. Moreover, several of the copula families used for two or three variables do not extend easily to these situations or produce unduly restrictive dependence structures.

New model building strategies for high-dimensional data are being actively developed. Among them, the so-called vine copula construction, rooted in the work of Harry Joe (1997), is quite popular at present (Kurowicka and Joe, 2010). For recent books that look at vine copula constructions and their simulation in a financial context, see Ma (2010) and Mai and Scherer (2012). More broadly, specialized conferences are regularly being held to stimulate research on copula modeling, including recent workshops at the Centre de recherches mathématiques in Montréal and at the Banff International Research Station.

Acknowledgments

This research was supported by the Canada Research Chairs Program and grants from the Natural Sciences and Engineering Research Council of Canada and the Fonds de recherche du Québec – Nature et technologies.

About the Authors

Christian Genest is the director of the Institut des sciences mathématiques du Québec and a professor of statistics at McGill University, where he holds a Canada Research Chair in Stochastic Dependence Modeling. He received a BSpSc from the Université du Québec à Chicoutimi, an MSc from the Université de Montréal, and a PhD from the University of British Columbia. He has served as editor of *The Canadian Journal of Statistics* and as president of the Statistical Society of Canada. He was the first winner of the CRM–SSC Prize and was awarded the 2011 SSC Gold Medal for research. He is a fellow of the American Statistical Association and the Institute of Mathematical Statistics.

Johanna G. Nešlehová is an associate professor of statistics at McGill University. She studied mathematics and statistics at Univerzita Karlova v Praze, Universität Hamburg and Carl von Ossietzky Universität Oldenburg. Before moving to Canada, she was at ETH Zürich for five years, as a postdoctoral fellow at RiskLab Switzerland and subsequently as Heinz Hopf Lecturer. Her primary research interests are multivariate extreme-value theory, dependence modeling, and applications of statistics to risk management and health. She was elected a member of the International Statistical Institute in 2011.

Bibliography

Cherubini, U., Luciano, E., and Vecchiato, W. (2004). *Copula Methods in Finance.* Wiley, New York.

Cook, R. J., Lawless, J. F., and Lee, K.-A. (2010). A copula-based mixed Poisson model for bivariate recurrent events under event-dependent censoring. *Statistics in Medicine*, 29:694–707.

Cox, D. R. and Small, N. J. H. (1978). Testing multivariate normality. *Biometrika*, 65:263–272.

Deheuvels, P. (1979). La fonction de dépendance empirique et ses propriétés. Un test non paramétrique d'indépendance. *Académie Royale de Belgique, Bulletin de la Classe des Sciences, 5e Série*, 65:274–292.

Denuit, M., Dhaene, J., Goovaerts, M. J., and Kaas, R. (2005). *Actuarial Theory for Dependent Risk: Measures, Orders and Models*. Wiley, New York.

Embrechts, P., McNeil, A. J., and Straumann, D. (2002). Correlation and dependence in risk management: Properties and pitfalls. In *Risk Management: Value at Risk and Beyond (Cambridge, 1998)*, pp. 176–223. Cambridge University Press, Cambridge.

Frees, E. W. and Valdez, E. A. (1998). Understanding relationships using copulas. *North American Actuarial Journal*, 2:1–25.

Genest, C. and Favre, A.-C. (2007). Everything you always wanted to know about copula modeling but were afraid to ask. *Journal of Hydrologic Engineering*, 12:347–368.

Genest, C., Gendron, M., and Bourdeau-Brien, M. (2009a). The advent of copulas in finance. *The European Journal of Finance*, 15:609–618.

Genest, C., Ghoudi, K., and Rivest, L.-P. (1995). A semiparametric estimation procedure of dependence parameters in multivariate families of distributions. *Biometrika*, 82:543–552.

Genest, C. and MacKay, R. J. (1986a). Copules archimédiennes et familles de lois bi-dimensionnelles dont les marges sont données. *The Canadian Journal of Statistics*, 14:145–159.

Genest, C. and MacKay, R. J. (1986b). The joy of copulas: Bivariate distributions with uniform marginals. *The American Statistician*, 40:280–283.

Genest, C. and Nešlehová, J. (2007). A primer on copulas for count data. *ASTIN Bulletin*, 37:475–515.

Genest, C. and Nešlehová, J. G. (2013). Assessing and modeling asymmetry in bivariate continuous data. In *Copulae in Mathematical and Quantitative Finance, Proceedings of the Workshop Held in Cracow, 10–11 July 2012*, pp. 91–114. Springer, Berlin.

Genest, C. and Rémillard, B. (2008). Validity of the parametric bootstrap for goodness-of-fit testing in semiparametric models. *Annales de l'Institut Henri Poincaré: Probabilités et Statistiques*, 44:1096–1127.

Genest, C., Rémillard, B., and Beaudoin, D. (2009b). Omnibus goodness-of-fit tests for copulas: A review and a power study. *Insurance: Mathematics & Economics*, 44:199–213.

Ghoudi, K. and Rémillard, B. (1998). Empirical processes based on pseudo-observations. In *Asymptotic Methods in Probability and Statistics (Ottawa, ON, 1997)*, pp. 171–197. North-Holland, Amsterdam.

Hoeffding, W. (1940). Maßstabinvariante Korrelationstheorie für diskontinuierliche Verteilungen. *Archiv für mathematische Wirtschafts- und Sozialforschung*, 7:4–70.

Joe, H. (1997). *Multivariate Models and Dependence Concepts*. Chapman & Hall, London.

Joe, H. (2005). Asymptotic efficiency of the two-stage estimation method for copula-based models. *Journal of Multivariate Analysis*, 94:401–419.

Kendall, M. G. (1938). A new measure of rank correlation. *Biometrika*, 30:81–89.

Kim, G., Silvapulle, M. J., and Silvapulle, P. (2007). Comparison of semiparametric and parametric methods for estimating copulas. *Computational Statistics & Data Analysis*, 51:2836–2850.

Kojadinovic, I., Yan, J., and Holmes, M. (2011). Fast large-sample goodness-of-fit tests for copulas. *Statistica Sinica*, 21:841–871.

Kolev, N. and Paiva, D. (2009). Copula-based regression models: A survey. *Journal of Statistical Planning and Inference*, 139:3847–3856.

Kurowicka, D. and Joe, H. (2010). *Dependence Modeling: Handbook on Vine Copulae*. World Scientific Publishing, Singapore/SG.

Lehmann, E. L. (1966). Some concepts of dependence. *The Annals of Mathematical Statistics*, 37:1137–1153.

Li, D. X. (2000). On default correlation: A copula function approach. *Journal of Fixed Income*, 9(4):43–54.

Ma, J. (2010). *Higher-dimensional Copula Models and Their Application: Bayesian Inference for D-vine Pair-copula Constructions Based on Different Bivariate Families*. VDM Publishing.

Mai, J.-F. and Scherer, M. (2012). *Simulating Copulas: Stochastic Models, Sampling Algorithms and Applications*. Imperial College Press, London.

McNeil, A. J. (1997). Estimating the tails of loss severity distributions using extreme value theory. *ASTIN Bulletin*, 27:117–137.

McNeil, A. J., Frey, R., and Embrechts, P. (2005). *Quantitative Risk Management: Concepts, Techniques and Tools*. Princeton University Press, Princeton, NJ.

Meissner, G. (2008). *The Definitive Guide to CDOs*. Risk Books, London.

Müller, A. and Stoyan, D. (2002). *Comparison Methods for Stochastic Models and Risks*. Wiley, Chichester.

Nelsen, R. B. (2006). *An Introduction to Copulas,* Second Edition. Springer, New York.

Patton, A. J. (2012). A review of copula models for economic time series. *Journal of Multivariate Analysis,* 110:4–18.

Rémillard, B. and Scaillet, O. (2009). Testing for equality between two copulas. *Journal of Multivariate Analysis,* 100:377–386.

Rüschendorf, L. (1976). Asymptotic distributions of multivariate rank order statistics. *The Annals of Statistics,* 4:912–923.

Salmon, F. (2012). The formula that killed Wall Street. *Significance,* 9(1):16–20.

Salvadori, G., De Michele, C., Kottegoda, N. T., and Rosso, R. (2007). *Extremes in Nature: An Approach Using Copulas.* Springer, Dordrecht.

Schweizer, B. and Wolff, E. F. (1981). On nonparametric measures of dependence for random variables. *The Annals of Statistics,* 9:879–885.

Shapiro, S. S. and Wilk, M. B. (1965). An analysis of variance test for normality: Complete samples. *Biometrika,* 52:591–611.

Shih, J. H. and Louis, T. A. (1995). Inferences on the association parameter in copula models for bivariate survival data. *Biometrics,* 51:1384–1399.

Sklar, A. (1959). Fonctions de répartition à n dimensions et leurs marges. *Publications de l'Institut de statistique de l'Université de Paris,* 8:229–231.

Spearman, C. E. (1904). The proof and measurement of association between two things. *American Journal of Psychology,* 15:72–101.

5

Lasso and Sparsity in Statistics

Robert J. Tibshirani

Stanford University, Stanford, CA

In this chapter, I discuss the lasso and sparsity, in the area of supervised learning that has been the focus of my research and that of many other statisticians. This area can be described as follows. Many statistical problems involve modeling important variables or outcomes as functions of predictor variables. One objective is to produce models that allow predictions of outcomes that are as accurate as possible. Another is to develop an understanding of which variables in a set of potential predictors are strongly related to the outcome variable. For example, the outcome of interest might be the price of a company's stock in a week's time, and potential predictor variables might include information about the company, the sector of the economy it operates in, recent fluctuations in the stock's price, and other economic factors. With technological development and the advent of huge amounts of data, we are frequently faced with very large numbers of potential predictor variables.

5.1 Sparsity, ℓ_1 Penalties and the Lasso

The most basic statistical method for what is called supervised learning relates an outcome variable Y to a linear predictor variables $x_1, \ldots, x_p$, viz.

$$Y = \beta_0 + \sum_{j=1}^{p} x_j \beta_j + \epsilon, \tag{5.1}$$

where ϵ is an error term that represents the fact that knowing $x_1, \ldots, x_p$ does not normally tell us exactly what Y will be. We often refer to the right-hand side of (5.1), minus ϵ, as the predicted outcome. These are referred to as linear regression models. If we have data on the outcome y_i and the predictor variables x_{ij} for each in a group of N individuals or scenarios, the method of least squares chooses a model by minimizing the sum of squared errors between the outcome and the predicted outcome, over the parameters (or regression coefficients) $\beta_0, \ldots, \beta_p$.

Linear regression is one of the oldest and most useful tools for data analysis. It provides a simple yet powerful method for modeling the effect of a set of predictors (or features) on an outcome variable. With a moderate or large number of predictors, we don't typically want to include all the predictors in the model. Hence one major challenge in regression is variable selection: choosing the most informative predictors to use in the model. Traditional variable selection methods search through all combinations of predictors and take too long to compute when the number of predictors is roughly larger than 30; see, e.g., Chapter 3 of Hastie et al. (2008) for details.

Penalized regression methods facilitate the application of linear regression to large problems with many predictors. The lasso uses ℓ_1 or absolute value penalties for penalized regression. In particular, it provides a powerful method for doing variable selection with a large number of predictors. In the end it delivers a sparse solution, i.e., a set of estimated regression coefficients in which only a small number are non-zero. Sparsity is important both for predictive accuracy, and for interpretation of the final model.

Given a linear regression with predictors x_{ij} and response values y_i for $i = 1, \ldots, N$ and $j = 1, \ldots, p$, the lasso solves the ℓ_1-penalized regression so as to minimize

$$\frac{1}{2} \sum_{i=1}^{N} \left(y_i - \beta_0 - \sum_{j=1}^{p} x_{ij}\beta_j \right)^2 + \lambda \sum_{j=1}^{p} |\beta_j|, \qquad (5.2)$$

for the unknown parameters $\beta_0, \ldots, \beta_p$. The second term above is called a penalty function; it balances the fit of the model with its complexity. The non-negative parameter λ governs that tradeoff. The larger λ, the more sparse the final solution vector $\hat{\beta}$. The statistician chooses the value of λ, or uses a method like cross-validation, to estimate it.

The lasso problem (5.2) is equivalent to minimizing the sum of squares with constraint

$$\sum_{j=1}^{p} |\beta_j| \leq s.$$

For every λ in (5.2), there is a bound parameter s yielding the same solution. Note that choosing $\lambda = 0$ or equivalently a sufficiently large value of s, yields the usual least squares solution. Lasso regression is similar to ridge regression, which has constraint

$$\sum_{j=1}^{p} \beta_j^2 \leq s.$$

Because of the form of the ℓ_1 penalty, as shown in Figure 5.1, the lasso does variable selection and shrinkage, while ridge regression, in contrast, only

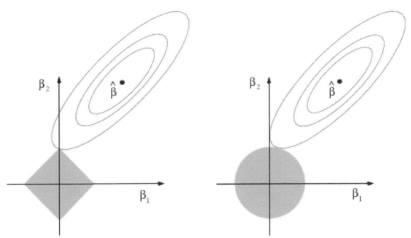

FIGURE 5.1: Estimation picture for the lasso (left) and ridge regression (right). Shown are contours of the error and constraint functions. The solid gray areas are the constraint regions $|\beta_1| + |\beta_2| \leq t$ and $\beta_1^2 + \beta_2^2 \leq t^2$, respectively, while the ellipses are the contours of the least squares error function, centered at the full least squares estimates $\hat{\beta}$. The sharp corners of the constraint region for the lasso regression yield sparse solutions. In high dimensions, sparsity arises from corners and edges of the constraint region.

shrinks. If we consider a more general penalty of the form

$$\left(\sum_{j=1}^{p} \beta_j^q \right)^{1/q},$$

then the lasso uses $q = 1$ and ridge regression has $q = 2$. Subset selection emerges as $q \to 0$, and the lasso corresponds to the smallest value of q (i.e., closest to subset selection) that yields a convex problem. [A convex problem is an optimization of a convex function over a convex set. If a function is strictly convex, the problem is guaranteed to have a unique global solution.]

Figure 5.1 gives a geometric view of the lasso and ridge regression. Figure 5.2 shows an example. The outcome is the value of the log PSA (prostate-specific antigen) for men whose prostate was removed during cancer surgery, modeled as a function of eight measurements such as age, cancer volume, tumor weight, etc. The figure shows the profiles of the lasso coefficients as the shrinkage factor s is varied. This factor is the bound on the total norm $|\hat{\beta}_1| + \cdots + |\hat{\beta}_p|$, and we have scaled it to lie between 0 and 1 for interpretability. The vertical dotted line is the value of s chosen by cross-validation: it yields a model with just three nonzero coefficients, lcavol, svi, and lweight. Recall that s is in one-to-one correspondence to the tuning parameter λ in (5.2): thus λ

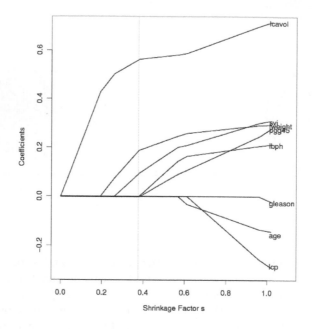

FIGURE 5.2: Profiles of the lasso coefficients for the prostate cancer example.

is large on the left of the plot forcing all estimates to be zero, and is zero on the right, yielding the least squares estimates.

5.2 Some Background

Lasso regression and ℓ_1 penalization have been the focus of a great deal of work in recent years. Table 5.1, adapted from Tibshirani (2011), gives examples of some this work.

My original lasso paper was motivated by an idea of Leo Breiman's called the garotte (Breiman, 1995). The garotte chooses $c_1, \ldots, c_p$ to minimize

$$\sum_{i=1}^{N} \left(y_i - \beta_0 - \sum_{j=1}^{p} c_j x_{ij} \hat{\beta}_j \right)^2 + \lambda \sum_{j=1}^{p} c_j, \tag{5.3}$$

where $\hat{\beta}_1, \ldots, \hat{\beta}_p$ are the usual least squares estimates and $c_j \geq 0$ for all $j \in \{1, \ldots, p\}$. Thus Leo's idea was to scale the least squares estimates by

TABLE 5.1: Some examples of generalizations of the lasso.

Method	Authors
Adaptive lasso	Zou (2006)
Compressive sensing	Donoho (2004), Candès (2006)
Dantzig selector	Candès and Tao (2007)
Elastic net	Zou and Hastie (2005)
Fused lasso	Tibshirani et al. (2005)
Generalized lasso	Tibshirani and Taylor (2011)
Graphical lasso	Yuan and Lin (2007b), Friedman et al. (2007)
Grouped lasso	Yuan and Lin (2007a)
Hierarchical interaction models	Bien et al. (2013)
Matrix completion	Candès and Tao (2009), Mazumder et al. (2010)
Multivariate methods	Joliffe et al. (2003), Witten et al. (2009)
Near-Isotonic regression	Tibshirani et al. (2011)

nonnegative constants, some of which might be zero. I noticed that the garotte wouldn't be defined for $p > N$, since the least squares estimates are not defined in that case. Hence I just simplified the method by removing the "middle man."

Not surprisingly, it turned out that using absolute value constraints in regression was not a completely new idea at the time. Around the same time, Chen, Donoho and Saunders proposed "Basis Pursuit" (Chen et al., 1998), which used absolute value constraints for signal processing. Earlier, Frank and Friedman (1993) had (briefly) discussed the "bridge" estimate, which proposed a family of penalties of the form $\sum |\beta_p|^q$ for some q.

The lasso paper was published in 1996, but did not get much attention at the time. This may have been in part the due to the relatively limited computational power that was available to the average statistician, and also the relatively small size of datasets (compared to today). Now the lasso and ℓ_1 constraint-based methods are a hot area, not only in statistics but in machine learning, engineering and computer science.

The original motivation for the lasso was for interpretability: it is an alternative to subset regression for obtaining a sparse (or parsimonious) model. In the past 20 years some unforeseen advantages of convex ℓ_1-penalized approaches emerged: statistical and computational efficiency.

On the statistical side, there has also been a great deal of interesting work on the mathematical aspects of the lasso, examining its ability recover a true underlying (sparse) model and to produce a model with minimal prediction error. Many researchers have contributed to this work, including Peter Bühlmann, Emmanuel Candès, David Donoho, Eitan Greenshtein, Iain Johnstone, Nicolai Meinshausen, Ya'acov Ritov, Martin Wainwright, Bin Yu, and many others. In describing some of this work, Hastie et al. (2001) coined the informal "Bet on Sparsity" principle. ℓ_1 methods assume that the truth is sparse, in some basis. If the assumption holds true, then the parameters can be efficiently estimated using ℓ_1 penalties. If the assumption does not hold

— so that the truth is dense — then no method will be able to recover the underlying model without a large amount of data per parameter. This is typically not the case when the number of predictors, p, is much larger than the sample size, N, a commonly occurring scenario.

On the computational side, the convexity of the problem and sparsity of the final solution can be used to great advantage. Parameters whose estimates are zero in the solution can be handled with minimal cost in the search. Powerful and scalable techniques for convex optimization can be applied to the problem, allowing the solution of very large problems. One particularly effective approach is coordinate descent (Fu, 1998; Friedman et al., 2007, 2010), a simple one-at-a-time method that is well-suited to the separable lasso penalty. This method is simple and flexible, and can also be applied to many other ℓ_1- penalized generalized linear models, including multinomial, Poisson and Cox's proportional hazards model for survival data. Coordinate descent is implemented in the **glmnet** package in the R statistical language, written by Jerome Friedman, Trevor Hastie, Noah Simon and myself.

Here is the basic idea of coordinate descent. Suppose we had only one predictor and wished to solve the lasso problem, i.e., minimize

$$\sum_{i=1}^{N}(y_i - x_i\beta)^2 + \lambda|\beta|.$$

Then the solution is easily shown to be the soft-thresholded estimate

$$\mathrm{sign}(\hat{\beta})(|\hat{\beta}| - \lambda)_+,$$

where $\hat{\beta}$ is usual least squares estimate, and the $+$ indicates positive part. The idea then, with multiple predictors, is to cycle through each predictor in turn, solving for the each estimate, using this method. We compute residuals

$$r_i = y_i - \sum_{j \neq k} x_{ij}\hat{\beta}_k$$

and apply univariate soft-thresholding, pretending that our data is (x_{ij}, r_i). We cycle through the predictors $j = 1, \ldots, p$ several times until convergence. Coordinate descent is like skiing to the bottom of a hill: but rather than pointing your skis towards the bottom of the hill, you go as far down as you can in the north-south direction, then east-west, then north-south, etc., until you (hopefully) reach the bottom.

5.3   A History for Coordinate Descent for the Lasso

This history is interesting, and shows the haphazard way in which science sometimes progresses. In 1997 Weijiang Fu, a graduate student of mine at the

University of Toronto wrote his thesis on lasso-related topics, and proposed the "shooting method" for computation of the lasso estimates. I read and signed it, but in retrospect apparently didn't understand it very well. And I basically forgot about the work, which was later published. Then by 2002 I had moved to Stanford and Ingrid Daubechies — an applied mathematician at Stanford — discussed a theorem about a coordinate-wise method for computing solutions to convex problems. Trevor Hastie and I went to the talk, took notes, and then programmed the proposed method ourselves in the S language. We made a mistake in the implementation: trying to exploit the efficient vector operations in Splus, we changed each parameter not one-at-a-time, but at the end of each loop of p updates. This turned out to be a fatal mistake, as the method did not even converge and so we just "wrote it off."

Then in 2006 our colleague Jerry Friedman was the external examiner at the PhD oral of Anita van der Kooij (in Leiden) who used the coordinate descent idea for the elastic net, a generalization of the lasso. Friedman showed us the idea and together we wondered whether this method would work for the lasso. Jerome, Trevor and I started working on this problem, and using some clever implementation ideas by Friedman, we produced some very fast code (`glmnet` in the R language). It was only then that I realized that Weijiang Fu had the same basic idea almost 10 years earlier! Now coordinate descent is a considered a state-of-the-art method for the lasso — one of the best methods around — and remarkable in its simplicity.

For a long time, even the convex optimization community did not seem to take coordinate descent very seriously. For example, my good friend Stephen Boyd's standard book on the topic (Boyd and Vandenberghe, 2004) does not even mention it. The only work I could find on the coordinate descent for convex problems was that of Paul Tseng, a Canadian at the University of Washington who proved (in the 1980s) some beautiful results showing the convergence a coordinate descent for separable problems (Tseng, 1988). These include the lasso, as a special case. When the problem is not separable, coordinate descent may not converge: this may explain the lack of interest in the method in the convex optimization world.

I never met Paul, but we corresponded by email and he was happy that his work was proving to be so important for the lasso. Sadly, in 2009 he went missing while kayaking in the Yangtze River in China and is now presumed dead. His seminal work provides the underpinning for the application of coordinate descent in the lasso and many related problems.

5.4   An Example in Medicine

I am currently working on a cancer diagnosis project with coworkers at Stanford. They have collected samples of tissue from a number of patients undergo-

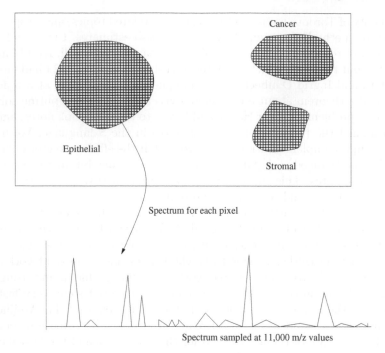

FIGURE 5.3: Schematic of the cancer diagnosis problem. Each pixel in each of the three regions labeled by the pathologist is analyzed by mass spectrometry. This gives a feature vector of 11,000 intensities for each pixel (bottom panel), from which we try to predict the class of that pixel.

ing surgery for cancer. We are working to build a classifier that can distinguish three kinds of tissue: normal epithelial, stromal, and cancer. Such a classifier could be used to assist surgeons in determining, in real time, whether they had successfully removed all of the tumor. It could also yield insights into the cancer process itself. The data are in the form of images, as sketched in Figure 5.3. A pathologist has labeled each region (and hence the pixels inside a region) as epithelial, stromal or cancer. At each pixel in the image, the intensity of metabolites is measured by a type of mass spectrometry, with the peaks in the spectrum representing different metabolites. The spectrum has been finely sampled, with the intensity measured at about $11,000$ sites (frequencies) across the spectrum. Thus, the task is to build a prediction model to classify each pixel into one of the three classes, based on the $11,000$ features. There are about 8000 pixels in all.

For this problem, I have applied an ℓ_1-regularized multinomial model. A multinomial model is one which predicts whether a tissue sample (pixel) is of type 1(epithelial), 2 (stromal) or 3 (cancer). For each class $k \in \{1, 2, 3\}$, the model has a vector of parameters $\beta_{1k}, \ldots, \beta_{pk}$, representing the weight given to each feature in that class. I used the `glmnet` package for fitting the model: it computes the entire path of solutions for all values of the regularization parameter λ, using cross-validation to estimate the best value of λ (I left one patient out at a time). The entire computation required just a few minutes on a standard Linux server.

The results so far are promising. The classifier shows 90–97% accuracy in the three classes, using only around 100 features. This means that when the model is used to predict the tissue type of a pixel, it is correct 90–97% of the time. These features could yield insights about the metabolites that are important in stomach cancer. The power of this approach, both its prediction accuracy and interpretability, are not shared by competing methods such as support vector machines or decision trees. For example, this method is based on the multinomial probability model and so we obtain not only class predictions but estimated probabilities for each class (unlike support vector machines). Thus for example we can create a "I don't know" category, and assign a pixel to that category if the gap between the two largest class probabilities in small (say 10%). There is much more work to be done — collecting more data, and refining and testing the model on more difficult cases. But this shows the potential of ℓ_1-penalized models in an important and challenging scientific problem.

5.5 Nearly Isotonic Regression

Another recent example of the use of ℓ_1 constraints is nearly isotonic regression (Tibshirani et al., 2011). Unlike the regression problem, here we have no predictors but just a sequence of outcome values $y_1, \ldots, y_N$ which we wish to approximate. Given this sequence, the method of isotonic regression solves

$$\text{minimize} \sum (y_i - \hat{y}_i)^2 \quad \text{subject to } \hat{y}_1 \leq \hat{y}_2 \leq \cdots$$

This assumes a monotone non-decreasing approximation, with an analogous definition for the monotone non-increasing case. The solution can be computed via the well-known Pool Adjacent Violators (PAVA) algorithm; see, e.g., Barlow et al. (1972). In nearly isotonic regression we solve

$$\text{minimize} \frac{1}{2} \sum_{i=1}^{N} (y_i - \beta_i)^2 + \lambda \sum_{i=1}^{n-1} (\beta_i - \beta_{i+1})_+,$$

with x_+ indicating the positive part, $x_+ = x\, \mathbf{1}(x > 0)$. The solutions $\hat{\beta}_i$ are the values $\hat{y}_i$ that we seek. This is a convex problem; with $\hat{\beta}_i = y_i$ at $\lambda = 0$ and culminating in the usual isotonic regression as $\lambda \to \infty$. Along the way it gives nearly monotone approximations. A toy example is given in Figure 5.4.

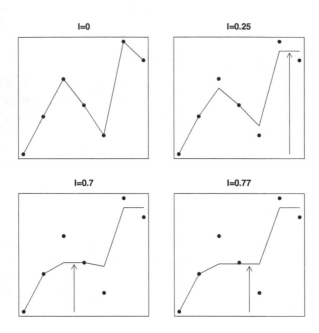

FIGURE 5.4: Illustration of nearly isotonic fits for a toy example. An interpolating function is shown in the top left panel. There are three joining events (indicated by the arrows) shown in the remaining panels, with the usual isotonic regression appearing in the bottom right panel.

Note that $(\beta_i - \beta_{i+1})_+$ is "half" of an ℓ_1 penalty on differences, penalizing dips but not increases in the sequence. This procedure allows one to assess the assumption of monotonicity by comparing nearly monotone approximations to the best monotone approximation. Tibshirani et al. (2011) provide a simple algorithm that computes the entire path of solutions, a kind of modified version of the PAVA procedure. They also show that the number of degrees of freedom is the number of unique values of $\hat{y}_i$ in the solution, using results from Tibshirani and Taylor (2010).

This kind of approach can be extended to higher order differences, and is also known as ℓ_1-trend filtering (Kim et al., 2009). For example a second-order difference penalty (without the positive part) yields a piecewise linear function estimate.

5.6 Conclusion

In this chapter I hope that I have conveyed my excitement for some recent developments in statistics, both in its theory and practice. These methods are already widely used in many areas, including business, finance and numerous scientific areas. The area of medical imaging may be greatly advanced by the advent of compressed sensing, a clever method based on ℓ_1 penalties (Candès and Tao, 2005; Donoho, 2006). I predict that sparsity and convex optimization will play an increasingly important role in the development of statistical methodology and in the applications of statistical methods to challenging problems in science and industry.

One particularly promising area is that of inference, where the covariance test recently proposed by Lockhart et al. (2014) provides a simple way to assess the significance of a predictor, while accounting for the adaptive nature of the fitting. In essence, the exponential distribution that arises in this new work is the analog of the usual chi-squared for the F-test for fixed (non-adaptive) regression. It appears that this new test will have broad applications in other problems such as principal components, clustering and graphical models. See Tibshirani (2014) for a brief overview.

About the Author

Robert J. Tibshirani is a professor of statistics, health research and policy at Stanford University; he was affiliated to the University of Toronto from 1985 to 1998. He received a BMath from the University of Waterloo, an MSc from the University of Toronto and a PhD from Stanford. His research interests include statistical theory, statistical learning, and a broad range of scientific areas. He received a Steacie Award, the 1996 COPSS Award, the 2000 CRM–SSC Award, and the 2012 SSC Gold Medal for research. He was elected to the Royal Society of Canada in 2001; he is a fellow of the American Statistical Association and the Institute of Mathematical Statistics.

Bibliography

Barlow, R. E., Bartholomew, D. J., Bremner, J. M., and Brunk, H. D. (1972). *Statistical Inference Under Order Restrictions: The Theory and Application of Isotonic Regression*. Wiley, New York.

Bien, J., Taylor, J., and Tibshirani, R. J. (2013). A lasso for hierarchical interactions. *The Annals of Statistics*, 41:1111–1141.

Boyd, S. and Vandenberghe, L. (2004). *Convex Optimization*. Cambridge University Press, Cambridge.

Breiman, L. (1995). Better subset selection using the non-negative garotte. *Technometrics*, 37:738–754.

Candès, E. J. (2006). Compressive sampling. In *Proceedings of the International Congress of Mathematicians, Madrid, Spain*.

Candès, E. J. and Tao, T. (2005). Decoding by linear programming. *IEEE Transactions on Information Theory*, 51:4203–4215.

Candès, E. J. and Tao, T. (2007). The Dantzig selector: Statistical estimation when p is much larger than n. *The Annals of Statistics*, 35:2313–2351.

Candès, E. J. and Tao, T. (2009). The power of convex relaxation: Near-optimal matrix completion. *IEEE Transactions on Information Theory*, 56:2053–2080.

Chen, S., Donoho, D. L., and Saunders, M. (1998). Atomic decomposition for basis pursuit. *SIAM Journal on Scientific Computing*, 20:33–61.

Donoho, D. L. (2004). *Compressed Sensing*. Technical Report, Statistics Department, Stanford University, Stanford, CA.

Donoho, D. L. (2006). Compressed sensing. *IEEE Transactions for Information Theory*, 52:1289–1306.

Frank, I. and Friedman, J. (1993). A statistical view of some chemometrics regression tools (with discussion). *Technometrics*, 35:109–148.

Friedman, J., Hastie, T., and Tibshirani, R. J. (2007). Pathwise coordinate optimization. *The Annals of Applied Statistics*, 1:302–332.

Friedman, J., Hastie, T., and Tibshirani, R. J. (2010). Regularization paths for generalized linear models via coordinate descent. *Journal of Statistical Software*, 33:Article 1.

Fu, W. J. (1998). Penalized regressions: The bridge versus the lasso. *Journal of Computational and Graphical Statistics*, 7:397–416.

Hastie, T., Tibshirani, R. J., and Friedman, J. (2001). *The Elements of Statistical Learning: Data Mining, Inference and Prediction*. Springer, New York.

Hastie, T., Tibshirani, R. J., and Friedman, J. (2008). *The Elements of Statistical Learning: Data Mining, Inference and Prediction*, Second Edition. Springer, New York.

Joliffe, I. T., Trendafilov, N. T., and Uddin, M. (2003). A modified principal component technique based on the lasso. *Journal of Computational and Graphical Statistics*, 12:531–547.

Kim, S.-J., Koh, K., Boyd, S., and Gorinevsky, D. (2009). ℓ_1 trend filtering. *SIAM Review, Problems and Techniques Section*, 51:339–360.

Lockhart, R. A., Taylor, J., Tibshirani, R. J., and Tibshirani, R. J. (2014). A significance test for the lasso (with discussion). *The Annals of Statistics*, in press.

Mazumder, R., Hastie, T., and Tibshirani, R. J. (2010). Spectral regularization algorithms for learning large incomplete matrices. *Journal of Machine Learning Research*, 11:2287–2322.

Tibshirani, R. J. (2011). Regression shrinkage and selection via the lasso: A retrospective. *Journal of the Royal Statistical Society, Series B*, 73:273–282.

Tibshirani, R. J. (2014). In praise of sparsity and convexity. In *Past, Present, and Future of Statistical Science*, pp. 497–505. Chapman & Hall, London.

Tibshirani, R. J., Hoefling, H., and Tibshirani, R. J. (2011). Nearly-isotonic regression. *Technometrics*, 53:54–61.

Tibshirani, R. J., Saunders, M., Rosset, S., Zhu, J., and Knight, K. (2005). Sparsity and smoothness via the fused lasso. *Journal of the Royal Statistical Society, Series B*, 67:91–108.

Tibshirani, R. J. and Taylor, J. (2010). *The Solution Path of the Generalized Lasso*. Technical Report, Stanford University, Stanford, CA.

Tibshirani, R. J. and Taylor, J. (2011). The solution path of the generalized lasso. *The Annals of Statistics*, 39:1335–1371.

Tseng, P. (1988). Coordinate ascent for maximizing nondifferentiable concave functions. Technical Report LIDS-P; 1840, Massachusetts Institute of Technology, Boston, MA.

Witten, D., Tibshirani, R. J., and Hastie, T. (2009). A penalized matrix decomposition, with applications to sparse principal components and canonical correlation analysis. *Biometrika*, 10:515–534.

Yuan, M. and Lin, Y. (2007a). Model selection and estimation in regression with grouped variables. *Journal of the Royal Statistical Society, Series B*, 68:49–67.

Yuan, M. and Lin, Y. (2007b). Model selection and estimation in the Gaussian graphical model. *Biometrika*, 94:19–35.

Zou, H. (2006). The adaptive lasso and its oracle properties. *Journal of the American Statistical Association*, 101:1418–1429.

Zou, H. and Hastie, T. (2005). Regularization and variable selection via the elastic net. *Journal of the Royal Statistical Society, Series B*, 67:301–320.

6

Optimizing and Adapting the Metropolis Algorithm

Jeffrey S. Rosenthal

University of Toronto, Toronto, ON

6.1 Introduction

Many modern scientific questions involve high-dimensional data and complicated statistical models. For example, data on weather consist of huge numbers of measurements across spatial grids, over a period of time. Even in simpler settings, data can be complex: for example, Bartolucci et al. (2007) consider recurrence rates for melanoma (skin cancer) patients after surgery. The probability of recurrence for an individual may depend on physical or biological characteristics of their cancerous lesion, as well as other factors.

Typically, a statistical model for high-dimensional data not only involves a large number of variables but also a correspondingly large number of parameters, which are often represented by a vector θ of some dimension d. To assess the relevance of specific variables for disease recurrence, and to build models that give a risk of recurrence for any given individual, researchers often use Bayesian analysis; see, e.g., Box and Tiao (1973), Gelman et al. (2003) and Carlin and Louis (2009). In this framework, the parameter vector is assumed to follow some probability distribution (of dimension d), and the challenge is to combine a "prior" distribution for θ (typically based on background information about the scientific area) with data that are collected, so as to produce a "posterior" distribution for θ. This probability distribution, call it $\pi(\theta)$, can then be used to answer important scientific questions (e.g., is the size of a cancerous lesion related to the risk of recurrence after surgery?) and to calculate specific probabilities (e.g., this person has a 20% probability of a recurrence within the next five years).

One challenge for Bayesian analysis in situations where the data and parameter vectors are high dimensional is that it is difficult or impossible to compute probabilities based on the posterior distribution. If there is some outcome A of interest (e.g., the outcome that a specific individual's cancer will recur), then its probability is given by an integral over A, written

$$\Pi(A) = \int_A \pi(x)\, dx. \tag{6.1}$$

Also, the expected (average) value of any particular quantity h (e.g., the size of the lesion if the cancer does recur) is also given by an integral (now over the state space $\mathcal{X}$ of all possible vectors), namely

$$E_\pi(h) = \int_{\mathcal{X}} h(x)\, \pi(x)\, dx. \tag{6.2}$$

So, to draw conclusions from a Bayesian statistical model, "all" we have to do is compute integrals like (6.1) and (6.2).

Unfortunately, integrals like (6.1) and (6.2) are sometimes very difficult to compute. For example, one commonly-used posterior density (corresponding to a "variance components model" in which individuals are divided into groups) is given by the formula

$$\pi(V, W, \mu, \theta_1, \ldots, \theta_K) = C\, e^{-b_1/V} V^{-a_1-1}$$
$$\times\, e^{-b_2/W} W^{-a_2-1} e^{-(\mu-a_3)^2/2b_3} V^{-K/2} W^{-\frac{1}{2}\sum_{i=1}^{K} J_i}$$
$$\times \exp\left\{ -\sum_{i=1}^{K}(\theta_i - \mu)^2/2V - \sum_{i=1}^{K}\sum_{j=1}^{J_i}(Y_{ij} - \theta_i)^2/2W \right\}.$$

Here K is the number of different groups (i.e., the number of different values θ_i), V, W and μ are additional unknown parameters of the model, while the a_i, b_i and Y_{ij} are known constants. A typical application might have, say, $K = 19$, so that π is a 22-dimensional function. In such cases, direct computation of integrals like (6.1) and (6.2), using calculus tricks or numerical integration or anything else, appears to be impossible, and alternative methods must be sought (Evans and Swartz, 2000). What can be done to compute estimates of quantities like $E_\pi(h)$ in such complicated cases?

6.2   Monte Carlo Algorithms

One answer to this question is provided by Monte Carlo algorithms. These algorithms, named after the famous casino in Monaco, use randomness to estimate quantities like (6.1) and (6.2). Surprisingly, this turns out to be very helpful!

The most basic ("classical") form of Monaco requires us to sample from π, i.e., generate a sequence of independent d-dimensional random vectors (variables) $X_1, \ldots, X_M$ which each follow the density π, i.e., each have probabilities given by $\Pr(X_i \in A) = \int_A \pi(x)\,\mathrm{d}x$ for all (measurable) $A \subset \mathcal{X}$. We can then use this random sample to estimate quantities like (6.2), by

$$\mathrm{E}_\pi(h) \approx \frac{1}{M} \sum_{i=1}^{M} h(X_i). \tag{6.3}$$

If the Monte Carlo sample size M is sufficiently large, then by the Law of Large Numbers the estimate (6.3) will be close to the true expected value (6.2).

Classical Monte Carlo can be very effective, and is widely used for lots of different applications. However, in many cases (such as the above variance components example), it is infeasible to directly sample from π in this sense, i.e., there is no known way to run a computer program which will produce the required sequence of vectors X_i. This state of affairs presented a serious limitation to the use of Monte Carlo algorithms in Bayesian statistical inference problems, until it was largely solved with the introduction of Markov chain Monte Carlo algorithms, as we now discuss.

6.3 Markov Chain Monte Carlo (MCMC)

Markov chain Monte Carlo (MCMC) algorithms were first developed in the statistical physics community by Metropolis et al. (1953), and later expanded by statisticians such as Hastings (1970) before being introduced into the wider statistical community by Gelfand and Smith (1990). They do not use or require a sequence of independent vectors X_i as above. Instead, they define a dependent sequence, more precisely a Markov chain, with each new vector X_{i+1} constructed using the previously-constructed vector X_i.

Suppose we can define a sequence of vectors $X_0, X_1, \ldots$, where each X_{i+1} is constructed using the previous X_i, such that for large n, X_n is approximately a sample (i.e., observation) from π. That is, for large n,

$$\Pr(X_n \in A) \approx \int_A \pi(x)\,\mathrm{d}x$$

for all $A \subset \mathcal{X}$. This situation is worse than for classical Monte Carlo as in (6.3), since it is just approximate, and furthermore the random vectors X_n are no longer independent but rather are each constructed sequentially using the previous vector. Nevertheless, it is sometimes true that if M is sufficiently large, then we can still approximate $\mathrm{E}_\pi(h)$ by

$$\mathrm{E}_\pi(h) \approx \frac{1}{M - B} \sum_{i=B+1}^{M} h(X_i). \tag{6.4}$$

That is, we still average many values together, similar to (6.3), even though they are no longer independent. Also, by convention we drop the first B "burn-in" observations since they might be too heavily influenced by our initial values X_0 and hence bias our estimate.

So when is an approximation like (6.4) valid? It turns out that this approximation holds as $M \to \infty$ provided that the Markov chain is ergodic for π, i.e., that the chain's probabilities converge to those of π in the sense that, for all $A \subset \mathcal{X}$,

$$\lim_{n \to \infty} \Pr(X_n \in A) = \int_A \pi(x)\,\mathrm{d}x.$$

This raises the question of how to define a computationally simple Markov chain (X_n) which guarantees that, as $n \to \infty$,

$$\Pr(X_n \in A) \to \int_A \pi(x)\,\mathrm{d}x.$$

By standard Markov chain theory, this will hold if the updating rules for (X_n) are irreducible (i.e., the chain can eventually get to anywhere in $\mathcal{X}$), and aperiodic (i.e., the chain doesn't have any forced cyclical behavior), and leave the density π stationary (i.e., if it starts distributed according to π, then it will remain distributed according to π at all future times too). How can we ensure that those conditions hold?

6.4   Metropolis Algorithm

An answer to this question was developed in the physics community over fifty years ago (Metropolis et al., 1953). Specifically, given a (possibly important and complicated and high-dimensional) target density π on some state space $\mathcal{X} \subset \mathbb{R}^d$ (with $\pi(x) = 0$ for $x \in \mathbb{R}^d \setminus \mathcal{X}$), the original Metropolis algorithm proceeds as follows.

First, choose a symmetric d-dimensional increment distribution Q; the most common choice is $Q = \mathcal{N}(0, \Sigma)$ for some fixed covariance matrix Σ like $\Sigma = c\,I_d$. Here $\mathcal{N}(0, \Sigma)$ is a d-dimensional Normal (Gaussian) distribution, and I_d is the $d \times d$ identity matrix.

Also, choose some initial vector X_0. Then, iteratively for $n = 1, 2, \ldots$, compute X_n from X_{n-1} as follows:

1. Let $Y_n = X_{n-1} + Z_n$, where $(Z_n) \sim Q$ are independent and identically distributed ("proposal").

2. Let $\alpha = \pi(Y_n)/\pi(X_{n-1})$ if this ratio is smaller than 1; otherwise, set $\alpha = 1$ ("acceptance probability").

3a. With probability α, let $X_n = Y_n$ ("accept").

3b. Otherwise, with probability $1 - \alpha$, let $X_n = X_{n-1}$ ("reject").

Steps 2, 3a, and 3b can all be accomplished by drawing a random variable U_n uniformly between 0 and 1, and then setting

$$
X_n = \begin{cases} Y_n & \text{if } U_n \leq \pi(Y_n)/\pi(X_{n-1}), \\[2mm] X_{n-1} & \text{if } U_n > \pi(Y_n)/\pi(X_{n-1}). \end{cases}
$$

Intuitively, the above acceptance probabilities α are useful because they encourage the algorithm to accept more moves toward larger values of π. More precisely, the formula for α turns out to be exactly the right one to ensure that the Markov chain (X_n) leaves the density π stationary, a key property for convergence (as discussed at the end of the previous section). Furthermore, the irreducibility property will almost always hold, e.g., it is guaranteed if Q has an everywhere-positive density like $\mathcal{N}(0, \Sigma)$. And, the aperiodicity property is essentially never a problem since the algorithm will eventually reject and thus avoid cyclic behavior. So, this simple algorithm has all the right properties to guarantee that, as $n \to \infty$, $\Pr(X_n \in A) \to \pi(A)$. It follows that we can use this algorithm to estimate $E_\pi(h)$ as in (6.4). Good!

The only problem is that sometimes the Metropolis algorithm will be too inefficient, i.e., it will take far too long (i.e., require too many iterations) to provide a decent approximation to π, which is a very important consideration; see, e.g., Rosenthal (1995). In some cases, even running the algorithm for months on the world's fastest computers would not provide a remotely reasonable approximation to π. Overcoming such problems has often necessitated new and more complicated MCMC algorithms; see, e.g., Bélisle et al. (1993), Neal (2003), Jain and Neal (2004), and Hamze and de Freitas (2012). In a different direction, detecting convergence of MCMC to π is so challenging that some authors have developed perfect sampling algorithms which guarantee complete convergence at the expense of a more complicated algorithm; see, e.g., Propp and Wilson (1996), Murdoch and Green (1998) or Fill et al. (2000). However, such perfect sampling algorithms are often infeasible to run, so we do not discuss them further here.

All of this raises the question of how to improve or optimize the speed of convergence of the Metropolis algorithm, for example by modifying the increment distribution Q, as we discuss next.

6.5 Goldilocks Principle

To illustrate the Metropolis optimization issues, consider the very simple case where $\pi = \mathcal{N}(0, 1)$, i.e., where the target density is just the standard Normal

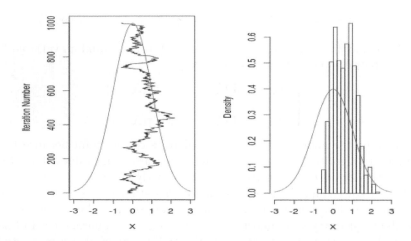

FIGURE 6.1: The trace plot (left) and histogram (right), along with the mound-shaped target density for a one-dimensional Metropolis algorithm with too small a proposal scaling σ, showing slow mixing and poor convergence.

distribution. (Of course we wouldn't actually need to use MCMC in such a simple case.) Assume that the proposal distribution is given by $Q = \mathcal{N}(0, \sigma^2)$. Our question of interest is, how should we choose σ?

As a first try, let's choose a small value of σ, say $\sigma = .1$, and run the Metropolis algorithm for 1000 iterations with that σ. The corresponding trace plot, graphing the values of the Markov chain (horizontal axis) at each iteration n (vertical axis), is shown in Figure 6.1 (left panel). Looking at this trace plot, we can see that the chain moves very slowly. It starts at the value zero, and takes many hundreds of iterations before it moves appreciably away from zero. In particular, it does not do a very good job of exploring the superposed mound-shaped target density. This is also illustrated by the histogram of those 1000 iterations, in Figure 6.1 (right panel), which does not match up very well with the target density.

As a second try, let's choose a large value of σ, say $\sigma = 25$, and again run the algorithm for 1000 iterations. The trace plot in this case is shown in Figure 6.2 (left panel). In this case, when the chain finally accepts a move, it jumps quite far, which is good. However, since it proposes such large moves, it hardly ever accepts them. (Indeed, it accepted just 5.4% of the proposed moves, compared to 97.7% when $\sigma = .1$.) So, this chain doesn't perform very well either, as illustrated by the histogram in Figure 6.2 (right panel), which again does not match up very well with the target density.

As a third try, let's choose a compromise value of σ, say $\sigma = 2.38$, and again run the algorithm for 1000 iterations. In this case, the chain performs very well. It accepts a medium fraction of its proposals (44.5%), and moves reasonably

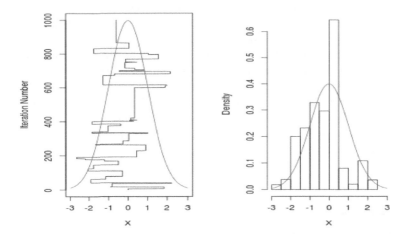

FIGURE 6.2: The trace plot (left) and histogram (right), along with the mound-shaped target density for a one-dimensional Metropolis algorithm with too large a proposal scaling σ, again showing slow mixing and poor convergence.

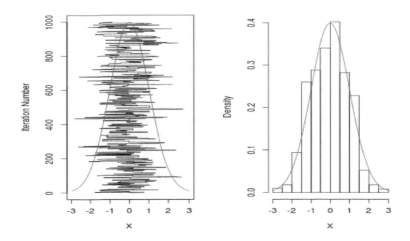

FIGURE 6.3: The trace plot (left) and histogram (right), along with the mound-shaped target density for a one-dimensional Metropolis algorithm with a good choice of proposal scaling σ, showing much better mixing and convergence properties.

far when it does accept. It thus explores the target density efficiently and well, as illustrated by the trace plot in Figure 6.3 (left panel). Furthermore it now provides fairly good samples from the target distribution, as illustrated by the histogram in Figure 6.3 (right panel).

We learn from this that it is best to choose values of the proposal increment scaling σ which are between the two extremes, i.e., not too small and not too big, but rather "just right" (as the little girl Goldilocks says in the classic children's fairy tale *The Three Bears*; http://w8r.com/the-colorful-story-book/the-three-bears). Correspondingly, the acceptance rate (i.e., the percentage of proposed moves which are accepted) should be far from 0% but also far from 100%.

6.6 Optimal Scaling

The above intuition was made more precise in a pioneering paper by Roberts et al. (1997). They decided to consider a Metropolis algorithm (X_n) in dimension d, with increment distribution

$$Q = \mathcal{N}\left(0, \frac{\ell^2}{d} I_d\right)$$

for some fixed scaling constant $\ell > 0$, and take the limit as $d \to \infty$. The beauty of their approach was that they could compute the speed of the algorithm in this limit, as an explicit (but messy and uncomputable) function $h(\ell)$ of the scaling constant ℓ. They then argued that the best choice of ℓ is the one which maximizes the limiting speed $h(\ell)$. This provided a clear standard for how to optimize the Metropolis algorithm.

The story gets even better. Roberts and his collaborators also considered the asymptotic acceptance rate, i.e., the fraction of Metropolis proposals that would be accepted in the limit as $d \to \infty$, and they computed an explicit function $A(\ell)$ for this as well. They then showed that the limiting speed $h(\ell)$ has a simple relation to the asymptotic acceptance rate $A(\ell)$. This in turn allowed them to compute that if ℓ_{opt} is the value of ℓ which maximizes the speed $h(\ell)$, then $A(\ell_{\text{opt}}) \approx .234 = 23.4\%$, a specific number that does not depend on any unknown quantities about π or anything else. This means that, at least under their (strong) assumptions, the optimal acceptance probability is 23.4%, which leads to the fastest limiting speed regardless of the target density π.

This provides a clear, simple rule for tuning Metropolis algorithms: adjust the proposal scaling ℓ so that the acceptance rate is roughly 23.4%. This rule appears to be quite robust, i.e., .234 is often a nearly optimal acceptance rate even if the theorem's formal assumptions are not satisfied. It has been implemented in numerous applied papers and software, including the hugely popular `WinBUGS` computer package (Lunn et al., 2000).

A number of authors have attempted to weaken and generalize the original strong assumptions; see e.g., Bédard (2007, 2008), Bédard and Rosenthal (2008), Beskos et al. (2009), and Sherlock and Roberts (2009). Corresponding results have been developed for Langevin MCMC algorithms (Roberts and Rosenthal, 1998), and for simulated tempering algorithms (Atchadé et al., 2011; Roberts and Rosenthal, 2013). It is also known that any acceptance rate between about 15% and 50% is still reasonably efficient (though 23.4% is best); see, e.g., Figure 3 in Roberts and Rosenthal (2001). Over all, this school of research has been very influential in guiding both applied usage and theoretical investigations for MCMC.

6.7   Proposal Shape

Despite the .234 rule's great success, an algorithm's acceptance rate is just a single scalar quantity which does not completely control the algorithm's efficiency. To illustrate this, consider the 20-dimensional target density $\pi = \mathcal{N}(0, \Sigma_*)$, where Σ_* is a 20-dimensional covariance matrix generated randomly as $\Sigma_* = M^\top M$ with M being a 20×20 matrix consisting of independent and identically distributed $\mathcal{N}(0, 1)$ entries, which shall remain fixed throughout the remainder of this chapter, and $M^\top$ is the transpose of M. We shall try running Metropolis algorithms on this density.

Based on the discussion in Section 6.6, we use an increment distribution of the form $Q = \mathcal{N}(0, \sigma^2 I_{20})$, where we try to adjust σ so that the resulting acceptance rate is approximately .234. After some experimenting, we take $\sigma = .5$, leading to an acceptance rate of .228 (close enough). Figure 6.4 shows the trace plot (left panel) and histogram (right panel) of the first coordinate of this run. The mixing and convergence aren't so bad, given the relatively large dimension, but they aren't great either. Indeed, the performance appears to be similar to our earlier one-dimensional example's first attempt with $\sigma = .1$: it only slowly explores the support of π.

Next, we instead try the increment distribution

$$Q = \mathcal{N}\left(0, \frac{(2.38)^2}{20} \Sigma_*\right),$$

which we shall justify later. This leads to an acceptance rate of .252 (again, close enough). Figure 6.5 shows the trace plot (left panel) and histogram (right panel) of the first coordinate of this run. Direct inspection (as well as more precise measurements like squared jumping distance and functional variance, not discussed here) indicate that this choice of Q, despite leading to a very similar acceptance rate (approximately .234), actually performs much

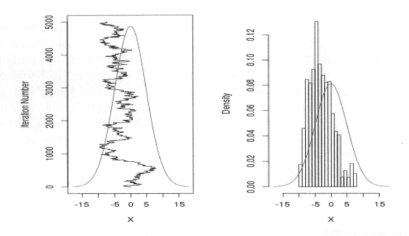

FIGURE 6.4: The trace plot (left) and histogram (right), along with the mound-shaped target density of the first coordinate of a 20-dimensional Metropolis algorithm with proposal scaling proportional to the identity matrix, showing slow mixing and poor convergence.

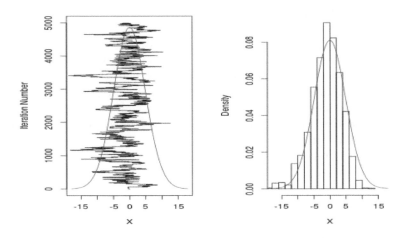

FIGURE 6.5: The trace plot (left) and histogram (right), along with the mound-shaped target density of the first coordinate of a 20-dimensional Metropolis algorithm with proposal scaling proportional to the target covariance matrix Σ_*, showing much faster mixing and much better convergence properties.

better, exploring the support of π and converging to the correct probabilities much more efficiently. This confirms that there is more to the story than just acceptance rate, and indeed that the shape of the proposal distribution (determined by the proposal covariance matrix, in this case $(2.38)^2\Sigma_*/20$) is also very important.

This concept was formalized by Roberts and Rosenthal (2001). They proved that under strong assumptions (similar to before), the optimal Gaussian proposal distribution is given (to three significant figures) by

$$Q = \mathcal{N}\left(0, \frac{(2.38)^2}{d}\Sigma_\pi\right),$$

(6.5)

where Σ_π is the $d \times d$ covariance matrix of the target density π, rather than $Q = \mathcal{N}(0, \sigma^2 I_d)$ for some σ^2. Furthermore, with this choice, the asymptotic acceptance rate will again be approximately .234. And, as before, this result appears to be robust in the sense of being nearly optimal even when the strong assumptions do not hold.

6.8 Adaptive MCMC

The optimization result of the previous section requires us to know and use the target covariance matrix Σ_π. Now, in most realistic situations, Σ_π would not be known in advance, and indeed would be at least as difficult to estimate as the quantity $E_\pi(h)$ of ultimate interest. How can we optimize the Metropolis algorithm in such situations?

In a pioneering paper, Haario et al. (2001) proposed to optimize the algorithm adaptively. That is, even if we don't know the optimal algorithm at the beginning, the computer can learn it during the run, and update the algorithm "on the fly" as it proceeds.

In its simplest form, this algorithm finds an approximately optimal increment distribution by replacing the unknown target $d \times d$ covariance matrix Σ_π by the sample $d \times d$ covariance $\Sigma_n = \text{cov}(X_0, \dots, X_{n-1})$ of the vectors visited so far during the run. If those X_i are indeed good approximate samples from π, then Σ_n will be a good approximation to Σ_π, and hence

$$Q = \mathcal{N}\left(0, \frac{(2.38)^2}{d}\Sigma_n\right)$$

can be used (after an initial phase, e.g., for $n \geq 40$ only) as a good approximation to the optimal proposal (6.5). If not, then the algorithm will not work well initially, but it will hopefully improve as it goes.

Now, Σ_n is easily computed, so this algorithm is quite feasible to run in practice. Running it on the above 20-dimensional target density $\pi = \mathcal{N}(0, \Sigma_*)$,

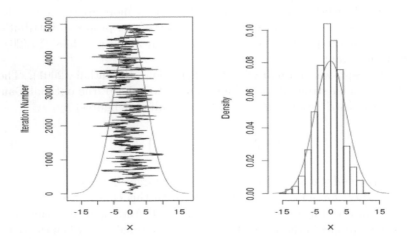

FIGURE 6.6: The trace plot (left) and histogram (right), along with the mound-shaped target density of the first coordinate of a 20-dimensional adaptive Metropolis algorithm with proposal scaling computed from previous iterations, showing fairly rapid mixing and good convergence properties, especially in later iterations (top of left plot).

the resulting trace plot and histogram of the first coordinate are shown in Figure 6.6. Direct inspection, and precise measurements, both indicate that this algorithm performs poorly during the initial phase (bottom of left plot), but then performs very well later on (top of left plot), exploring the support of π nearly as efficiently as the optimal algorithm presented above, even though it does not require any prior knowledge about Σ_π.

Adaptive MCMC algorithms have recently been used in a number of different statistical applications, and often lead to significant speed-ups, even in hundreds of dimensions; see, e.g., Roberts and Rosenthal (2009), Craiu et al. (2009), Giordani and Kohn (2010) and Richardson et al. (2011). On the other hand, adaptive MCMC algorithms use previous iterations to determine their future transitions, so they violate the Markov property which provides the justification for conventional MCMC. This raises the question of whether adaptive MCMC algorithms are valid, i.e., whether they converge (asymptotically, at least) to the target density π.

The answer to this question is "no" in general. However it is "yes" under various conditions; see, e.g., Andrieu and Moulines (2006), Haario et al. (2001), Atchadé and Rosenthal (2005), Roberts and Rosenthal (2007), Saksman and Vihola (2010), Fort et al. (2012) and Latuszyński et al. (2013). In particular, Roberts and Rosenthal (2007) show that adaptive algorithms will still converge to the target density π provided they satisfy two fairly mild conditions: "Diminishing Adaptation" (the algorithm adapts by less and less as time goes on), and "Containment" (the chain never gets too lost, in the sense that it remains bounded in probability). Conditions such as these have been used to formally justify adaptive algorithms in many examples; see, e.g., Roberts and Rosenthal (2009) and Richardson et al. (2011). Adaptive MCMC appears to hold great promise for improving statistical computation in many application areas in the years ahead.

6.9   Summary

We summarize the points made in this chapter as follows.

a) The Metropolis algorithm is very important in many applications.

b) This algorithm sometimes runs so slowly that computations are infeasible; thus, optimization of the algorithm can be crucial.

c) The simplest optimization result is the Goldilocks Principle that the acceptance rate should be far from 0, but also far from 1.

d) A more detailed result says that the optimal acceptance rate is .234 under certain strong assumptions (though the conclusion appears to be fairly robust even when the assumptions are not satisfied).

e) Another theorem says that the optimal increment distribution is $\mathcal{N}\big(0, \ (2.38)^2 \Sigma_\pi / d\big)$, again under strong assumptions (though again with a fairly robust conclusion).

f) When certain key information is unknown (e.g., Σ_π), it may still be possible to adapt toward the optimal algorithm. Such adaption is not valid in general, but is valid under various conditions such as "Diminishing Adaptation" and "Containment." It can eventually lead to tremendous speed-ups, even in very high dimensions.

g) In short, to improve statistical computation in important applied areas, optimization and adaption may well be worth the trouble!

Acknowledgments

This work was financed in part by the Natural Sciences and Engineering Research Council of Canada. The author thanks the editors and reviewers for many detailed comments.

About the Author

Jeffrey S. Rosenthal is a professor of statistics at the University of Toronto. He received his BSc from the University of Toronto and his PhD in Mathematics from Harvard University. His research interests include Markov Chain Monte Carlo algorithms and interdisciplinary applications. He has also published two textbooks, and a general-interest book about probability. He has received the 2006 CRM–SSC Prize, the 2007 COPSS Award, and the 2013 SSC Gold Medal for research. He is a fellow of the Institute of Mathematical Statistics and was elected a fellow of the Royal Society of Canada in 2012.

Bibliography

Andrieu, C. and Moulines, É. (2006). On the ergodicity properties of some adaptive Markov Chain Monte Carlo algorithms. *The Annals of Applied Probability*, 16:1462–1505.

Atchadé, Y. F., Roberts, G. O., and Rosenthal, J. S. (2011). Towards optimal scaling of Metropolis-Coupled Markov Chain Monte Carlo. *Statistics and Computing*, 21:555–568.

Atchadé, Y. F. and Rosenthal, J. S. (2005). On adaptive Markov Chain Monte Carlo algorithms. *Bernoulli*, 11:815–828.

Bartolucci, A. A., Singh, K. P., and Bae, S. (2007). Analyzing clinical trial data via the Bayesian multiple logistic random effects model. In *MODSIM 2007 conference proceedings, Modelling and Simulation Society of Australia and New Zealand*, pp. 2861–2866.

Bédard, M. (2007). Weak convergence of Metropolis algorithms for non-i.i.d. target distributions. *The Annals of Applied Probability*, 17:1222–1244.

Bédard, M. (2008). Optimal acceptance rates for Metropolis algorithms: Moving beyond 0.234. *Stochastic Processes and their Applications*, 118:2198–2222.

Bédard, M. and Rosenthal, J. S. (2008). Optimal scaling of Metropolis algorithms: Heading toward general target distributions. *The Canadian Journal of Statistics*, 36:483–503.

Bélisle, C., Romeijn, H. E., and Smith, R. L. (1993). Hit-and-run algorithms for generating multivariate distributions. *Mathematics of Operations Research*, 18:255–266.

Beskos, A., Roberts, G., and Stuart, A. (2009). Optimal scalings for local Metropolis–Hastings chains on nonproduct targets in high dimensions. *The Annals of Applied Probability*, 19:863–898.

Box, G. E. P. and Tiao, G. C. (1973). *Bayesian Inference in Statistical Analysis.* Addison-Wesley, Reading, MA.

Carlin, B. P. and Louis, T. A. (2009). *Bayesian Methods for Data Analysis,* Third Edition. CRC Press, Boca Raton, FL.

Craiu, R. V., Rosenthal, J. S., and Yang, C. (2009). Learn from thy neighbor: Parallel-chain and regional adaptive MCMC. *Journal of the American Statistical Association,* 104:1454–1466.

Evans, M. J. and Swartz, T. B. (2000). *Approximating Integrals via Monte Carlo and Deterministic Methods.* Oxford University Press, Oxford.

Fill, J. A., Machida, M., Murdoch, D. J., and Rosenthal, J. S. (2000). Extension of Fill's perfect rejection sampling algorithm to general chains. In *Proceedings of the Ninth International Conference "Random Structures and Algorithms" (Poznan, 1999),* volume 17, pp. 290–316.

Fort, G., Moulines, É., and Priouret, P. (2012). Convergence of adaptive and interacting Markov Chain Monte Carlo algorithms. *The Annals of Statistics,* 39:3262–3289.

Gelfand, A. E. and Smith, A. F. M. (1990). Sampling-based approaches to calculating marginal densities. *Journal of the American Statistical Association,* 85:398–409.

Gelman, A., Carlin, J. B., Stern, H. S., and Rubin, D. B. (2003). *Bayesian Data Analysis,* Second Edition. Chapman & Hall, London.

Giordani, P. and Kohn, R. (2010). Adaptive independent Metropolis–Hastings by fast estimation of mixtures of normals. *Journal of Computational and Graphical Statistics,* 19:243–259.

Haario, H., Saksman, E., and Tamminen, J. (2001). An adaptive Metropolis algorithm. *Bernoulli,* 7:223–242.

Hamze, F. and de Freitas, N. (2012). Intracluster moves for constrained discrete-space MCMC. *CoRR,* abs/1203.3484.

Hastings, W. (1970). Monte Carlo sampling methods using Markov Chains and their applications. *Biometrika,* 57:97–109.

Jain, S. and Neal, R. M. (2004). A split-merge Markov Chain Monte Carlo procedure for the Dirichlet process mixture model. *Journal of Computational and Graphical Statistics,* 13:158–182.

Łatuszyński, K., Roberts, G. O., and Rosenthal, J. S. (2013). Adaptive Gibbs samplers and related MCMC methods. *The Annals of Applied Probability,* 23:66–98.

Lunn, D., Thomas, A., Best, N., and Spiegelhalter, D. (2000). WinBUGS: A Bayesian modelling framework: Concepts, structure, and extensibility. *Statistics and Computing,* 10:325–337.

Metropolis, N., Rosenbluth, A. W., Rosenbluth, M. N., Teller, A. H., and Teller, E. (1953). Equations of state calculations by fast computing machines. *The Journal of Chemical Physics*, 21:1087–1092.

Murdoch, D. J. and Green, P. J. (1998). Exact sampling from a continuous state space. *Scandinavian Journal of Statistics*, 25:483–502.

Neal, R. M. (2003). Slice sampling (with discussion). *The Annals of Statistics*, 31:705–767.

Propp, J. G. and Wilson, D. B. (1996). Exact sampling with coupled Markov Chains and applications to statistical mechanics. In *Proceedings of the Seventh International Conference on Random Structures and Algorithms (Atlanta, GA, 1995)*, 9:223–252.

Richardson, S., Bottolo, L., and Rosenthal, J. S. (2011). Bayesian models for sparse regression analysis of high dimensional data. In *Bayesian Statistics 9*, pp. 539–560. Oxford University Press.

Roberts, G. O., Gelman, A., and Gilks, W. R. (1997). Weak convergence and optimal scaling of random walk Metropolis algorithms. *The Annals of Applied Probability*, 7:110–120.

Roberts, G. O. and Rosenthal, J. S. (1998). Optimal scaling of discrete approximations to Langevin diffusions. *Journal of the Royal Statistical Society, Series B*, 60:255–268.

Roberts, G. O. and Rosenthal, J. S. (2001). Optimal scaling for various Metropolis–Hastings algorithms. *Statistical Science*, 16:351–367.

Roberts, G. O. and Rosenthal, J. S. (2007). Coupling and ergodicity of adaptive MCMC. *Journal of Applied Probability*, 44:458–475.

Roberts, G. O. and Rosenthal, J. S. (2009). Examples of adaptive MCMC. *Journal of Computational and Graphical Statistics*, 18:349–367.

Roberts, G. O. and Rosenthal, J. S. (2013). Minimising MCMC variance via diffusion limits, with an application to simulated tempering. *The Annals of Applied Probability*, in press.

Rosenthal, J. S. (1995). Minorization conditions and convergence rates for Markov Chain Monte Carlo. *Journal of the American Statistical Association*, 90:558–566. [Correction: p. 1136].

Saksman, E. and Vihola, M. (2010). On the ergodicity of the adaptive Metropolis algorithm on unbounded domains. *The Annals of Applied Probability*, 20:2178–2203.

Sherlock, C. and Roberts, G. O. (2009). Optimal scaling of the random walk Metropolis on elliptically symmetric unimodal targets. *Bernoulli*, 15:774–798.

7

Design of Computer Experiments for Optimization, Estimation of Function Contours, and Related Objectives

Derek Bingham
Simon Fraser University, Burnaby, BC

Pritam Ranjan
Acadia University, Wolfville, NS

William J. Welch
University of British Columbia, Vancouver, BC

7.1 Introduction

A computer code or simulator is a mathematical representation of a physical system, for example a set of differential equations. Such a simulator takes a set of input values or conditions, $\mathbf{x}$, and from them produces an output value, $y(\mathbf{x})$, or several such outputs. For instance, one application we use for illustration simulates the average tidal power, y, generated as a function of a turbine location, $\mathbf{x} = (x_1, x_2)$, in the Bay of Fundy, Nova Scotia (Ranjan et al., 2011). Performing scientific or engineering experiments via such a computer code (i.e., a computer experiment) is often more time and cost effective than running a physical experiment or collecting data directly.

A computer experiment may have objectives similar to those of a physical experiment. For example, computer experiments are often used in manufacturing or process development. If y is a quality measure for a product or process, an experiment could aim to optimize y with respect to $\mathbf{x}$. Similarly, an experiment might aim to find sets or contours of $\mathbf{x}$ values that make y equal a specified target value — a type of inverse problem. Such scientific and engineering objectives are naturally and efficiently achieved via so-called data-adaptive sequential design, which we describe below. Essentially, each new run (i.e., new set of input values) is chosen based on the analysis of the data so far, to make the best expected improvement in the objective. In a computer experiment, choosing new experimental runs, restarting the experiment, etc. pose only minor logistical challenges if these decisions are also computer-controlled, a distinct advantage relative to a physical experiment.

Choosing new runs sequentially for optimization, moving y to a target, etc. has been formalized using the concept of expected improvement (Jones et al., 1998). The next experimental run is made where the expected improvement in the function of interest is largest. This expectation is with respect to the predictive distribution of y from a statistical model relating y to $\mathbf{x}$. By considering a set of possible inputs $\mathbf{x}$ for the new run, we can choose that which gives the largest expectation.

We illustrate this idea with two examples in Section 7.2. Then we describe formulations of improvement functions and their expectations in Section 7.3. Expectation implies a statistical model, and in Section 7.4 we outline the use of Gaussian process models for fast emulation of computer codes. In Section 7.5 we describe some extensions to other, more complex scientific objectives.

7.2   Expected Improvement and Sequential Design: Basic Ideas

The ideas behind expected improvement and data-adaptive sequential design are illustrated via two examples. The first, a tidal-power application, shows the use of expected improvement in sequential optimization. In the second example, we use a simulator of volcanic pyroclastic flow to illustrate how to map out a contour of a function.

7.2.1   Optimization

Ranjan et al. (2011) described output from a 2D computer-model simulation of the power produced by a tidal turbine in the Minas Passage of the Bay of Fundy, Nova Scotia. In this simplified version of the problem there are just two inputs for the location of a turbine. Originally, the input space was defined by latitude-longitude coordinates for a rectangular region in the Minas Passage (see Figure 5 of Ranjan et al., 2011). The coordinates were transformed so that x_1 is in the direction of the flow and x_2 is perpendicular to the flow. Furthermore, only an interesting part of the Minas Passage was considered, with $x_1 \in [.75, .95]$ and $x_2 \in [.2, .8]$. The computational model generates an output, y, the extractable power in MW, averaged over a tidal cycle for inputs (x_1, x_2). For the simplified demonstration here, y was computed for 533 runs on a 13×41 grid of x_1 and x_2 values, which produced the contour plot of Figure 7.1(a).

We now demonstrate how the turbine location optimizing the power (i.e., $\max y(x_1, x_2)$) can be found with far fewer than 533 runs of the computer code. Such an approach would be essential for the computer experiment of

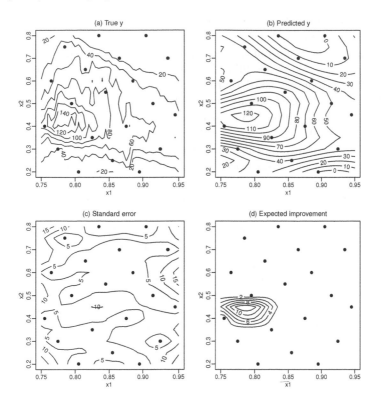

FIGURE 7.1: Initial 20-run design and analysis for the tidal-power application: (a) true power, y, in MW; (b) predicted power, $\hat{y}(\mathbf{x})$; (c) standard error of prediction, $s(\mathbf{x})$; and (d) expected improvement, $\mathrm{E}\{I(\mathbf{x})\}$. The design points from an initial 20-run maximin Latin hypercube are shown as filled circles. All plots are functions of the two input variables, x_1 and x_2, which are transformed longitude and latitude.

ultimate interest. A more realistic computer model has a grid resolution 10 times finer in each coordinate and introduces vertical layers in a 3D code. The running time would be increased by several orders of magnitude. Moreover, the final aim is to position several turbines, which would interfere with each other, and so the optimization space (or input space) is larger than two or three dimensions. Thus, the ultimate goal is to optimize a high-dimensional function with a limited number of expensive computer model runs. Inevitably, much of the input space cannot be explicitly explored, and a statistical approach to predict outcomes (extractable power) along with an uncertainty measure is required to decide where to make runs and when to stop. The expected improvement criterion addresses these two requirements.

Consider a more limited computer experiment with just 20 runs, as shown by the points in Figure 7.1. The experimental design (i.e., the locations of the 20 points) is a maximin Latin hypercube (Morris and Mitchell, 1995), a stratified scheme that is "space-filling" even in higher dimensions. The choice of 20 runs is based on the heuristic rule that an initial computer experiment has $n = 10d$ observations (Loeppky et al., 2009), where d is the input dimension; here $d = 2$. Among the 20 initial runs, the largest y observed, denoted by $y_{max}^{(20)}$, is 109.7 MW at $(x_1, x_2) = (.755, .4)$. The expected improvement algorithm tries to improve on the best value found so far, as new runs are added.

At each iteration of the computer experiment we obtain a predictive distribution for $y(\mathbf{x})$ conditional on the runs so far. This allows prediction of the function at input vectors $\mathbf{x}$ where the code has not been run. A Gaussian process (GP) statistical model is commonly used for prediction, as outlined in Section 7.4, though this is not essential. A GP model was fit here to the data from the first 20 runs, giving the point-wise predictions, $\hat{y}(\mathbf{x})$, of $y(\mathbf{x})$ in Figure 7.1(b) along with standard errors of prediction, $s(\mathbf{x})$, in Figure 7.1(c). The standard error is a statistical measure of uncertainty concerning the closeness of the predicted value to the actual true value of $y(\mathbf{x})$. We show the predicted values and standard errors through contours in Figures 7.1(b) and 7.1(c).

Figures 7.1(b) and (c) are informative about regions in the input space that are promising versus unpromising for further runs of the code. While the $\hat{y}(\mathbf{x})$ prediction surface is nonlinear, it suggests there is a single, global optimum. Moreover, the $s(\mathbf{x})$ surface is uniformly below about 15 MW: For much of the input space, $\hat{y}(\mathbf{x})$ is so much smaller than $y_{max}^{(20)}$ relative to $s(\mathbf{x})$ that a new run is expected to make virtually zero improvement.

The expected improvement (EI) for a candidate new run at any $\mathbf{x}$ is computed from the predictive distribution of $y(\mathbf{x})$; see Section 7.3.1 for the formal definition of EI. Figure 7.1(d) shows the EI surface based on predictive distributions from a GP that was fit to the data from the initial 20 runs of the tidal-power code. By the definition in Section 7.3.1, the improvement can never be negative (if the output from the new run does beat the current optimum, the current optimum stands). Consequently, the EI is always non-negative too. Figure 7.1(d) indicates that for most of the input space EI is near zero and a new run would be wasted, but there is a sub-region where EI is more than 12 MW. Evaluating EI over the 13×41 grid shows that the maximum EI is 13.9 MW at $\mathbf{x} = (.785, .45)$. In other words, a new code run to evaluate $y(.785, .45)$ is expected to beat $y_{max}^{(20)} = 109.7$ MW by 13.9 MW.

Thus, run 21 of the sequential design for the computer experiment is at $\mathbf{x}^{(21)} = (.785, .45)$. The actual power obtained from the simulator is $y = 159.7$ MW, so the best y found after 21 runs is $y_{max}^{(21)} = 159.7$ MW, and this is the value to beat at the next iteration. Note that the actual improvement in the optimum from the new run is $159.7 - 109.7 = 50.0$ MW, compared with an expectation of about 13.9 MW.

The new run raises concerns about the statistical model. Before making the new run, the predictive distribution of $y(\mathbf{x}^{(21)})$ is approximately Normal, specifically $\mathcal{N}(123.5, 5.67^2)$, an implausible distribution given the large value of the standardized residual

$$\frac{y(\mathbf{x}^{(21)}) - \hat{y}(\mathbf{x}^{(21)})}{s(\mathbf{x}^{(21)})} = \frac{159.7 - 123.5}{5.67} = 6.4.$$

One deficiency is that $s(\mathbf{x})$ may not reflect all sources of uncertainty in estimation of the parameters of the GP (see Section 7.4). A more important reason here, however, is that the new observation successfully finds a peak in the input space, a sub-region where the output function is growing rapidly and uncertainty is larger. In contrast, the first 20 runs were at locations where the function is flatter and easier to model. The GP model fit to the initial runs underestimated the uncertainty of prediction in a more difficult part of the input space.

Careful consideration of the properties of a GP model and the possible need for transformations is particularly relevant for sequential methods based on predictive distributions. Uncertainty of prediction is a key component of the EI methodology, so checking that a model has plausible standard errors of prediction is critical.

One way of improving the statistical emulator of the tidal-power code is to consider transformation of the output. This is described in the context of the volcano example of Section 7.2.2, where transformation is essential. For the tidal-power example, persisting with the original model will show that it adapts to give more plausible standard errors with a few more runs.

The GP model and predictive distributions are next updated to use the data from all 21 runs now available. Figure 7.2(a) shows the location of the new run as a "+" and the updated $\hat{y}(\mathbf{x})$. Similarly, Figure 7.2(b) gives the updated $s(\mathbf{x})$. A property of the GP fit is that $s(\mathbf{x})$ must be zero at any point $\mathbf{x}$, where $y(\mathbf{x})$ is in the dataset for the fit; see Jones et al. (1998) for a derivation of this result. Thus, $s(\mathbf{x})$ is zero at the new run, and Figure 7.2(b) shows it is less than 5 MW near the new run. Comparing with Figure 7.1(c), it is seen that $s(\mathbf{x})$ was 5 MW or more in this neighborhood for the GP fit before the new run.

On the other hand, comparison of Figures 7.1(c) and 7.2(b) shows that $s(\mathbf{x})$ has *increased* outside the neighborhood of the new run. For example, at the right edge of Figure 7.1(c), $s(\mathbf{x})$ barely reaches 15 MW, yet $s(\mathbf{x})$ often exceeds 15 MW or even 20 MW at the same locations in Figure 7.2(b). The 21-run GP fit has adapted to reflect the observed greater sensitivity of the output to x_1 and x_2. (For instance, the estimate of the GP variance parameter σ^2, defined in Section 7.4, increased.) Thus, the model has at least partially self corrected and we continue with it.

The EI contour plot in Figure 7.2(c) suggests that there is little further improvement to be had from a further run anywhere. If a run number 22 is made, however, it is not located where $\hat{y}(\mathbf{x})$ is maximized; that location co-

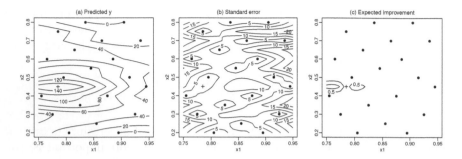

FIGURE 7.2: Analysis of the tidal-power application after 21 runs: (a) predicted power, $\hat{y}$; (b) standard error of prediction, $s(\mathbf{x})$; and (c) expected improvement, $E\{I(\mathbf{x})\}$. The new design point is shown as a "+."

incides with run 21 and there would be no gain. Rather, the maximum EI of about 1.3 MW occurs at a moderate distance from the location providing maximum $\hat{y}(\mathbf{x})$. As we move away from run 21, the standard error of prediction increases from zero until it is large enough to allow a modest expected improvement. Thus, this iteration illustrates that EI trades off local search (evaluate where $\hat{y}(\mathbf{x})$ is optimized) and global search (evaluate where uncertainty concerning fitted versus actual output values, characterized by $s(\mathbf{x})$, is optimized).

With this approach, EI typically indicates smaller potential gains as the number of iterations increases. Eventually, the best EI is deemed small enough to stop. It turns out that run 21 found the global maximum for extractable power on the 13×41 grid of locations.

7.2.2   Contour Estimation

We illustrate sequential design for mapping out a specified contour of a computer-model function using TITAN2D computer model runs provided by Elaine Spiller. They relate to the Colima volcano in Mexico. Again for ease of illustration, there are two input variables: x_1 is the pyroclastic flow volume (log base 10 of m^3) of fluidized gas and rock fragments from the eruption; and x_2 is the basal friction angle in degrees, defined as the minimum slope for the volcanic material to slide. The output z is the maximum flow height (m) at a single, critical location. As is often the case, the code produces functional output, here flow heights over a 2D grid on the earth's surface, but the output for each run is reduced to a scalar quantity of interest, the height at the critical location.

Following Bayarri et al. (2009), the scientific objective is to find the values of x_1 and x_2 where $z = 1$, a contour delimiting a "catastrophic" region. Bayarri et al. (2009) used the same TITAN2D code but for a different volcano. They

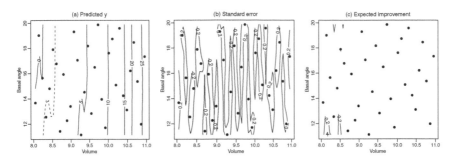

FIGURE 7.3: Analysis of the initial 32-run design for the volcano application: (a) predicted height, $\hat{y}(\mathbf{x})$, where $y = \sqrt{z}$; (b) standard error of prediction, $s(\mathbf{x})$; and (c) expected improvement, $E\{I(\mathbf{x})\}$. The design points of the initial 32-run design are shown as filled circles. The new design point chosen by the EI criterion is shown as a "+" in the lower left corner of (c).

also conducted their sequential experiment in a less formal way than in our illustration of the use of EI.

There are 32 initial runs of the TITAN2D code. They are located at the points shown in Figure 7.3(a). The predicted flow height surface also shown in Figure 7.3(a) relates to a GP model fit to the transformed simulator output $y = \sqrt{z}$. This choice was made by trying GP models on three different scales: the z untransformed height; $\log(z + 1)$, as chosen by Bayarri et al. (2009); and $\sqrt{z}$. Our final choice of $y = \sqrt{z}$ results from inspection of standard cross-validation diagnostics for GP models (Jones et al., 1998).

The dashed curve in Figure 7.3(a) shows the contour where $\hat{y}(\mathbf{x}) = 1$. This maps out the contour of interest in the (x_1, x_2) input space, but it is based on predictions subject to error. The standard errors in Figure 7.3(b) are substantial, and sequential design via EI aims to improve the accuracy of the estimate of the true $y(\mathbf{x}) = 1$ contour.

The EI criterion adapted for the contouring objective is defined in Section 7.3.2. It is computed for the initial design of the volcano example in Figure 7.3(c). EI suggests improving the accuracy of the contour by taking the next run at $(x_1, x_2) = (8.2, 11.1)$. Inspection of Figures 7.3(a) and 7.3(b) show that this location is intuitively reasonable. It is in the vicinity of the predicted $\hat{y}(\mathbf{x}) = 1$ contour and has a relatively large predictive standard error. Reducing substantial uncertainty in the vicinity of the estimated contour is the dominant aspect of the EI measure of Equation (7.4).

The tidal-flow and volcano applications both have a 2D input space for ease of exposition, but the same approaches apply to higher dimensions, where choosing runs in a sequential design would be more problematic with ad hoc methods.

7.3 Expected Improvement Criteria

In this section, we briefly define the improvement and EI in general. We then review two implementations, specific to global optimization and contour estimation, respectively.

Let $I(\mathbf{x})$ be an improvement function defined for any $\mathbf{x}$ in the input space, χ. The form of the improvement depends on the scientific objective, such as improving the largest y found so far in maximization. In general, it is formulated for efficient estimation of a pre-specified computer-model output feature, $\psi(y)$. Typically, before taking another run, $I(\mathbf{x})$ is an unobserved function of $\mathbf{x}$, the unknown computer-model output $y(\mathbf{x})$, the predictive distribution of $y(\mathbf{x})$, and the best estimate so far of $\psi(y)$.

Given a definition of $I(\mathbf{x})$, as its name suggests, the corresponding EI criterion is given by the expectation of $I(\mathbf{x})$, viz.

$$\mathrm{E}\{I(\mathbf{x})\} = \int I(\mathbf{x}) f(y|\mathbf{x}) \, dy.$$

Here expectation is with respect to the predictive distribution of $y(\mathbf{x})$ conditional on all runs so far, $f(y|\mathbf{x})$. Assuming the sequential design scheme selects one new input point at a time, the location of the new point, $\mathbf{x}^{\mathrm{new}}$, is the global maximizer of $\mathrm{E}\{I(\mathbf{x})\}$ over $\mathbf{x} \in \chi$.

7.3.1 EI for Global Optimization

Finding the global minimum, $\psi(y) = \min\{y(\mathbf{x}) : x \in \chi\}$, of a function which is expensive to evaluate is an extensively investigated optimization problem. (Finding the maximum is reformulated as $\min -y(\mathbf{x})$, and the following results apply.) Jones et al. (1998) proposed an efficient sequential solution via the improvement function to assess the gain if a new evaluation is made at $\mathbf{x}$. The improvement function is

$$I(\mathbf{x}) = \max\{y_{\mathrm{min}}^{(n)} - y(\mathbf{x}), 0\},$$

where $y_{\mathrm{min}}^{(n)}$ is the minimum value of y found so far with n runs. The objective is improved by $y_{\mathrm{min}}^{(n)} - y(\mathbf{x})$ if $y_{\mathrm{min}}^{(n)} > y(\mathbf{x})$, otherwise there is no improvement.

The GP statistical model outlined in Section 7.4 leads to a Gaussian predictive distribution for $f(y|\mathbf{x})$, i.e., $y(\mathbf{x}) \sim \mathcal{N}[\hat{y}(\mathbf{x}), s^2(\mathbf{x})]$. The Gaussian predictive model leads to a simple, closed form for the expected improvement,

$$\mathrm{E}\{I(\mathbf{x})\} = s(\mathbf{x})\phi(u) + \{y_{\mathrm{min}}^{(n)} - \hat{y}(\mathbf{x})\}\Phi(u), \tag{7.1}$$

where

$$u = \frac{y_{\mathrm{min}}^{(n)} - \hat{y}(\mathbf{x})}{s(\mathbf{x})},$$

while ϕ and Φ denote the standard Normal probability density function (pdf) and cumulative distribution function (cdf), respectively.

Large values of the first term in (7.1) support global exploration in regions of the input space sparsely sampled so far, where $s(\mathbf{x})$ is large. The second term favors search where $\hat{y}(\mathbf{x})$ is small, which is often close to the location giving $y_{\min}^{(n)}$, i.e., local search. This trade-off between local and global search makes EI-based sequential design very efficient, and it often requires relatively few computer-model evaluations to achieve a desired accuracy in estimating min y.

For instance, in the tidal-power application, the EI surface displayed in Figure 7.1(d) indicates that the first follow-up run is at the location giving the maximum predicted power; see Figure 7.2(a). Thus, the local-search component dominates. Conversely, the suggested location for the second follow-up run is in an unsampled region near the maximum predicted power; see Figure 7.2(c).

Attempts have been made to control this local versus global trade-off for faster convergence (i.e., using as few runs as possible) to the true global minimum. For instance, Schonlau et al. (1998) proposed an exponentiated improvement function, $I^g(\mathbf{x})$, for $g \geq 1$. With $g > 1$, there is more weight on larger improvements when expectation is taken to compute EI. Such large improvements will have a non-trivial probability even if $\hat{y}(\mathbf{x})$ is unfavorable, provided $s(\mathbf{x})$ is sufficiently large. Hence, global exploration of high-uncertainty regions can receive more attention with this adaptation. Similarly, Sóbester et al. (2005) developed a weighted expected improvement function (WEIF) by introducing a user-defined weight parameter $w \in [0, 1]$ in the EI criterion of Jones et al. (1998), and Ponweiser et al. (2008) proposed clustered multiple generalized expected improvement.

7.3.2   EI for Contour Estimation

Ranjan et al. (2008) developed an EI criterion specific to estimating a threshold (or contour) of y. They applied it to a 2-queue 1-server computer network simulator that models the average delay in a queue for service.

Let the feature of interest $\psi(y)$ be the set of input vectors $\mathbf{x}$ defining the contour at level a, viz.

$$S(a) = \{x : y(\mathbf{x}) = a\}. \tag{7.2}$$

The improvement function proposed by Ranjan et al. (2008) is

$$I(\mathbf{x}) = \epsilon^2(\mathbf{x}) - \min[\{y(\mathbf{x}) - a\}^2, \epsilon^2(\mathbf{x})], \tag{7.3}$$

where $\epsilon(\mathbf{x}) = \alpha s(\mathbf{x})$ for a positive constant α (e.g., $\alpha = 1.96$, corresponding to 95% confidence/credibility under approximate Normality). This improvement function defines a limited region of interest around $S(a)$ for further experimentation. Point-wise, the extent of the region depends on the uncertainty $s(\mathbf{x})$ and hence the tolerance $\epsilon(\mathbf{x})$.

Under a Normal predictive distribution, $y(\mathbf{x}) \sim \mathcal{N}[\hat{y}(\mathbf{x}), s^2(\mathbf{x})]$, the expectation of $I(\mathbf{x})$ can again be written in closed form, namely

$$
\begin{aligned}
\mathrm{E}\{I(\mathbf{x})\} \;=\; & \left[\epsilon^2(\mathbf{x}) - \{\hat{y}(\mathbf{x}) - a\}^2\right]\{\Phi(u_2) - \Phi(u_1)\} \\
& + s^2(\mathbf{x})\left[\{u_2\phi(u_2) - u_1\phi(u_1)\} - \{\Phi(u_2) - \Phi(u_1)\}\right] \\
& + 2\{\hat{y}(\mathbf{x}) - a\}s(\mathbf{x})\{\phi(u_2) - \phi(u_1)\},
\end{aligned}
\tag{7.4}
$$

where

$$
u_1 = \frac{a - \hat{y}(\mathbf{x}) - \epsilon(\mathbf{x})}{s(\mathbf{x})} \quad \text{and} \quad u_2 = \frac{a - \hat{y}(\mathbf{x}) + \epsilon(\mathbf{x})}{s(\mathbf{x})}.
$$

Like EI for optimization, the EI criterion in (7.4) trades off the twin aims of local search near the predicted contour of interest and global exploration. The first term on the right of (7.4) recommends an input location with a large $s(\mathbf{x})$ in the vicinity of the predicted contour. When it dominates, the follow-up point is often essentially the maximizer of $\epsilon^2(\mathbf{x}) - \{\hat{y}(\mathbf{x}) - a\}^2$. This consideration led to the new point in Figure 7.3(c) of the volcano application, for instance. The last term in (7.4) gives weight to points far away from the predicted contour with large uncertainties. The second term is often dominated by the other two terms in the EI criterion.

The EI criterion in (7.4) can easily be extended to related aims. For simultaneous estimation of k contours $S(a_1), \ldots, S(a_k)$, with $S(\cdot)$ defined in (7.2), the improvement function becomes

$$
I(\mathbf{x}) = \epsilon^2(\mathbf{x}) - \min\left[\{y(\mathbf{x}) - a_1\}^2, \ldots, \{y(\mathbf{x}) - a_k\}^2, \epsilon^2(\mathbf{x})\right],
$$

and the corresponding EI can also be written in a closed form. Roy (2008) considered estimation of the contour defining a given percentile of the output distribution when inputs to the computer model are random (described further in Section 7.5).

7.4 Gaussian Process Models and Predictive Distributions

Evaluation of an EI criterion requires the computation of the expectation of $I(\mathbf{x})$ with respect to the predictive distribution of $y(\mathbf{x})$. In principle, any predictive distribution can be used, but for the method to be useful, it should faithfully reflect the data obtained up to the run in question. In practice, treating the data from the computer-model runs as a realization of a GP is nearly ubiquitous in computer experiments. A GP model leads to a Gaussian predictive distribution, which in turn leads to the closed form expressions in (7.1) and (7.4) and easy interpretation of the trade-off between local and global search.

A GP model is a computationally inexpensive statistical emulator of a computer code. A key feature of many codes is that they are deterministic: rerunning the computer model with the same values for all input variables will give the same output values. Such a deterministic function is placed within a statistical framework by considering a given computer-model input-output relationship as the realization of a stochastic process, $Z(\mathbf{x})$, indexed by the input vector. A single realization of the process is non-random, hence the relevance for a deterministic computer code. For a continuous function, the process is usually assumed to be Gaussian, possibly after transformation, as was done for the volcano application.

This GP or Gaussian Stochastic Process (GaSP) paradigm for modeling a computer code dates back to Sacks et al. (1989a,b), Currin et al. (1991), and O'Hagan (1992). Specifically, the code output function, $y(\mathbf{x})$, is treated as a realization of $Y(\mathbf{x}) = \mu(\mathbf{x}) + Z(\mathbf{x})$, where $\mu(\mathbf{x})$ is a mean (regression) function in $\mathbf{x}$, and $Z(\mathbf{x})$ is a Gaussian process with mean 0 and variance σ^2.

Crucial to this approach is the assumed correlation structure of $Z(\mathbf{x})$. For two configurations of the d-dimensional input vector, $\mathbf{x} = (x_1, \ldots, x_d)$ and $\mathbf{x}' = (x_1', \ldots, x_d')$, the correlation between $Z(\mathbf{x})$ and $Z(\mathbf{x}')$ is denoted by $R(\mathbf{x}, \mathbf{x}')$. Here, $R(\cdot, \cdot)$ is usually a parametric family of functions, for which there are many choices (e.g., Santner et al., 2003, Section 2.3). The computations for the applications in Section 7.2 were based on a constant (intercept) regression and a stationary power-exponential correlation function,

$$R(\mathbf{x}, \mathbf{x}') = \exp\left(-\sum_{j=1}^{d} \theta_j |x_j - x_j'|^{p_j}\right).$$

Here, θ_j (with $\theta_j \geq 0$) and p_j (with $1 \leq p_j \leq 2$) control the properties of the effect of input variable j on the output. A larger value of θ_j implies greater sensitivity (activity) of y with respect to x_j, whereas a larger value of p_j implies smoother behavior of y as a function of x_j.

Under this model the output values from n runs of the code, $Y_1, \ldots, Y_n$, have a joint multivariate Normal distribution. If the parameters in the statistical model — in the mean function, in the correlation function, and σ^2 — are treated as known, the predictive distribution of Y at a new $\mathbf{x}$ has a Normal distribution, $\mathcal{N}[\hat{y}(\mathbf{x}), s^2(\mathbf{x})]$, where $\hat{y}(\mathbf{x})$ is the conditional mean of $Y(\mathbf{x})$ given $Y_1, \ldots, Y_n$, and $s^2(\mathbf{x})$ is the conditional variance. Without assuming Normality, $\hat{y}(\mathbf{x})$ can also be interpreted as the best linear unbiased predictor, and $s^2(\mathbf{x})$ is the associated mean squared error. In practice, the unknown parameters have to be estimated, usually by maximum likelihood or Bayesian methods. The predictive distribution is then only approximately Normal, and $\hat{y}(\mathbf{x})$ and $s(\mathbf{x})$ are computed from estimated parameter values. Moreover, Bayesian estimation of the correlation parameters may be necessary to capture all sources of uncertainty in the predictive distribution.

As mentioned already, neither a GP model nor a Normal predictive distribution are essential for sequential design with an EI criterion. For instance,

Chipman et al. (2012) used the optimization improvement function of Jones et al. (1998) with Bayesian additive regression trees (BART). Thus, the emulator was a non-parametric ensemble of tree models.

7.5   Other EI-Based Criteria

Over the last two decades, a plethora of EI-based criteria have been proposed for other scientific and engineering objectives.

Applications can involve several outputs of interest. For instance, constrained optimization problems arise where the code generating the objective function $y(\mathbf{x})$ or another code gives values for a constraint function, $c(\mathbf{x})$, (or several functions). For a feasible solution, $c(\mathbf{x})$ must lie in $[a, b]$. If $c(\mathbf{x})$ is also expensive to compute, one can build an emulator, $\hat{c}(\mathbf{x})$, for it too. The predictive distribution for $c(\mathbf{x})$ leads to an estimate of the probability that $a < c(\mathbf{x}) < b$ for any new run $\mathbf{x}$ under consideration. EI in (7.1) is multiplied by this probability of feasibility to steer the search to locations where EI for the objective $y(\mathbf{x})$ is large and $c(\mathbf{x})$ is likely to be feasible (Schonlau et al., 1998). For a code with multivariate output, Henkenjohann and Kunert (2007) proposed an EI criterion for estimating the global maximum of the desirability scores of simulator outputs.

In a computer experiment, each simulator run requires a fixed, "known" set of values for the $\mathbf{x}$ input vector, but in the real world the input variables may vary randomly, say due to different environmental conditions. Thus, there is an induced output distribution, and its properties can be of interest. Roy (2008) estimated the contour in the input space defining the pth percentile, ν_p, of the output distribution for given $0 < p < 1$. At any iteration of the sequential search based on a limited number of simulator runs, the distribution of the inputs, $\mathbf{x}$, is propagated through the predictive model, giving a distribution of the output variable. In practice, this is achieved by taking a large random sample from the known distribution of the inputs and evaluating $\hat{y}(\mathbf{x})$ for each sampled value. This Monte Carlo sample gives an estimate of the output distribution and hence an estimate $\hat{\nu}_p$ of ν_p. The improvement function for the next point is

$$I^g(\mathbf{x}) = \epsilon^g(\mathbf{x}) - \min[\{y(\mathbf{x}) - \hat{\nu}_p\}^g, \epsilon^g(\mathbf{x})].$$

For $g = 2$ this is the improvement function in (7.3) with $a = \hat{\nu}_p$. Hence, the contour of interest, a, is not fixed throughout the sequential design procedure of Ranjan et al. (2008) but adapts as $\hat{\nu}_p$ changes. Bichon et al. (2009) adapted this criterion to estimate the probability of rare events and system failure in reliability-based design optimization.

Sometimes the simulator input variables partition into those that are controllable in reality and those that are uncontrollable. In a manufacturing process context, for instance, these two types of input could be nominal engineering dimensions and random environmental conditions, respectively. Lehman et al. (2004) developed improvement functions for finding M- and V-robust designs for optimization for such a process. For a given configuration of the control variables, $\mathbf{x}_c$, let $\mu(\mathbf{x}_c)$ and $\sigma^2(\mathbf{x}_c)$ be the unknown mean and variance of the simulator output y with respect to the distribution of the uncontrollable variables. An M-robust engineering design minimizes $\mu(\mathbf{x}_c)$ with respect to $\mathbf{x}_c$ subject to a constraint on $\sigma^2(x_c)$, whereas V-robust engineering design minimizes $\sigma^2(x_c)$ subject to a constraint on $\mu(x_c)$.

The inclusion of measurement error (or equivalently, considering a non-deterministic simulator) is becoming more popular in computer experiments, often due to unavoidable simulator biases and inaccurate modeling assumptions. It is undesirable to ignore the measurement error when minimizing a noisy output response, and Ranjan (2013) recommended minimizing a lower quantile, $q(\mathbf{x})$, via an estimate $\hat{q}(\mathbf{x})$ from the predictive distribution, e.g., $\hat{q}(\mathbf{x}) = \hat{y}(\mathbf{x}) - 1.96s(\mathbf{x})$ under a Normal predictive distribution. The proposed improvement function is

$$I(\mathbf{x}) = \max\{0, \hat{q}_{\min}^{(n)} - q(\mathbf{x})\},$$

where $\hat{q}_{\min}^{(n)}$ is the minimum $\hat{q}(\mathbf{x})$ from n runs so far, and $q(\mathbf{x}) = y(\mathbf{x}) - 1.96s(\mathbf{x})$ is an unobservable random quantity. Treating $s(\mathbf{x})$ as non-stochastic and assuming $y(\mathbf{x}) \sim \mathcal{N}[\hat{y}(\mathbf{x}), s^2(\mathbf{x})]$, the corresponding EI criterion is

$$\mathrm{E}\{I(\mathbf{x})\} = s(\mathbf{x})\phi(u) + \{\hat{q}_{\min}^{(n)} - \hat{y}(\mathbf{x}) + 1.96s(\mathbf{x})\}\Phi(u), \qquad (7.5)$$

where

$$u = \frac{\hat{q}_{\min}^{(n)} - \hat{y}(\mathbf{x}) + 1.96s(\mathbf{x})}{s(\mathbf{x})}.$$

Like the EI criterion in (7.1), EI in (7.5) facilitates the trade-off between local and global search. One can easily generalize this EI criterion to $\mathrm{E}\{I^q(\mathbf{x})\}$ (as in Schonlau et al., 1998) or introduce a user specified weight (as in Sóbester et al., 2005).

For complex physical phenomena like climate and tidal power, multiple computer simulators with different computational demands are often available for experimentation. For instance, there are 2D and 3D codes for the tidal-power application; the 3D version is a higher-fidelity representation of reality but is much more expensive to run. They can be combined to obtain more informed prediction, and Huang et al. (2006) proposed augmented expected improvement for finding the global minimum of the highest-fidelity process, subject to noise.

7.6 Summary

The essence of these approaches for sequential computer experiments is to formulate the scientific objective through an improvement function. Following some initial runs, the next run is chosen to maximize the expected improvement. In contrast to physical experiments, sequential design is convenient, with the computer handling the logistics of iterating analysis of the data so far, choice of the next run, and making the new run.

With objectives like optimization and contouring in high-dimensional applications, sequential strategies are efficient in terms of solving the problem with a relatively small number of runs. For these reasons, we expect this area of the design of computer experiments will continue to receive considerable research attention from methodologists and users.

Of course, the usefulness of this strategy depends on having a computer model that provides a satisfactory description of the physical process of interest. Such models have to be checked by reference to real data from the physical process (Bayarri et al., 2007). However, once a model has been validated it provides an efficient route to achieving the goals discussed here.

Acknowledgments

The authors thank the editor and an associate editor for their insightful comments leading to clarifications. Funding in partial support of this work was provided by the Canada Research Chairs Program and the Natural Sciences and Engineering Research Council of Canada.

About the Authors

Derek Bingham is a professor of statistics and actuarial science at Simon Fraser University and a Canada Research Chair in industrial statistics. He holds degrees in mathematics and statistics from Concordia, Carleton, and Simon Fraser University. His primary research interests lie in the design and analysis of physical and computer experiments. He was the 2013 recipient of the CRM–SSC Prize, offered jointly by the Centre de recherches mathématiques (in Montréal) and the Statistical Society of Canada.

Pritam Ranjan is an associate professor in the Department of Mathematics and Statistics at Acadia University, Wolfville, Nova Scotia. He obtained BStat and MStat degrees from the Indian Statistical Institute in Kolkata, and a PhD degree in statistics from Simon Fraser University. His research interests include design and analysis of computer experiments, statistical models for computer simulators, sequential designs for feature estimation, and fractional factorial designs with randomization restrictions.

William J. Welch joined the Department of Statistics at the University of British Columbia as a professor in 2003 and was head from 2003 to 2008. Prior to that he was at the University of Waterloo for 16 years. He received his PhD from Imperial College London. His research has concentrated on computer-aided design of experiments, quality improvement, the design and analysis of computer experiments, and drug-discovery from high-throughput screening data. He won the American Statistical Association's Statistics in Chemistry Prize and is an associate director of the Canadian Statistical Sciences Institute.

Bibliography

Bayarri, M. J., Berger, J. O., Calder, E. S., Dalbey, K., Lunagomez, S., Patra, A. K., Pitman, E. B., Spiller, E. T., and Wolpert, R. L. (2009). Using statistical and computer models to quantify volcanic hazards. *Technometrics*, 51:402–413.

Bayarri, M. J., Berger, J. O., Paulo, R., Sacks, J., Cafeo, J. A., Cavendish, J., Lin, C.-H., and Tu, J. (2007). A framework for validation of computer models. *Technometrics*, 49:138–154.

Bichon, B. J., Mahadevan, S., and Eldred, M. S. (2009). Reliability-based design optimization using efficient global reliability analysis. In *50th AIAA/ASME/ASCE/AHS/ASC Structures, Structural Dynamics, and Materials Conference*, AIAA 2009–2261, Palm Springs, CA. American Institute of Aeronautics and Astronautics.

Chipman, H. A., Ranjan, P., and Wang, W. (2012). Sequential design for computer experiments with a flexible Bayesian additive model. *The Canadian Journal of Statistics*, 40:663–678.

Currin, C., Mitchell, T., Morris, M., and Ylvisaker, D. (1991). Bayesian prediction of deterministic functions, with applications to the design and analysis of computer experiments. *Journal of the American Statistical Association*, 86:953–963.

Henkenjohann, N. and Kunert, J. (2007). An efficient sequential optimization approach based on the multivariate expected improvement criterion. *Quality Engineering*, 19:267–280.

Huang, D., Allen, T. T., Notz, W. I., and Miller, R. A. (2006). Sequential kriging optimization using multiple-fidelity evaluations. *Structural and Multidisciplinary Optimization*, 32:369–382.

Jones, D. R., Schonlau, M., and Welch, W. J. (1998). Efficient global optimization of expensive black-box functions. *Journal of Global Optimization*, 13:455–492.

Lehman, J. S., Santner, T. J., and Notz, W. I. (2004). Designing computer experiments to determine robust control variables. *Statistica Sinica*, 14:571–590.

Loeppky, J. L., Sacks, J., and Welch, W. J. (2009). Choosing the sample size of a computer experiment: A practical guide. *Technometrics*, 51:366–376.

Morris, M. D. and Mitchell, T. J. (1995). Exploratory designs for computational experiments. *Journal of Statistical Planning and Inference*, 43:381–402.

O'Hagan, A. (1992). Some Bayesian numerical analysis. In *Bayesian Statistics 4*, pp. 345–363. Oxford University Press.

Ponweiser, W., Wagner, T., and Vincze, M. (2008). Clustered multiple generalized expected improvement: A novel infill sampling criterion for surrogate models. In *2008 IEEE Congress on Evolutionary Computation (CEC 2008)*, pp. 3515–3522.

Ranjan, P. (2013). Comment: EI criteria for noisy computer simulators. *Technometrics*, 55:24–28.

Ranjan, P., Bingham, D., and Michailidis, G. (2008). Sequential experiment design for contour estimation from complex computer codes. *Technometrics*, 50:527–541.

Ranjan, P., Haynes, R., and Karsten, R. (2011). A computationally stable approach to Gaussian process interpolation of deterministic computer simulation data. *Technometrics*, 53:366–378.

Roy, S. (2008). *Sequential-Adaptive Design of Computer Experiments for the Estimation of Percentiles*. Doctoral dissertation, Ohio State University, Columbus, OH.

Sacks, J., Schiller, S. B., and Welch, W. J. (1989a). Designs for computer experiments. *Technometrics*, 31:41–47.

Sacks, J., Welch, W. J., Mitchell, T. J., and Wynn, H. P. (1989b). Design and analysis of computer experiments (with discussion). *Statistical Science*, 4:409–435.

Santner, T. J., Williams, B. J., and Notz, W. I. (2003). *The Design and Analysis of Computer Experiments*. Springer, New York.

Schonlau, M., Welch, W. J., and Jones, D. R. (1998). Global versus local search in constrained optimization of computer models. In *New Developments and Applications in Experimental Design*, volume 34, pp. 11–25. Institute of Mathematical Statistics, Hayward, CA.

Sóbester, A., Leary, S. J., and Keane, A. J. (2005). On the design of optimization strategies based on global response surface approximation models. *Journal of Global Optimization*, 33:31–59.

8

Statistical Genetic Modeling and Analysis of Complex Traits

Shelley B. Bull

*Lunenfeld–Tanenbaum Research Institute
and University of Toronto, Toronto, ON*

Jinko Graham

Simon Fraser University, Burnaby, BC

Celia M. T. Greenwood

McGill University, Montréal, QC

8.1 Introduction and Overview

The past 15 years have witnessed remarkable developments in the nature and volume of genetic variation data that are available for statistical genetic analysis. Genetic variation generally refers to differences between individuals in the DNA that is inherited from parents. Normally, identical DNA is contained in each of our cells and does not change during a lifetime; it is organized as a string of paired nucleotides for each of 22 chromosomes (autosomes), plus the X and Y sex-chromosomes. A base-pair refers to a pair of nucleotides at a specific position. Simply speaking, a gene can be defined as a set of specific DNA instructions that code an RNA or protein product. These in turn can affect the development of physical features as well as the production of proteins and metabolites that have biological consequences and may eventually play a role in disease causation and physiological variation. The genome refers to all of a person's nucleotides across the chromosomes, and is comprised of 3 billion nucleotides in total, indexed by base-pair position, with roughly 2% within the coding regions of genes, known as exons. As a first step toward discovering and characterizing the role of genes, genetic analysis of DNA variation investigates relationships of specific DNA variants with measurable human traits.

In response to technological advances in measuring DNA variation across the genome of an individual, new ways have arisen to investigate the role of genetic factors in complex traits and this has led to new methods of statistical modeling and analysis. Our objective in this chapter is to describe some contributions of Canadian researchers to statistical methods in human genetics. We will begin in Section 8.2 by giving a short description of genetic studies that involve either families or groups of unrelated individuals, with an outline of some essentials of the methods. Then in Section 8.3 we will discuss advances in two areas of research, highlighting their applications and impact in disease studies. Section 8.4 will comment on some other current and emerging areas of study. Before doing this, we will first describe some of the ways that information on a person's genome is obtained.

In the investigation of the genetics of complex traits and diseases, in which trait variation and disease risk arise from a combination of multiple genetic and environmental factors, the ultimate scientific objective is to understand the underlying genetic and biological mechanisms. This process often begins by scanning (i.e., screening) the entire genomes of specific individuals to identify a chromosomal region that may harbor a genetic variant that is a risk factor for the disease or explains variation in the trait of interest. The two primary methods of analysis used in such genome-wide studies are genetic linkage and genetic association analysis. Genome-wide studies involve measurement of a large number of genetic markers, each with a known genomic position (locus) on a specific chromosome. At an observed genetic locus, the measured marker can take on different values or variants, known as alleles.

As summarized in Table 8.1, early genome scans used sets of markers known as microsatellites that can take on many different values (referred to as polymorphisms), but these were subsequently replaced with cost-effective arrays of single nucleotide polymorphisms (SNPs) which are markers that take on only two values (e.g., alleles A and a) at a single base-pair position on a chromosome. The design of these arrays exploits information about local dependence between neighboring SNP markers (known as linkage disequilibrium, LD) to choose so-called tagSNPs which serve to represent the genetic information in a small region, saving the effort of measuring every SNP therein. A tagSNP usually has no biological role, but can indirectly detect variation at an unobserved SNP that is directly involved in disease expression. SNP arrays are designed to comprehensively assess the kind of genetic variation across the genome that occurs reasonably often in human populations, both within genes and in regions outside genes. The density of SNPs measured per chromosome has steadily increased with each improvement in array technology, so that standard arrays can measure more than one million SNP markers genome-wide, with estimates of another 2 million by genetic imputation based on LD information available from external population sources through the International HapMap Project. Next generation sequencing (NGS) technology, now emerging, aims to directly measure variation at every base-pair position with sufficient accuracy for near complete characterization of an individual's genome, including variants that occur very rarely in a population.

TABLE 8.1: Summary of technologies for measurement of genetic variation. Microsatellite markers have been widely used in genetic analysis of study designs involving pedigrees, affected relatives, and case-parent trios. SNP arrays now predominate in most study designs, including those with unrelated individuals, particularly case-control studies, and use of next generation sequencing is increasing.

High-Throughput Technology	Type of Variant	Number of Markers/ Variants	Genotyping Accuracy
Microsatellite Markers	Highly polymorphic	Hundreds	High
SNP Arrays	Binary: common	1 Million	Moderate to High
Next Generation Sequencing	Binary: Rare, low frequency and common	3 Billion	Variable

8.2 Essentials of Statistical Genetic Methods

By examining patterns of inheritance in families, genetic linkage analysis aims to detect chromosomal regions containing genes that influence the risk of specific inherited diseases or traits. We need to know the pedigree (relationships among family members), as well as disease or trait information and genetic typing for at least some of the family members. Two genetic loci are said to be linked when the parental alleles transmitted to a child at one locus are not independent of the parental alleles at the other locus. Figure 8.1 illustrates the genetic transmission of chromosomal material to the children of two parents drawn from a randomly mating population (i.e., within a nuclear family).

In Figure 8.1(a), the patterned vertical bars represent a pair of chromosomes for each of six individuals in a population. The horizontal tick marks represent chromosome positions at which a genetic marker or variant can be measured; one allele occurs on each chromosome. As a simple example, suppose that the dark diagonal and light dotted chromosomes each carry the A allele at a marker locus on this chromosome, and the other chromosomes carry the a allele at the same locus. A genotype for an individual is composed of the pair of unordered alleles observed at the marker locus, for example AA, Aa,

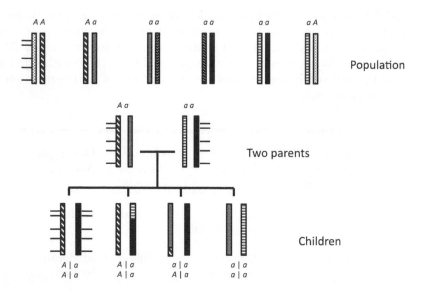

FIGURE 8.1: Illustration of (a) the genetic variability in a population, as well as (b) genetic transmission from parents to offspring in a nuclear family (two parents and their children) and identical-by-descent (IBD) sharing among siblings.

or *aa*. Genotyping methods provide information about the two alleles at the particular location, but not the parental source (so that *Aa* is not distinguishable from *aA*). In the nuclear family depicted in Figure 8.1(b), which consists of parents and their children, the parents drawn from the population produce four children. A chromosome composed of two patterns represents a recombinant chromosome, while a chromosome of one pattern is non-recombinant. A recombinant chromosome can be created by the occurrence of crossovers during gamete formation (either the egg or the sperm) in which chromosomal material is exchanged between the two chromosomes of one parent, although exchange does not always happen. For each gamete, such crossover events occur independently with varying probabilities and locations. Each child then inherits one chromosome from their mother, and likewise, one from their father. When a genetic marker is close to a disease gene, it is more likely that a parental allele at the marker locus is jointly transmitted with a specific allele from the same parent at the disease locus. Informative parental genotyping, here represented by differently-patterned chromosomes, is usually required for precise inference about which parent transmitted a particular allele.

One type of genetic linkage analysis, usually referred to as model-based or parametric linkage analysis, requires explicit model assumptions about the relationship between the unknown gene variants and the disease or trait, and the pattern of disease inheritance in a family. Another type of linkage analysis, known as model-free analysis, does not require specification of a disease inheritance model; see Xu et al. (2012) for a comprehensive summary. Rather, patterns of genetic similarity among affected relatives at a marker locus are compared to what one would expect at a location distant from a disease gene. Genetic similarity is typically assessed by allele sharing, which is discussed in the following subsection. Genetic linkage tests have been used as a first step in the search for disease susceptibility genes, and a genome-wide scan for linkage may simply identify broad regions that harbor such genes. These regions can then be examined more closely by analysis with alternative methods or by collection and analysis of higher density genetic typing data.

Genetic association analysis also aims to detect chromosomal regions that harbor genes for complex traits, but proceeds by testing the null hypothesis of no association between a SNP genotype and disease status, or trait value. Case-control studies of unrelated individuals that compare the distribution of genotypes in those with and without disease (i.e., "affected" and "unaffected") have predominated. Nevertheless, statistical methods for genetic association analysis involving families have remained of considerable interest. Family-based designs for assessing genetic association in families include the case-parent trio design, in which an affected child and his or her parents are genotyped. Under this design, one examines whether an affected child inherits a parental allele more often than expected under Mendelian inheritance, or in other words, whether allele transmission from the parents in a sample of trios exceeds the expected proportion of 1/2. Family-based designs have been extended to the analysis of disease status and trait values in nuclear families and large pedigrees, combining comparisons between unrelated individuals with comparisons between related individuals from the same family (Mirea et al., 2012).

Recently and with the advent of next generation sequencing (NGS), there is renewed interest in classical genetic linkage analysis methods. In families carrying rare genetic mutations that nearly always produce disease, linkage analysis can efficiently narrow down the genomic regions likely to carry the causal variant. Consequently, there is a substantial increase in the accuracy of inferring which of the genetic alterations identified by NGS is most likely to be a disease-causing mutation.

8.2.1 Modeling Genetic Sharing in Sibships and Relative Pairs

Here, we focus our discussion on genetic linkage analysis for a binary disease state (affected/unaffected), and we assume that sufficient genetic data are available so that it is possible to quite accurately infer whether two related individuals share 0, 1 or 2 copies of a chosen chromosomal region. Such sharing implies that the chromosomal region was inherited from a common ancestor, and this is termed identical-by-descent (IBD) sharing of chromosomal regions.

In the nuclear family illustrated in Figure 8.1, the first and the second child share the dark diagonal chromosome inherited from one parent but share only the lower (solid dark) part of the chromosome inherited from the other parent, so IBD sharing is equal to 1 in the upper part of the chromosome and equal to 2 in the lower part. The second and third child have IBD sharing equal to 0 in the upper part of the chromosome, IBD sharing equal to 1 in the middle part, and IBD sharing equal to 2 at the lower tip. In contrast, the first and the last child have IBD sharing equal to 0 across the entire chromosome. An observation in this family that the first, second, and third child were all affected with disease, whereas the last was not, would be consistent with a disease gene locus in the lower tip of the chromosome.

Mendel's law says that a child is equally likely to inherit one half or the other half of a parent's chromosomal material. In sib pairs generally (i.e., not selected according to their disease status), or in affected sib pairs at markers that are distant from a disease gene locus, we expect IBD sharing proportions of $(1/4, 1/2, 1/4)$ for IBD sharing of $(0, 1, 2)$ copies, respectively. (At a specific location, for example at the top of the chromosomes in Figure 8.1(b), there are four possible genotypes that are equally likely to be inherited by a child. Then for a pair of siblings, there are 16 possible combinations of two genotypes, and among these, there are four that share IBD $= 0$, eight that share IBD $= 1$, and four that share IBD $= 2$). However at a marker close to a disease susceptibility gene, we expect to observe excess IBD in a sample of affected sib pairs, with a higher proportion of IBD $= 2$ and a lower proportion of IBD $= 0$.

Conceptually, the goal of genetic linkage analysis is to find chromosomal regions with excess IBD sharing among related individuals. Such a chromosomal region is then likely to contain a gene or genetic variant that alters disease risk. In the absence of any association with disease, the probability distribution of IBD sharing follows from Mendel's laws of inheritance. We denote the probability that a pair of related individuals share k alleles identical by descent as $z_k = \Pr(\text{share } k \text{ copies})$. For a randomly chosen chromosomal region (i.e., a region in which no disease gene is located), the IBD sharing probabilities are $(z_0, z_1, z_2) = (1/4, 1/2, 1/4)$ for siblings, and $(3/4, 1/4, 0)$ for first cousins, for example. When multiple markers within a chromosome region are examined for linkage in a sample of affected relatives, a location with sufficiently large excess IBD sharing is inferred to be close to a potential disease gene locus, while more distant loci exhibit lesser amounts of excess sharing.

Figure 8.2 illustrates the characteristic reduction in average IBD sharing in affected siblings with increasing distance from a linkage peak at which IBD sharing is maximal. Marker locations under the linkage curve of excess sharing constitute a linkage region.

Statistical estimation and testing of hypotheses is often done using likelihood functions (LF). A likelihood function associated with a particular dataset is obtained by considering the probability of the observed data under a hypothesized model; it thus is a function of the parameters that specify the model. Here the z_k are the parameters in the likelihood for a simple allele

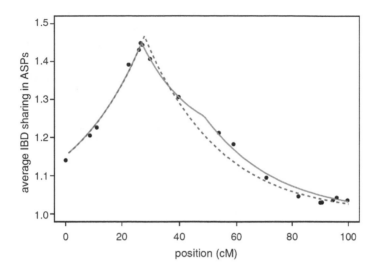

FIGURE 8.2: Patterns of identical-by-descent (IBD) sharing in a region of linkage on chromosome 6 in a sample of affected sib pairs (ASPs) with diabetes (taken from Biernacka et al., 2005). The solid circles indicate average IBD sharing values in the ASPs at each of 18 markers. The two curves show the expected IBD sharing for a one-locus model (dashed line) and for a two-locus model (solid line); cM is a measure of chromosome marker position commonly used in linkage analysis.

sharing linkage model, and they can be estimated and tested using the LF. Formally, one way to test for linkage is by a likelihood ratio (LR) statistic where the likelihood at the estimated IBD proportions $z = (\hat{z}_0, \hat{z}_1, \hat{z}_2)$ is compared to the likelihood under the null hypothesis of no excess sharing. The LR statistic can be maximized with respect to the z_k parameters, and the null hypothesis of no linkage is rejected when the maximized LR statistic is large in comparison to a test criterion. The maximum likelihood estimation is constrained such that the estimated IBD proportions sum to 1. In affected sibling pairs, the power of the LR test to detect linkage can be improved by evaluating the likelihood at a different point z^*, where z^* is constrained to lie within the so-called plausible triangle of z_k values ($z_1 \leq 1/2$ and $z_1 \geq 2z_0$) that are consistent with the underlying genetics of IBD allele sharing (Holmans, 1993; Feng et al., 2005).

8.2.2 Models for Genetic Transmission in Families

Family-based designs for genetic association are also useful in the analysis of disease status in families, and analysis can be conducted in a nuclear family

TABLE 8.2: Formulation of allele transmission from parents to an affected child, for family-based association tests of excess transmission under a case-parent trio design: n_{Aa} is the observed count of Aa parents that transmit A and not a, n_{aA} is the observed count of aA parents that transmit a and not A, n_{AA} is the observed count of homozygous AA parents, and n_{aa} is the observed count of homozygous aa parents.

	Not Transmitted	
Transmitted	A	a
A	n_{AA}	n_{Aa}
a	n_{aA}	n_{aa}

with only one affected child. Returning to Figure 8.1, now we consider that child 1 is affected and with two parents together constitutes a case-parent trio. Suppose that the dark diagonal chromosome carries the A allele at a marker locus on this chromosome, and the other chromosomes carry the a allele at the same locus. When we genotype a marker locus on the chromosome, we find that the first parent transmitted their diagonal chromosome A allele and did not transmit their gray chromosome a allele to the child, whereas the second parent transmitted their solid dark a allele and not their striped a allele. More generally, assuming n trios with $2n$ parents, we can construct a 2×2 table as given in Table 8.2 in which n_{Aa} is the observed count of parents that transmit A and not a, n_{aA} is the observed count that transmit a and not A, and so on. In our simple example, parent 1 would contribute to the n_{Aa} count, but parent 2 is homozygous for a and contributes to the n_{aa} count. If the marker allele is unrelated to being affected with disease then the probability that a parent transmits one allele or the other is $1/2$, and among parents with the Aa genotype, no difference is expected in the proportions of A and a alleles that are transmitted. In contrast, excess transmission of the A allele with n_{Aa} greater than n_{aA}, for example, would provide family-based evidence for association of that allele with disease. As in the allele-sharing analysis, the success of this approach is dependent on the presence of linkage between the marker locus and the disease locus.

In the case of a biallelic genetic variant such as a SNP (e.g., alleles A and a) and a single affected child, a test statistic to detect excess transmission can be constructed from the off-diagonal cell counts in the 2×2 table, where the margins correspond to transmitted and untransmitted alleles for the parents, and each parent contributes one observation to the table (Table 8.2). Among several model formulations developed for this kind of pattern of genetic inheritance, known as transmission disequilibrium, one that lends itself well to generalization is the conditional on parental genotypes (CPG) likelihood (Schaid and Sommer, 1993). Following the notation of Mirea et al. (2012) for the case-parent trio design, this likelihood models $\Pr(G \mid G_m, G_f, D = 1)$, the probability of the child's genotype G conditional on the child being a case ($D = 1$) and the parental genotypes (G_m for the mother and G_f for the father), in terms of the genotype relative risk parameters

$$\psi_{AA} = \frac{\Pr(D = 1 \mid G = AA, G_m, G_f)}{\Pr(D = 1 \mid G = aa, G_m, G_f)}$$

and

$$\psi_{Aa} = \frac{\Pr(D = 1 \mid G = Aa, G_m, G_f)}{\Pr(D = 1 \mid G = aa, G_m, G_f)}.$$

Assuming a gene-dose model where

$$\psi_{Aa} = \psi \quad \text{and} \quad \psi_{AA} = \psi_{Aa}^2 = \psi^2,$$

the maximum likelihood estimate is $\hat{\psi} = n_{Aa}/n_{aA}$. Families in which both parents are homozygous (e.g., AA or aa) at a marker locus are uninformative. The null hypothesis specifying no excess transmissions, $\mathcal{H}_0 : \psi = 1$, can be formally tested using a LR statistic.

8.2.3   Genetic Association in Unrelated Individuals

The era of genome-wide association studies (GWAS) has been largely driven by the availability of inexpensive SNP-based technology to systematically examine the entire genome. Statistical analysis in GWAS is characterized by the application of conventional methods for hypothesis testing of each one of a very large number of SNPs genotyped using arrays (see Chapter 9 by Craiu and Sun for discussion of approaches to statistical inference in GWAS). The problem of high false positive error rates generated by conducting millions of tests has been addressed by requiring strict significance thresholds in the discovery study and replication in an independent study. Because stringent criteria for genome-wide statistical significance are necessary to reduce the number of false positive associations, very large sample sizes are required, and case-control study designs have predominated; these studies compare the genotypes of unrelated individuals affected with disease (i.e., cases) to those of healthy individuals from the general population or individuals known to be unaffected with the disease (i.e., controls).

As illustrated in Table 8.3, assuming n cases and m controls have been genotyped at a SNP locus, we can construct a 2×3 table to compare their genotype frequencies. We denote the genotype probabilities for cases and controls by (p_2, p_1, p_0) and (q_2, q_1, q_0) respectively, according to the number of copies of the less frequent allele in the genotype (e.g., the number of copies of A in AA, Aa or aA, aa). Assuming a gene-dose model in which the probability

TABLE 8.3: Formulation of genetic association in unrelated individuals under a case-control design: n_2 is the observed count of cases with genotype AA, n_1 is the observed count of cases with genotype Aa, n_0 is the observed count of cases with genotype aa, and m_x is the observed count of controls with x copies of allele A.

	Individual Genotype		
	AA	Aa	aa
Case	n_2	n_1	n_0
Control	m_2	m_1	m_0

of being affected depends on the number of copies of the allele, a genotypic odds ratio parameter, OR, can be defined such that

$$OR = \frac{p_2/q_2}{p_1/q_1} = \frac{p_1/q_1}{p_0/q_0},$$

and estimated from the observed genotype counts using likelihood methods.

GWAS provide a cost-effective platform to detect associations, and can often identify a narrower chromosome region than linkage analysis. However, association analysis has been largely limited to SNP variants that occur in more that 5% of individuals, and a disease-associated GWAS tagSNP is not usually causal itself but is close to a causal variant. GWAS typically have not provided sufficient refinement at the base-pair level to identify potential disease-causing variants. Next generation sequencing (NGS) and genetic imputation to augment the set of genotyped SNPs can be implemented in GWAS samples to better determine the base-pair location of the genetic variant that is possibly relevant. In some studies, only the regions surrounding the significant tag SNPs are sequenced. In other studies, the entire genome or exome is imputed or sequenced. The ability to comprehensively examine all genetic variation, including that occurring rarely or at low frequency in a population, is the main attraction of NGS for genetic association studies, but in this chapter we leave aside discussion of methods for the analysis of NGS rare variants, which is currently a very active area of research.

8.3   Advances in Statistical Genetic Methods

8.3.1   Linkage with Covariate Data

Allele-sharing models such as the one introduced in Section 8.2.1 have been widely used for the study of genetic linkage in complex traits and diseases. When the etiology of a disease is complex (i.e., when there are many factors coming together to cause disease), heterogeneity can occur because of differences in genetic or environmental factors that cause disease, or because individuals with disease subtypes having different causes appear to be affected with the same disease. In this situation, families may exhibit excess IBD sharing at several marker loci and subsets of the families may be linked to different markers. Likewise, if an environmental exposure as well as a high-risk gene variant need to co-occur in order to increase risk of disease, then unexposed families will not exhibit excess sharing even for a marker close to the disease gene locus. Classifying families according to covariate data (i.e., data on factors that may be disease-related) can often help to reduce this heterogeneity, and improve the sensitivity of linkage analysis. The inherent challenge in considering covariate effects on genetic linkage is that a covariate is defined for an individual, but linkage is defined by examining sharing in pairs or sets of individuals. It is therefore necessary to construct covariates that apply to a pair or a set. For example, in a pair of diseased siblings, it might be of interest to know if they were both diagnosed at a young age, with the idea that genetic factors are more important in early onset than in adult onset disease. The concept of heterogeneity in linkage evidence was instrumental in a study of genetic factors influencing inflammatory bowel disease. Canadian families containing at least two individuals diagnosed with either Crohn's disease (CD) or ulcerative colitis (UC) were recruited and a linkage analysis based on IBD allele sharing in 158 families identified a region of interest on chromosome 5 (Rioux et al., 2000). Exploration of covariate associations in affected relative pairs in an overlapping but slightly larger set of families (167 families and 199 affected sibling pairs) found evidence for differences in allele sharing that depended on diagnostic subtype and age at diagnosis (Bull et al., 2002). The sharing patterns in the families were associated more strongly with CD than with UC, and particularly with families containing at least one individual with a young age at diagnosis.

For simplicity, the discussion here assumes that one chromosomal region is being examined and that the genetic data for an individual at that location are represented by G. If the probability of being diagnosed with disease D depends on a covariate, x, such as age or an environmental exposure, as well as the genotype G, this can be thought of as a modification of the conditional probabilities $\Pr(D|G)$ by the covariate; i.e.,

$$\Pr(D|G, x_1) \neq \Pr(D|G, x_2),$$

where x_1 and x_2 are different covariate values. Given this assumption, it can then be shown that the probability distribution of IBD genetic sharing between a pair of diseased, related individuals also depends on the covariate values of the pair; calculations for these probabilities essentially use Bayes rule. Let x_P represent a covariate that applies to a pair of affected relatives, and $k = 0, 1, 2$ enumerate the IBD states. In one of the first studies in this direction (Greenwood and Bull, 1999a), the dependence of the IBD proportions on covariates in affected sibling pairs was parameterized by a multinomial model,

$$z_k(x_P) = \frac{\exp(\beta_k x_P)}{1 + \exp(\beta_1 x_P) + \exp(\beta_2 x_P)}$$

in which the parameters β_1 and β_2 correspond to differences in IBD sharing that depend on covariates. In the study of CD and UC families noted above, both family-level and pair-level covariates were defined (i.e., ethnic background: Jewish versus non-Jewish; diagnostic subtype: all sibs CD, all sibs UC, or both CD and UC sibs; and age at diagnosis: one sib diagnosed by age 16 years versus both sibs diagnosed after age 16). Although inference about linkage in the absence of covariate effects can be improved by the triangle constraints introduced in Section 8.2.1, their use in linkage models with covariates is more complicated, since the constraints may no longer apply for all sibling pairs, particularly for pairs who have different values for the covariate. The most attractive option may be to implement partial constraints, for example within covariate-defined subgroups, which although not optimal, may nevertheless improve power of the likelihood ratio test in most situations (Greenwood and Bull, 1999a).

Subsequent development of approaches to test for linkage in the presence of covariates beyond the affected sibpair analysis we have described here allow many types of relative pairs to be analyzed simultaneously. For example, because the expected IBD sharing proportions depend on the type of relative pair, Xu et al. (2006) extended the model of Olson (1999) to include relative-pair-level covariates. In this framework, the likelihood contributions of different relative pair types are unified by writing the probabilities in terms of the increased disease risk for IBD $= k$ ($k = 1$ or 2) relative to IBD $= 0$. The resulting covariate-dependent relative risk formulation could then be incorporated into a tree-based algorithm designed to select the important covariates; this method identified clinical covariates altering evidence for linkage at a candidate region in families including cases of bipolar affective disorder (Xu et al., 2006). It is also possible to include individual-level covariates in IBD-sharing-based linkage analysis through judicious weighting of each individual's contribution to the test of linkage (Whittemore and Halpern, 2006). Finally, for linkage studies that examine multiple pairs of relatives in the same family, practical methods to account for the dependence between the pairs provide accurate assessment of statistical significance (Greenwood and Bull, 1999b; Schaid et al., 2007).

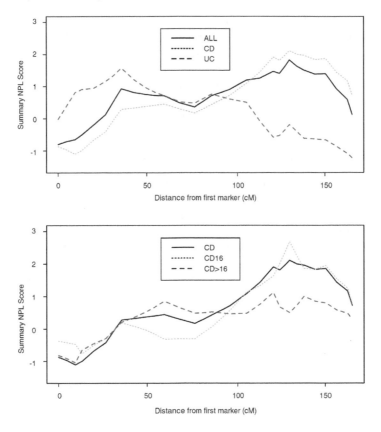

FIGURE 8.3: Summary nonparametric linkage (NPL) scores for family sub-groups with inflammatory bowel disease in a region of linkage on chromosome 5 (taken From Mirea et al., 2004). The solid line in the upper figure applies to all families, CD and UC denote diagnostic subgroups, while CD 16 and CD>16 distinguishes subgroups defined by minimum age at onset $\leq$ 16 years in the family. The summary NPL score is an average of family-specific linkage statistics calculated using 17 markers. cM is a measure of chromosome marker position commonly used in linkage analysis.

Further analysis of the inflammatory bowel disease families at the chromo-some 5 locus using family-level tests of covariate-based heterogeneity (Mirea et al., 2004) also demonstrated some evidence for heterogeneity in linkage depending on the pedigree minimum age of onset (Xu et al., 2012). The non-parametric linkage (NPL) statistic used in Figure 8.3 is a summary of excess IBD sharing among all the affected relatives in a family. We see differences between CD and UC subtypes in the family-specific NPL score, and according to early age at diagnosis in the CD subtype. The prominent genomic location in the latter has shown consistent association with disease risk in subsequent studies (Weizman and Silverberg, 2012), but unfortunately, identifying exactly which DNA changes are responsible for the increased disease risk has proven extremely elusive. Investigation of the genetics of inflammatory bowel disease continues, with a recent large meta-GWAS study of over 75,000 cases and controls, yielding a cumulative total of 163 loci that meet genome-wide significance thresholds (Jostins et al., 2012).

8.3.2 Effect Estimation in Genome-Wide Analysis

Whether conducted by linkage or association analysis, genome-wide investigations are fundamentally discovery studies in that they aim to comprehensively examine all regions of the genome and discover regions, genes, or variants that will be evaluated in subsequent studies. It is important for scientific acceptance that such potential effects also be seen in one or more independent "replication" studies of individuals or families that did not participate in the discovery study. When the same observations are used for both gene discovery and effect estimation, and a genetic effect is estimated only when the test for linkage or association at that locus meets genome-wide statistical significance criteria, the effect estimate is on average larger in magnitude than the true value (Göring et al., 2001). Bias in genetic effect estimates, a phenomenon also known as the winner's curse or the Beavis effect (Xu, 2003), can occur in both genome-wide linkage and genome-wide association studies. The bias introduced in this way is a form of selection bias (Efron, 2011), and the observation of a smaller effect in an independent replication sample is common. This phenomenon is well-illustrated by a case-control GWAS of psoriasis that reported odds ratios for genetic association calculated in independent discovery and replication samples (Nair et al., 2009). All of the 10 most significant SNPs identified in the discovery sample of 2759 individuals had smaller odds ratios in the replication sample of 10,099 individuals, with the percentage reduction ranging from 5% to 20%.

Such bias is particularly critical in replication study design. When an estimate from a discovery study is used to calculate the sample size required for a replication study, a biased effect estimate will produce a sample size estimate that is too small, leaving the investigators with a study under-powered to replicate a true association. If the association is not replicated, it may be assumed the original finding was a false positive association, when in fact it is a true association that was not replicated due to inadequate sample size. In addition, an accurate estimate of the genetic effect is important for estimation of the proportion of heritability explained by significant genetic associations, for meta-analyses from multiple studies, and for interpretation of detected associations.

To address the problem of selection bias, initially in the setting of allele-sharing linkage analysis, Sun and Bull (2005) developed a genome-wide "bootstrap" resampling method in which random samples drawn from the original study data are reanalyzed. Within each such bootstrap resample, which consists of a subset of the original study observations, repeating the entire genome-wide analysis imitates a "discovery" study, and captures the effects of the statistical significance criterion, as well as the correlation structure among the genetic loci. A genetic effect estimate for a locus with a test statistic that exceeds the genome-wide significance threshold in the bootstrap resample analysis, will be subject to the "winner's curse." Then to imitate an independent study, the genetic effect at the locus discovered in the bootstrap resample is estimated in the original study observations not included in the bootstrap resample. The difference between these two estimates reflects the winner's curse bias. This process is repeated in each of a large number of bootstrap resamples and the average difference is taken as an estimate of bias. Because the size of the bias depends on the ranking of the loci according to the size of the test statistic (i.e., whether it is the largest or the tenth largest, for example), the bias for the kth ranked locus detected in the original analysis is estimated with the bias estimates for the kth ranked locus in each of the bootstrap resamples. This is called a shrinkage estimator because the bias calculated in this way is used to reduce the genetic effect estimate for each of the loci identified in the original study.

This bootstrap implementation has proven valuable in several scientific settings, including genetic linkage analysis, GWAS and NGS (Wu et al., 2006; Yu et al., 2007; Faye et al., 2011, 2013). Wu et al. (2006) evaluated the bootstrap method for a quantitative trait linkage scan and, in the situation where multiple significant markers are detected, described how to estimate the proportion of genetic variance explained by a marker (a quantity known as locus-specific heritability). In GWAS, it can account for SNP selection according to statistical significance or relative ranking among SNPs, and a software package with an efficient implementation for GWAS is publicly available (`www.utstat.utoronto.ca/sun/Software/BR2/br2-web/br2.html`). Applications in genetic association studies of psoriasis, diabetic glycemia, and time to diabetic nephropathy have provided additional insight into the performance and prospects for replication studies (Al-Kateb et al., 2008; Paterson et al., 2010; Sun et al., 2011; Poirier et al., 2013). Additional aspects of selection bias have been studied in the NGS setting in which regions surrounding highly ranked GWAS SNPs are sequenced by NGS in the same sample used to detect the region (Faye and Bull, 2011). In addition, SNP genotyping is subject to both measurement error and to errors of imputation. Recent work has considered further the consequences of differential sequencing error and imputation error, and applied the bootstrap shrinkage estimator in an analytic re-ranking method that improves identification of potentially causal variants, for example, in locating sequence variants within a chromosome region associated with the risk of prostate cancer (Faye et al., 2013).

8.4   Commentary: Current and Emerging Issues

For complex traits, such as cancer and obesity, epidemiologic evidence implicates environment and lifestyle as major factors. On the other hand, familial clustering of disease suggests that genetics also plays a role. An emerging focus of complex-trait epidemiology is the role of the environment (E) together with genes (G), often referred to as $G \times E$, or gene-environment interaction. In early genome-wide association studies (GWAS), environmental and lifestyle factors were rarely considered, owing to challenges with the feasibility, cost and validity of covariate measurement. However, GWAS have explained very little of the variation in traits that is due to genetic differences (Manolio et al., 2009), and so investigators are taking a closer look at $G \times E$ (Ober and Vercelli, 2011), with some success. For instance, among smokers, colorectal cancer is strongly associated with exposure to carcinogens in well-done red meat, but only in genetically susceptible individuals (Le Marchand et al., 2001). As another example, obesity is more highly associated with fat intake in carriers than in non-carriers of the risk allele in the PPAR-γ gene (e.g., Garaulet et al., 2011). In fact, the concept of $G \times E$ was instrumental in a recent longitudinal study of body-mass index (Abarin et al., 2012). The study followed 1096 children from birth to 14 years of age and analyzed the variation in their BMI growth trajectories. Among carriers of the risk allele in the fat mass and obesity gene, *FTO*, the increase in BMI was attenuated by longer duration of exclusive breastfeeding. These $G \times E$ findings have public health relevance for preventing obesity in genetically susceptible individuals.

Despite the successes described in this chapter, statistical challenges remain. For example, in cancer and other rare diseases, the case-control study is a practical design, but the ability to detect $G \times E$ is low, even with large sample sizes (Greenland, 1983). When the study population is ethnically homogeneous and E is not genetically influenced, it is reasonable to assume that G and E are statistically independent; i.e., a person's genotype is unrelated to their environmental covariates. In these cases, power may be enhanced by enforcing G–E independence (e.g., Umbach and Weinberg, 1997; Chatterjee and Carroll, 2005; Shin et al., 2007), as implemented in the R software package LUCA (Shin et al., 2007), available from the Comprehensive R Archive Network (CRAN; cran.r-project.org). Outside genetics, these methods can also be applied to exploit the independence of randomized treatments and covariates in nested case-control studies (Breslow et al., 2013). One drawback is that even slight departures from the assumed form of G–E dependence can inflate the chance of false-positive results in tests of $G \times E$ (e.g., Shin et al., 2007). Statistical independence of G and E within families, rather than in the population, is a weaker assumption that can increase the power to detect $G \times E$ in case-parent study designs. These designs are frequently used for rare diseases of early onset, such as childhood cancers (e.g., Infante-Rivard et al., 2007) in which prenatal factors such as parental occupational exposures may be relevant. Shin (2012) developed a data-driven approach that incorporates G–E independence within families to visualize and test $G \times E$ in case-parent trios. The approach is implemented in the R software package trioGxE available on CRAN.

Measurement or misclassification error can affect the interpretation of results for $G \times E$ in ways that require more research in terms of identification and correction (Greenland, 1980). Development of approaches to appropriately deal with the misclassification of G (e.g., Weinberg et al., 2011; Shin et al., 2012) and with E (Thomas, 2010) is an important area of ongoing research. Recent commentaries (Thomas, 2012; Thomas et al., 2012) suggest that methods for the joint analysis of genes and environment are likely to take on increasing prominence, in part due to the development of molecular technologies and biomarkers that can broadly assess environmental exposures.

Acknowledgments

We would like to mention here, with thanks, our trainees, colleagues, and collaborators: David Andrews, Joanna Biernacka, Andy Boright, Laurent Briollais, Mary Corey, Gerarda Darlington, Linnea Duke, Laura Faye, Claire Infante-Rivard, Sophia Lee, Juan Pablo Lewinger, Brad McNeney, Lucia Mirea, Kenneth Morgan, Andrew Paterson, Thomas Schulze, Jean Shin, Mark Silverberg, John Spinelli, Lei Sun, Dave Tritchler, Longyang Wu, and Wei Xu. In addition, we wish to acknowledge the following sources of funding: Mathematics of Information Technology and Complex Systems, Natural Sciences and Engineering Research Council, Canadian Institutes of Health Research.

About the Authors

Shelley B. Bull is a senior investigator in the Lunenfeld–Tanenbaum Research Institute of Mount Sinai Hospital, and a professor of biostatistics in the Dalla Lana School of Public Health, University of Toronto, where she co-directs the CIHR Strategic Training Program in Advanced Genetic Epidemiology and Statistical Genetics. Her undergraduate degree in mathematics and master's degree in statistics are from the University of Waterloo, and her PhD in epidemiology and biostatistics is from the University of Western Ontario. She is the recipient of the Anthony Miller Award for Excellence in Research in Public Health and the International Genetic Epidemiology Leadership Award. She was program chair of the Statistical Society of Canada Annual Meeting in 2011.

Jinko Graham is an associate professor of statistics and Actuarial Science at Simon Fraser University. Her research is directed toward the development of statistical methods for inference from genomic data, with a focus on genetic association studies and applications of the coalescent. Her undergraduate degree in mathematics and master's degree in statistics are from the University of British Columbia, and her PhD in biostatistics is from the University of Washington. From 2005 to 2012, she served on the CIHR Institute of Genetics Priority and Planning Committee for Population Genetics, Genetic Epidemiology and Complex Diseases.

Celia M. T. Greenwood is a senior scientist at the Lady Davis Institute for Medical Research, Jewish General Hospital, Montréal, and an associate professor at McGill University, with affiliations to three departments: Oncology, Epidemiology, Biostatistics and Occupational Health, and Human Genetics. Her research interests include numerous aspects of methodology for the analysis of human genetic and genomic data. She has degrees from McGill, the University of Waterloo, and the University of Toronto. After several years at the Hospital for Sick Children and the University of Toronto, she was appointed Weekend to End Cancer Career Scientist at the Lady Davis Institute in 2010.

Bibliography

Abarin, T., Wu, Y. Y., Warrington, N., Lye, S., Pennell, C., and Briollais, L. (2012). The impact of breastfeeding on FTO-related BMI growth trajectories: An application to the Raine pregnancy cohort study. *International Journal of Epidemiology*, 41:1650–1660.

Al-Kateb, H., Boright, A. P., Mirea, L., Xie, X., Sutradhar, R., Mowjoodi, A., Bharaj, B., Liu, M., Bucksa, J. M., Arends, V. L., Steffes, M. W., Cleary, P. A., Sun, W., Lachin, J. M., Thorner, P. S., Ho, M., McKnight, A. J., Maxwell, A. P., Savage, D. A., Kidd, K. K., Kidd, J. R., Speed, W. C., Orchard, T. J., Miller, R. G., Sun, L., Bull, S. B., Paterson, A. D., and Diabetes Control and Complications Trial/Epidemiology of Diabetes Interventions and Complications Research Group (2008). Multiple SOD1/SFRS15 variants are associated with the development and progression of diabetic nephropathy: The DCCT/EDIC Genetics study. *Diabetes*, 57:218–228.

Biernacka, J. M., Sun, L., and Bull, S. B. (2005). Simultaneous localization of two linked disease susceptibility genes. *Genetic Epidemiology*, 28:33–47.

Breslow, N. E., Amorim, G., Pettinger, M. B., and Rossouw, J. (2013). Using the whole cohort in the analysis of case-control data. *Statistics in Biosciences*, 5:232–249.

Bull, S. B., Greenwood, C. M. T., Mirea, L., and Morgan, K. (2002). Regression models for allele sharing: Analysis of accumulating data in affected sib pair studies. *Statistics in Medicine*, 21:431–444.

Chatterjee, N. and Carroll, R. J. (2005). Semiparametric maximum likelihood estimation exploiting gene-environment independence in case-control studies. *Biometrika*, 92:399–418.

Efron, B. (2011). Tweedie's formula and selection bias. *Journal of the American Statistical Association*, 106:1602–1614.

Faye, L. L. and Bull, S. B. (2011). Two-stage study designs combining genome-wide association studies, tag single-nucleotide polymorphisms, and exome sequencing: Accuracy of genetic effect estimates. *BMC Proceedings*, 5 Suppl 9:S64.

Faye, L. L., Machiela, M. J., Kraft, P., Bull, S. B., and Sun, L. (2013). Re-ranking sequencing variants in the post-GWAS era for accurate causal variant identification. *PLoS Genetics*, 9(8):e1003609. doi:10.1371/journal.pgen.1003609.

Faye, L. L., Sun, L., Dimitromanolakis, A., and Bull, S. B. (2011). A flexible genome-wide bootstrap method that accounts for ranking and threshold-selection bias in GWAS interpretation and replication study design. *Statistics in Medicine*, 30:1898–1912.

Feng, Z. Z., Chen, J., and Thompson, M. E. (2005). The universal validity of the possible triangle constraint for affected sib pairs. *The Canadian Journal of Statistics*, 33:297–310.

Garaulet, M., Smith, C., Hernández-González, T., Lee, Y., and Ordovás, J. (2011). PPAR-γ Pro12Ala interacts with fat intake for obesity and weight loss in a behavioural treatment based on the Mediterranean diet. *Molecular Nutrition & Food Research*, 55:1771–1779.

Göring, H. H., Terwilliger, J. D., and Blangero, J. (2001). Large upward bias in estimation of locus-specific effects from genomewide scans. *American Journal of Human Genetics*, 69:1357–1369.

Greenland, S. (1980). The effect of misclassification in the presence of covariates. *American Journal of Epidemiology*, 112:564–569.

Greenland, S. (1983). Tests for interaction in epidemiologic studies: A review and a study of power. *Statistics in Medicine*, 2:243–251.

Greenwood, C. M. T. and Bull, S. B. (1999a). Analysis of affected sib pairs, with covariates–with and without constraints. *American Journal of Human Genetics*, 64:871–885.

Greenwood, C. M. T. and Bull, S. B. (1999b). Down-weighting of multiple affected sib pairs leads to biased likelihood-ratio tests, under the assumption of no linkage. *American Journal of Human Genetics*, 64:1248–1252.

Holmans, P. (1993). Asymptotic properties of affected-sib-pair linkage analysis. *American Journal of Human Genetics*, 52:362–374.

Infante-Rivard, C., Vermunt, J., and Weinberg, C. R. (2007). Excess transmission of the NAD(P)H:Quinone Oxidoreductase 1 (NQO1) C609T polymorphism in families of children with acute lymphoblastic leukemia. *American Journal of Epidemiology*, 165:1248–1254.

Jostins, L., Ripke, S., Weersma, R. K., Duerr, R. H., McGovern, D. P., Hui, K. Y., Lee, J. C., Schumm, L. P., Sharma, Y., Anderson, C. A., et al. (2012). Host-microbe interactions have shaped the genetic architecture of inflammatory bowel disease. *Nature*, 491:119–124.

Le Marchand, L., Hankin, J. H., Wilkens, L. R., Pierce, L. M., Franke, A., Kolonel, L. N., Seifried, A., Custer, L. J., Chang, W., Lum-Jones, A., and Donlon, T. (2001). Combined effects of well-done red meat, smoking, and rapid n-acetyltransferase 2 and CYP1A2 phenotypes in increasing colorectal cancer risk. *Cancer Epidemiology Biomarkers & Prevention*, 10:1259–1266.

Manolio, T. A., Collins, F. S., Cox, N. J., Goldstein, D. B., Hindorff, L. A., Hunter, D. J., McCarthy, M. I., Ramos, E. M., Cardon, L. R., Chakravarti, A., et al. (2009). Finding the missing heritability of complex diseases. *Nature*, 461:747–753.

Mirea, L., Briollais, L., and Bull, S. B. (2004). Tests for covariate-associated heterogeneity in IBD allele sharing of affected relatives. *Genetic Epidemiology*, 26:44–60.

Mirea, L., Infante-Rivard, C., Sun, L., and Bull, S. B. (2012). Strategies for genetic association analyses combining unrelated case-control individuals and family trios. *American Journal of Epidemiology*, 176:70–79.

Nair, R. P., Duffin, K. C., Helms, C., Ding, J., Stuart, P. E., Goldgar, D., Gudjonsson, J. E., Li, Y., Tejasvi, T., Feng, B.-J., Ruether, A., Schreiber, S., Weichenthal, M., Gladman, D., Rahman, P., Schrodi, S. J., Prahalad, S., Guthery, S. L., Fischer, J., Liao, W., Kwok, P.-Y., Menter, A., Lathrop, G. M., Wise, C. A., Begovich, A. B., Voorhees, J. J., Elder, J. T., Krueger, G. G., Bowcock, A. M., Abecasis, G. R., and Collaborative Association Study of Psoriasis (2009). Genome-wide scan reveals association of psoriasis with IL-23 and NF-kappaB pathways. *Nature Genetics*, 41:199–204.

Ober, C. and Vercelli, D. (2011). Gene-environment interactions in human disease: Nuisance or opportunity? *Trends in Genetics*, 27:107–115.

Olson, J. M. (1999). A general conditional-logistic model for affected-relative-pair linkage studies. *American Journal of Human Genetics*, 65:1760–1769.

Paterson, A. D., Waggott, D., Boright, A. P., Hosseini, S. M., Shen, E., Sylvestre, M.-P., Wong, I., Bharaj, B., Cleary, P. A., Lachin, J. M., MAGIC (Meta-Analyses of Glucose and Insulin-related traits Consortium), Below, J., Nicolae, D., Cox, N. J., Canty, A. J., Sun, L., Bull, S. B., and Diabetes Control and Complications Trial/Epidemiology of Diabetes Interventions and Complications Research Group (2010). A genome-wide association study identifies a novel major locus for glycemic control in type 1 diabetes, as measured by both HbA1C and glucose. *Diabetes*, 59:539–549.

Poirier, J., Faye, L. L., Dimitromanolakis, A., Paterson, A. D., Sun, L., and Bull, S. B. (2013). A general procedure to address the winner's curse in genetic association studies: Bias-reduction in analysis of time-to-event traits. *Submitted for publication.*

Rioux, J. D., Silverberg, M. S., Daly, M. J., Steinhart, A. H., McLeod, R. S., Griffiths, A. M., Green, T., Brettin, T. S., Stone, V., Bull, S. B., Bitton, A., Williams, C. N., Greenberg, G. R., Cohen, Z., Lander, E. S., Hudson, T. J., and Siminovitch, K. A. (2000). Genomewide search in Canadian families with inflammatory bowel disease reveals two novel susceptibility loci. *American Journal of Human Genetics*, 66:1863–1870.

Schaid, D. J., Sinnwell, J. P., and Thibodeau, S. N. (2007). Testing genetic linkage with relative pairs and covariates by quasi-likelihood score statistics. *Human Heredity*, 64:220–233.

Schaid, D. J. and Sommer, S. S. (1993). Genotype relative risks: Methods for design and analysis of candidate-gene association studies. *American Journal of Human Genetics*, 53:1114–1126.

Shin, J.-H. (2012). *Inferring gene-environment interaction from case-parent trio data: Evaluation of and adjustment for spurious $G \times E$ and development of a data-smoothing method to uncover true $G \times E$.* Doctoral Dissertation, Simon Fraser University, Burnaby, BC. Available at `http://summit.sfu.ca/item/12281\#310`.

Shin, J.-H., Infante-Rivard, C., Graham, J., and McNeney, B. (2012). Adjusting for spurious gene-by-environment interaction using case-parent triads. *Statistical Applications in Genetics and Molecular Biology*, 11:1–21.

Shin, J.-H., McNeney, B., and Graham, J. (2007). Case-control inference of interaction between genetic and nongenetic risk factors under assumptions on their distribution. *Statistical Applications in Genetics and Molecular Biology*, 6:Article 13.

Sun, L. and Bull, S. B. (2005). Reduction of selection bias in genomewide studies by resampling. *Genetic Epidemiology*, 28:352–367.

Sun, L., Dimitromanolakis, A., Faye, L. L., Paterson, A. D., Waggott, D., DCCT/EDIC Research Group, and Bull, S. B. (2011). BR-squared: A practical solution to the winner's curse in genome-wide scans. *Human Genetics*, 129:545–552.

Thomas, D. C. (2010). Methods for investigating gene-environment interactions in candidate pathway and genome-wide association studies. *Annual Review of Public Health*, 31:21–36.

Thomas, D. C. (2012). Genetic epidemiology with a capital E: Where will we be in another 10 years? *Genetic Epidemiology*, 36:179–182.

Thomas, D. C., Lewinger, J. P., Murcray, C. E., and Gauderman, W. J. (2012). Invited commentary: GE-Whiz! Ratcheting gene-environment studies up to the whole genome and the whole exposome. *American Journal of Epidemiology*, 175:203–7; discussion 208.

Umbach, D. M. and Weinberg, C. R. (1997). Designing and analysing case-control studies to exploit independence of genotype and exposure. *Statistics in Medicine*, 16:1731–1743.

Weinberg, C. R., Shi, M., and Umbach, D. M. (2011). A sibling-augmented case-only approach for assessing multiplicative gene-environment interactions. *American Journal of Epidemiology*, 174:1183–1189.

Weizman, A. V. and Silverberg, M. S. (2012). Have genomic discoveries in inflammatory bowel disease translated into clinical progress? *Current Gastroenterology Reports*, 14:139–145.

Whittemore, A. S. and Halpern, J. (2006). Nonparametric linkage analysis using person-specific covariates. *Genetic Epidemiology*, 30:369–379.

Wu, L. Y., Sun, L., and Bull, S. B. (2006). Locus-specific heritability estimation via the bootstrap in linkage scans for quantitative trait loci. *Human Heredity*, 62:84–96.

Xu, S. (2003). Theoretical basis of the Beavis effect. *Genetics*, 165:2259–2268.

Xu, W., Bull, S. B., Mirea, L., and Greenwood, C. M. T. (2012). Model-free linkage analysis of a binary trait. In *Statistical Human Genetics*, volume 850 of *Methods in Molecular Biology*, pp. 317–345. Humana Press, New York.

Xu, W., Schulze, T. G., DePaulo, J., Bull, S. B., McMahon, F., and Greenwood, C. M. T. (2006). A tree-based model for allele-sharing-based linkage analysis in human complex diseases. *Genetic Epidemiology*, 30:155–169.

Yu, K., Chatterjee, N., Wheeler, W., Li, Q., Wang, S., Rothman, N., and Wacholder, S. (2007). Flexible design for following up positive findings. *American Journal of Human Genetics*, 81:540–551.

9

Bayesian Methods in Fisher's Statistical Genetics World

Radu V. Craiu and Lei Sun

University of Toronto, Toronto, ON

9.1   Background and Introduction

Statistical genetics is a scientific discipline that covers any statistical analysis of genetic data. The interplay between statistics and genetics has a long history, dating back to the seminal work by Fisher almost a century ago to confirm the genetic theory of chromosomal inheritance (Piegorsch, 1990). Recent advancements in genotyping (i.e., collecting genetic data) technologies have produced vast amounts of data, offering statisticians great opportunities in methodological development, implementation and application. For example, the high-dimensional genome-wide association studies conducted in the last few years have led to the development of a catalog of novel statistical methods for dissecting the genetic architecture of complex human traits; see, e.g., Thomas et al. (2009) and Begum et al. (2012).

The types of genetic data available are quite diverse; they include microsatellites, single-nucleotide polymorphisms, copy number variations, DNA methylation, and gene expression. The corresponding statistical methodologies are equally diverse, as illustrated by Shelley Bull, Jinko Graham, and Celia Greenwood in Chapter 8. For clarity and a more focused discussion, this expository piece is centered around studies of genetic association between single-nucleotide polymorphisms and heritable human traits. In the following, we first provide relevant genetic terminology. We then formulate genetic association studies in terms of regression models in which inferences on the regression coefficients are of interest. Using a published genome-wide association study as an example, we first describe the commonly used frequentist approaches to achieve testing and estimation objectives, and we then discuss alternative Bayesian methods and associated advantages as well as challenges. We conclude with discussions of other recent developments in Bayesian statistical genetics, focusing on contributions made by Canadian statisticians, and comment on future directions.

9.1.1   Basic Genetic Terminology

The building blocks of the human genome are base pairs, which are part of the DNA in each person's chromosomes. There are about three billion base pairs and over 99% are identical between individuals. A base pair that varies in a population and has two variants (alleles) is called a single-nucleotide polymorphism (SNP).

Let A and a denote the two alleles, and without loss of generality, let a be the minor allele for which the population allele frequency, $p(a)$, is such that $p(a) \leq .5$. This frequency is called the minor allele frequency (MAF) and some terminology is based on this: a common SNP has MAF $\geq .05$, a low frequency SNP has $.01 < \text{MAF} < .05$, and a rare SNP or rare variant has MAF $\leq .01$.

Although the exact number is a moving target, it is believed that there are over 10 million common SNPs and even more rare variants (1000Genome-sProjectConsortium, 2012). Human chromosomes are paired, with one inherited from the mother and the other from the father. Therefore, at each SNP location along the genome, there are two alleles forming a person's genotype, which can be AA, Aa, or aa.

9.1.2   Statistical Set-Up of Genetic Association Studies: Two Intertwined Issues

Let Y be the trait (i.e., phenotype) of interest, e.g., the presence of Type 1 diabetes (T1D) or blood glucose level, and let X represent the genotype of a SNP under study. Genetic association studies assess whether the response variable Y varies between different levels of X.

For example, if individuals with genotype aa tend to have a higher risk of developing T1D than individuals with genotype AA, then there is an association between X and Y. To statistically assess the relationship between Y and X, a so-called simple linear regression model can be considered if the trait is (approximately) normally distributed (e.g., blood glucose level), viz.

$$Y_i = \alpha + \beta X_i + \epsilon_i, \quad \epsilon_i \sim \mathcal{N}(0, \sigma^2), \tag{9.1}$$

where Y_i is the trait value for individual i, X_i is the SNP genotype for this individual, and $\epsilon_1, \ldots, \epsilon_n$ represent independent random variations, which are assumed to follow a Normal distribution with mean 0 and variance σ^2.

In many applications, the three genotypes, AA, Aa and aa, are coded numerically (additively) as $X = 0$, 1 and 2 to represent the number of copies of the minor allele a. In that case, model (9.1) assumes that increasing X by one unit will increase the Y value, on average, by β. This is also called the linear additive model because the effect of two copies of the minor allele (genotype aa, $X = 2$) is assumed to be twice the effect of 1 copy (genotype Aa, $X = 1$). In some genetic settings, other genotype models may be used, which we discuss in Section 9.1.3.

If the trait of interest is binary (e.g., the presence of Type 1 diabetes, T1D), the classical logistic regression can be used (Agresti, 2002),

$$\text{logit}\{E(Y_i)\} = \alpha + \beta X_i, \tag{9.2}$$

where Y_i indicates whether individual i has the disease ($Y_i = 1$) or not ($Y_i = 0$), $E(Y_i) = \Pr(Y = 1)$ and

$$\text{logit}\{\Pr(Y = 1)\} = \log\left\{\frac{\Pr(Y = 1)}{1 - \Pr(Y = 1)}\right\}$$

is called the log odds. Model (9.2) implies that when increasing X by 1 unit we expect an increase in the log odds value of Y by β, and the interpretation of β is the well known log odds ratio (log OR). The log odds, rather than just $E(Y_i)$, is used in (9.2) because it tends to give a better description of the relationship between Y_i and X_i.

In either regression setting, the primary objective of an association study is to identify which SNPs are related to the trait; this is equivalent to comparing the two hypotheses for each SNP, viz.

$$\mathcal{H}_0 : \beta = 0, \quad \mathcal{H}_1 : \beta \neq 0.$$

It is in this context that we will describe in Section 9.2 the commonly used frequentist and Bayesian approaches to assess whether the evidence provided by the data supports $\mathcal{H}_0$ or $\mathcal{H}_1$.

To a statistician, this may seem like an exceedingly simple problem. However, significant complications arise in applications. For example:

a) additional variables (also known as covariates) such as sex and age may need to be included in the regression models (9.1) or (9.2);

b) measurement errors occur in both X and Y;

c) individuals' phenotypes and genotypes can be correlated;

d) individuals may come from different populations;

e) multiple (common or rare) SNPs need to be jointly analyzed to increase power; and

f) multiple (binary, continuous or both) traits can be of interest.

Proper statistical treatment for any of these issues requires experience in statistical genetics and genetic epidemiology. Since we do not assume here that the reader has such prior knowledge, we will focus on explaining some of the basic statistical techniques used for genetic association studies. We briefly discuss some more complex issues in Section 9.4.

Once a trait-associated SNP has been identified, it is of interest to report the corresponding genetic effect size such as log OR, β in (9.2), and to plan follow-up studies that seek to replicate the initial finding. An interesting statistical question arises here. Let us assume that β is the true log OR of an associated SNP, and without loss of generality, assume that the minor allele of the SNP increases the risk of the disease under study (i.e., OR > 1 and $\beta > 0$). Let $\hat{\beta}$ be an estimated value reported from a study based on (9.2). It is known that $\hat{\beta}$ is unbiased, which means that if we were to repeat the experiment over and over again by collecting a different sample from the same population and applying the same statistical analysis, then the average of all the values of $\hat{\beta}$ would be β. However, not all studies are successful in identifying the association and only studies with sufficiently large $\hat{\beta}$ values are significant (i.e., result in rejecting $\mathcal{H}_0$). Therefore, $\hat{\beta}$ reported from a significant study is on average bigger than the true value. How to correct for this upward bias requires non-trivial statistical remedies, which we discuss in Section 9.3.

9.1.3   GWAS and an Example

Many current "high-throughput" genetic studies, such as the genome-wide association studies (GWAS) or next generation sequencing (NGS) studies, comprehensively investigate the whole genome to identify trait-associated SNPs. In that case, the scientific objective is to look for association between a trait and genotypes of hundreds of thousands or millions of SNPs.

Let X_{ij} be the genotype for individual i at SNP j with $i \in \{1, \ldots, n\}$ and $j \in \{1, \ldots, p\}$. One of the key features of GWAS and NGS is $n \ll p$, where n is in the range of 1,000 to 10,000 while p is in the range of 1 million for GWAS (or 10 millions for NGS). Although joint analyses of multiple SNPs have recently been tackled using more advanced regression methods such as the Lasso (see Chapter 5 by Rob Tibshirani), single-SNP association analysis is still predominant due to its simplicity and interpretability. In that case, regression models (9.1) or (9.2) are fitted repeatedly for each SNP $j \in \{1, \ldots, p\}$, and the corresponding (frequentist) decision rule concerning $\mathcal{H}_0$ is based on a p-value, which is the probability, computed assuming that $\mathcal{H}_0$ is true, that the test statistic is as extreme as the one observed.

Intuitively, the smaller the p-value, the smaller the evidence in favor of $\mathcal{H}_0$ provided by the data. Traditionally, if the p-value is less than .05 we reject the null hypothesis $\mathcal{H}_0$. However, even if the SNP is not associated with the trait, there is a 5% probability of wrongly rejecting $\mathcal{H}_0$; this is known as the Type 1 error. In the GWAS setting, we perform millions of tests in one analysis and errors accumulate. In order to control the overall (i.e., genome-wide or family-wise) Type 1 error rate at 5%, a stringent criterion such as p-value $< 5 \times 10^{-8}$ is typically used for each SNP included in the study (Dudbridge and Gusnanto, 2008).

WTCCC (2007) by the Wellcome Trust Case Control Consortium is a landmark work in which a GWAS was conducted for each of the seven common diseases, including coronary heart disease, Type 1 diabetes (T1D), Type 2 diabetes, rheumatoid arthritis, Crohn's disease, bipolar disorder and hypertension, with a total sample size of 16,179. Here we focus on the T1D study for which the sample consisted of 1,926 individuals with T1D (cases) and 2,872 subjects without T1D (controls). Genotype data were collected for these individuals at $469,557$ SNPs (after data quality control). To assess the association evidence between T1D and each of the SNPs, WTCCC (2007) calculated p-values for testing $\mathcal{H}_0 : \beta_j = 0$ for $1 \leq j \leq 469,557$, and reported the findings in their Table 2 (five SNPs with p-values $< 5 \times 10^{-7}$) and Table 3 (seven additional SNPs with p-values $< 10^{-5}$). They also provided Bayes factors for each of the 12 SNPs (Section 9.2.2).

Section 9.2 describes the details of their frequentist and Bayesian association analyses. Section 9.3 discusses the upward bias inherent in their reported β_j estimates and reviews different approaches for reducing the estimation bias using only the summary statistics.

9.2 Identification of Trait-Associated SNPs

9.2.1 p-Value

To identify SNPs associated with T1D using as an example the WTCCC (2007) study, first consider the logistic regression model

$$\text{logit}\{\text{E}(Y_i)\} = \alpha + \beta_j X_{ij},$$

where Y_i indicates whether individual i in the sample has T1D ($Y_i = 1$) or not ($Y_i = 0$), and $X_{ij} = 0, 1$ or 2 if the genotype of SNP j for individual i is, respectively, AA, Aa or aa. This is the additive model as discussed in Section 9.1.2. The p-value for testing $\mathcal{H}_0 : \beta_j = 0$ is denoted as $p_{j,\text{add}}$. Covariates such as age at onset of T1D were not included in the WTCCC analysis.

To discover SNPs with non-additive genetic effects, it is also important to consider the more flexible genotypic model, where the genotype of a SNP is treated as a categorical variable with three levels. For the genotypic model,

$$\text{logit}\{\text{E}(Y_i)\} = \alpha + \beta_{1j} X1_{ij} + \beta_{2j} X2_{ij}$$

can be used, where $X1$ and $X2$ are the standard dummy variables: ($X1_{ij} = 0, X2_{ij} = 0$) for genotype AA, ($X1_{ij} = 1, X2_{ij} = 0$) for Aa and ($X1_{ij} = 0, X2_{ij} = 1$) for aa. The p-value for testing $\mathcal{H}_0 : \beta_{1j} = \beta_{2j} = 0$ is denoted as $p_{j,\text{geno}}$. WTCCC (2007) then defined SNPs with strong evidence of association with T1D as SNPs with $\min(p_{j,\text{add}}, p_{j,\text{geno}}) < 5 \times 10^{-7}$, and moderate evidence of association as $\min(p_{j,\text{add}}, p_{j,\text{geno}}) < 10^{-5}$. We summarize their key results in Table 9.1. Although the two models generally give similar results, SNP rs17166496 clearly did not follow the additive assumption and would have been missed without the consideration of the alternative genotypic model.

TABLE 9.1: Results of the WTCCC genome-wide association study (GWAS) of Type 1 diabetes (T1D). All values except the ranks are restated here based on Tables 2 and 3 of WTCCC (2007). Note that the ranks in the table are within these 12 SNPs not at the genome-wide level involving all 469,557 SNPs.

| Chromo-some | SNP | MAF | Additive Model | | | |
| | | | Frequentist | | Bayesian | |
			p-value	Rank	$\log_{10}$BF	Rank
6	rs9272346	.387	2.42 E-134	1	141.93	1
1	rs6679677	.096	1.17 E-26	2	23.07	2
12	rs17696736	.424	2.17 E-15	3	12.53	3
12	rs11171739	.423	1.14 E-11	4	8.89	4
16	rs12708716	.350	9.24 E-08	5	5.15	5
4	rs17388568	.260	5.00 E-07	6	4.42	6
18	rs2542151	.163	1.89 E-06	7	3.91	7
10	rs2104286	.286	7.97 E-06	8	3.31	9
5	rs2544677	.242	8.23 E-06	9	3.32	8
1	rs2639703	.276	8.46 E-06	10	3.25	10
12	rs11052552	.486	1.02 E-04	11	2.22	11
5	rs17166496	.391	6.06 E-01	12	$-.97$	12

| Chromo-some | SNP | MAF | Genotypic Model | | | |
| | | | Frequentist | | Bayesian | |
			p-value	Rank	$\log_{10}$BF	Rank
6	rs9272346	.387	5.47 E-134	1	139.77	1
1	rs6679677	.096	5.43 E-26	2	22.83	2
12	rs17696736	.424	1.51 E-14	3	11.56	3
12	rs11171739	.423	9.71 E-11	4	8.24	4
16	rs12708716	.350	4.92 E-07	5	4.71	5
4	rs17388568	.260	3.27 E-06	7	3.89	6
18	rs2542151	.163	1.16 E-05	9	3.52	8
10	rs2104286	.286	4.32 E-05	11	2.88	11
5	rs2544677	.242	4.43 E-05	12	2.70	12
1	rs2639703	.276	1.74 E-05	10	3.06	10
12	rs11052552	.486	7.24 E-07	6	3.80	7
5	rs17166496	.391	5.20 E-06	8	3.25	9

9.2.2   Bayes Factor

The previous section showed that alternative models may lead to different conclusions regarding the association of a SNP with a given trait. Combining analyses produced by different models is desirable when there is no clear evidence supporting a specific model. However, this is difficult under the frequentist framework and leads us to consider Bayesian alternatives. WTCCC (2007) were among the first who considered a Bayesian framework for GWAS, and in particular they relied on Bayes factors (Kass and Raftery, 1995) to rank SNPs. Other early Bayesian work in the GWAS context includes Marchini et al. (2007) and Wakefield (2007), and Stephens and Balding (2009) gave an excellent review on this topic.

The Bayesian approach treats the parameter of interest, θ, as a random variable for which a prior distribution, $p(\theta|M)$, is first defined under assumed model M. The prior distribution can be interpreted as a summary of our information about the parameter θ before the data is collected. Inference is developed based on the corresponding posterior density which is conditional on the observed data D and is calculated using

$$p(\theta|D, M) = \frac{f(D|\theta, M)p(\theta|M)}{p(D|M)} = \frac{f(D|\theta, M)p(\theta|M)}{\int_{\Theta} f(D|\theta, M)p(\theta|M)d\theta}. \quad (9.3)$$

The denominator of (9.3) is the probability of data under model M. We can interpret $p(D|M)$ as the weighted average probability of observing the data under the assumed model M, over all possible values of the parameter θ.

Let us revisit the logistic additive model (9.2). For a given SNP, let M_0 ($\mathcal{H}_0 : \beta = 0$) denote the null model of no association and M_1 ($\mathcal{H}_1 : \beta \neq 0$) the alternative model. To decide which of the two competing models, M_0 and M_1, is more suitable, a natural choice is to rely on the Bayes factor of M_1 against M_0, viz.

$$\mathrm{BF}_{10} = \frac{p(D|M_1)}{p(D|M_0)}.$$

In the WTCCC (2007) application, $\theta_0 = (\alpha, 0)$, $\theta_1 = (\alpha, \beta)$, $\alpha \sim \mathcal{N}(0, 1)$ and $\beta \sim \mathcal{N}(0, .2)$. The prior for β reflects the belief that genetic effects (expressed as the odds ratio, $\mathrm{OR} = \exp(\beta)$) of SNPs associated with complex human traits are in the range of .5 to 2 and most likely between .67 and 1.5; see the Supplementary Figure 1 of WTCCC (2007). The choice of prior is study specific and subjective, which contributes to scientists' reluctance to using Bayesian methods; see Chapter 10 by Paul Gustafson on Bayesian methods in observational epidemiology studies. However, techniques such as model-averaging (see below) can alleviate some of the concerns that different prior specifications might lead to different conclusions.

Let $\mathrm{BF}_{10}(\mathrm{add})$ denote the Bayes factor for SNP j under the additive model and $\mathrm{BF}_{10}(\mathrm{geno})$ under the genotypic model. In the latter case, $\theta_0 = (\alpha, 0, 0)$ and $\theta_1 = (\alpha, \beta_1, \beta_2)$, and prior distributions are specified for both β_1 and β_2; see WTCCC (2007) for more details on the choice of prior. For each of the 12 SNPs selected based on the p-value criterion as described in Section 9.2.1, the corresponding Bayes factors, $\mathrm{BF}_{10}(\mathrm{add})$ and $\mathrm{BF}_{10}(\mathrm{geno})$ were also calculated and are restated (on the $\log_{10}$ scale) in Table 9.1.

9.2.3 Additional Considerations

The immediate conclusion from Table 9.1 is that rankings of the SNPs are remarkably consistent between p-value and Bayes factor. This consistency has been steadily reported in the GWAS literature; see, e.g., Strömberg et al. (2009). It has been theoretically justified by Wakefield (2009) for case-control GWAS. However, such conclusions depend on several assumptions, an important one being that the MAFs are not too small. Although SNPs analyzed by GWAS are usually common SNPs, this is not the case for the emerging NGS

TABLE 9.2: Results of WTCCC GWAS of T1D after "Model Averaging." Here, $\min(p\text{-value}) = \min(p\text{-value}_{\text{add}}, p\text{-value}_{\text{geno}})$, $\log_{10} \text{BF}(.8, .2) = \log_{10}\{.8\,\text{BF}(\text{add}) + .2\,\text{BF}(\text{geno})\}$, $p\text{-value}_{\text{add}}$, $p\text{-value}_{\text{geno}}$, $\text{BF}(\text{add})$ and $\text{BF}(\text{geno})$ are from Table 9.1.

Chro-mosome	SNP	MAF	Frequentist min(p-value)	Rank	Bayesian $\log_{10} \text{BF}(.8, .2)$	Rank
6	rs9272346	.387	2.42 E-134	1	141.83	1
1	rs6679677	.096	1.17 E-26	2	23.03	2
12	rs17696736	.424	2.17 E-15	3	12.44	3
12	rs11171739	.423	1.14 E-11	4	8.82	4
16	rs12708716	.350	9.24 E-08	5	5.09	5
4	rs17388568	.260	5.00 E-07	6	4.35	6
18	rs2542151	.163	1.89 E-06	8	3.86	7
10	rs2104286	.286	7.97 E-06	10	3.25	8
5	rs2544677	.242	8.23 E-06	11	3.25	9
1	rs2639703	.276	8.46 E-06	12	3.22	10
12	rs11052552	.486	7.24 E-07	7	3.14	11
5	rs17166496	.391	5.20 E-06	9	2.55	12

area. For analysis of rare variants, it is unclear if the traditional frequentist approaches (e.g., Lee et al., 2012; Derkach et al., 2014) and Bayesian methods (e.g., Yi and Zhi, 2011) lead to similar rankings.

Let us reconsider SNP rs17166496 in Table 9.1 for which different genetic model assumptions (additive vs genotypic) resulted in strikingly different results. Let $p(M)$ denote the prior distribution for genetic model M, and for simplicity consider only two models, $M \in \{\text{add}, \text{geno}\}$, with user-specified prior probabilities

$$\begin{aligned} \Pr(M = \text{add}) &= p(\text{add}), \\ \Pr(M = \text{geno}) &= p(\text{geno}) = 1 - p(\text{add}). \end{aligned}$$

Under the Bayesian framework, we can define an overall Bayes factor as a weighted average of the individual Bayes factors, viz.

$$\text{BF}_{10} = p(\text{add}) \times \text{BF}_{10}(\text{add}) + p(\text{geno}) \times \text{BF}_{10}(\text{geno}).$$

For a more detailed discussion we refer the readers to Stephens and Balding (2009). Therefore, instead of selecting one "best" model and reporting the corresponding finding, the Bayesian analysis provides a principled way to combine the evidence from each model, weighted by the prior belief in that model. For the SNPs in Table 9.1, Table 9.2 provides the Bayes factors after model averaging (see Section 9.3.1), with more weight ($p(\text{add}) = .8$) given to the additive genetic model reflecting the common belief that most SNPs act in an (approximately) additive manner (Hill et al., 2008).

9.3   Replication of a Significant Finding

In most scientific studies, a negative result is usually not published and a positive finding needs to be replicated in an independent sample. For example, other investigators might want to validate the association between SNP rs2542151 and T1D found by WTCCC (2007), and they would typically use the reported genetic effect size (i.e., $\exp(\hat{\beta})$) to calculate the sample size needed for a replication study (e.g., with 80% power at the .05 level). However, as discussed in Section 9.1.2, the use of the same data first for testing and then for estimation leads to an upward-biased estimate of the effect size and, therefore, an under-powered replication study. This phenomenon is also known as the winner's curse and is ubiquitous in GWAS. In the following, we discuss the frequentist and Bayesian approaches to correcting such bias using only the summary statistics that are typically reported.

9.3.1   Conditional MLE vs. Bayesian Model Averaging

Consider $\hat{\beta}$, the estimator of β, $\widehat{SE}(\hat{\beta})$, the standard error of $\hat{\beta}$, and

$$T = \hat{\beta}/\widehat{SE}(\hat{\beta}),$$

the test statistic for testing $\mathcal{H}_0$ as defined in Section 9.1.2. To simplify the discussion, assume that $\beta > 0$ for a truly associated SNP. Since we focus attention on estimates for which the associated test was significant (i.e., $T > c$, where c is some specified value), $\hat{\beta}$ is not an (approximately) unbiased estimator of β; in fact, $\mathrm{E}(\hat{\beta}|T > c) \geq \beta$.

Ghosh et al. (2008) proposed to obtain the maximum likelihood estimate (MLE) based on the conditional likelihood that incorporates the fact that the observed test statistic exceeded the significance criterion. Specifically, under the Normal assumption that T follows the $\mathcal{N}[\beta/SE(\hat{\beta}), 1]$ distribution, the conditional MLE of β is the parameter value that maximizes the likelihood

$$L(\beta) = p(T|T > c) = \frac{\phi\{T - \beta/\widehat{SE}(\hat{\beta})\}}{\Phi\{-c + \beta/\widehat{SE}(\hat{\beta})\}}, \qquad (9.4)$$

where ϕ and Φ are respectively the density and cumulative distribution function of the standard Normal distribution, $\mathcal{N}(0, 1)$ (Ghosh et al., 2008). The $L(\beta)$ in (9.4) is a conditional likelihood because it quantifies the conditional probability of the observed data given that they yielded a significant finding for β. Conditioning on the significant result creates a more realistic statistical framework. Indeed, compared with the unconditional $\hat{\beta}$, the estimate obtained from this conditional MLE approach is, on average, closer to the true value β. However, it is difficult to incorporate prior information on β into (9.4) in settings where it might be available.

As an alternative, Xu et al. (2011) considered the Bayesian approach that specifies a prior distribution for the parameter of interest, $\theta = \beta$ (i.e., the log OR), and infers the posterior as in (9.3). In the same vein as the conditional likelihood approach, the Bayesian inference is performed conditional on the significant finding for β. Therefore, the prior specification can incorporate this information. In this setting, Xu et al. (2011) specified the prior distribution for β as a mixture of a discrete distribution that places all probability mass at 0 and a continuous distribution g with support on $\mathbb{R}$,

$$p(\beta) = \xi\delta_{\{0\}}(\beta) + (1 - \xi)g(\beta). \qquad (9.5)$$

Priors such as (9.5) are known as spike-and-slab priors and have been used repeatedly in Bayesian variable selection and shrinkage estimation; see, e.g., Kuo and Mallick (1998). In this application, the mixture parameter ξ allows the possibility that the observed significance is due to chance (i.e., $\beta = 0$); the $g(\beta)$ density function reflects the prior belief about the range and distribution of the effect of a truly associated SNP. The Bayesian estimator of β is the posterior mean $\mathrm{E}(\beta|D)$. The spike at 0 will shrink the posterior mean and reduce the posterior variance of β. The prior distribution $p(\xi)$ assigned to ξ allows us to quantify the uncertainty about a discovery being true or false and influence the amount of shrinkage in the posterior. For instance, if we use a Beta prior for ξ, denoted $\mathcal{B}(a, b)$, and if the sample size of the discovery study was small, we may choose large a and small b values (e.g., $a = 8$ and $b = .5$) to represent our initial skeptical view on the significance of any finding from such a study.

In practice, a researcher may have a difficult time deciding which particular prior to use for the analysis. In many cases, the analysis can be better served by considering several priors rather than choosing just one of them. Suppose M_1 is the model with $p(\xi) = \mathcal{B}(8, .5)$ and M_2 with $p(\xi) = \mathcal{B}(.5, 8)$, and let $p(M_1)$ and $p(M_2)$ be the prior probabilities for the two models; see Xu et al. (2011) for the specification of $p(M_1)$ and $p(M_2)$. Then the Bayesian model averaging (BMA) paradigm offers a consistent method to combine the two models by using

$$p(\theta|D) = \sum_{k=1}^{2} p(\theta|D, M_k)p(M_k|D), \qquad (9.6)$$

where $p(M_k|D)$ is the posterior probability of model M_k and is proportional to $p(M_k|D) \propto p(D|M_k)p(M_k)$. Formally, the Bayesian BMA estimator is the mean of the distribution $p(\theta|D)$, but in practice, the theoretical mean is not available in closed form and it is approximated via the Monte Carlo method as discussed in Section 9.3.2.

Using SNP rs2542151 as an example, the genetic effect reported by WTCCC (2007) was

$$\hat{\beta}_{\text{naive}} = .285, \quad \widehat{OR}_{\text{naive}} = 1.33.$$

However, estimates were substantially smaller after bias correction by both the frequentist approach,

$$\hat{\beta}_{\text{freq}} = .140, \quad \widehat{OR}_{\text{freq}} = 1.15,$$

and the Bayesian method,

$$\hat{\beta}_{\text{Bayes}} = .117, \quad \widehat{OR}_{\text{Bayes}} = 1.12.$$

This reduction in genetic effect estimate (OR from 1.33 to just over 1.1) is practically meaningful and important because a) the sample size needed for a successful replication study would not be underestimated; and b) the potential clinical importance of this SNP would not be overestimated. Between the two correction methods, Xu et al. (2011) have shown that the Bayesian estimator has better accuracy (as measured by the mean squared error) than the conditional MLE in studies with low power, where the winner's curse is more likely to impact the conclusions of the analyses.

9.3.2 Computational Considerations

Realistic Bayesian modeling and analysis have been made possible thanks to significant advances in computational algorithms, particularly Markov Chain Monte Carlo (MCMC) samplers. Note that the integral that appears in the denominator of (9.3) is often intractable. For example, the model considered here along with the prior (9.5) and a uniform prior on β (i.e., θ) yields a posterior distribution that cannot be studied analytically, e.g., one cannot compute in closed form the posterior mean, viz.

$$E(\theta|D, M) = \int \theta\, p(\theta|D, M)\mathrm{d}\theta.$$

However, we can still approximate $E(\theta|D, M)$ as long as we can draw from $p(\theta|D, M)$. For instance, if we can generate a sample $\theta_1, \ldots, \theta_K$ from the posterior $p(\theta|D, M)$ we can then estimate $E(\theta|D, M)$ by

$$\widehat{E(\theta|D, M)} = \frac{1}{K}\sum_{i=1}^{K} \theta_i.$$

Due to space constraints, we cannot get into the details of constructing the algorithms used to sample from posterior distributions, but we refer the reader to the review article of Craiu and Rosenthal (2014) and Jeff Rosenthal's review of Metropolis algorithms in Chapter 6.

The tight connection between Bayesian inference and computational algorithms has also been exploited in situations in which the likelihood cannot be calculated in closed form (e.g., Tavaré et al., 1997) or when it may be too expensive to compute; see, e.g., Wegmann et al. (2009) and Row et al. (2011).

9.4   Conclusion and Discussion

The collection of vast amounts of genetic data in recent years has provided statisticians with numerous challenges and opportunities. In the effort to understand the genetic architecture of complex human traits, one important area of research has focused on genetic association studies which are the central theme of the discussion here. In addition to the frequentist framework we have emphasized the value of the alternative Bayesian paradigm. For many, a major hurdle in a Bayesian analysis comes from the computational and methodological challenges involved in the study of the posterior distribution via MCMC sampling. However, the increase in computational expertise among statistical geneticists and the advent of user-friendly software has spurred renewed interest in Bayesian methods.

The frequentist and Bayesian bias-correction methods discussed in Section 9.3 assume that only the summary statistics are available and only for the reported significant SNPs. When genome-wide results are available, it might be beneficial to let the empirical distribution of the estimated effects influence the specification of the prior. Built upon the work by Efron (2011), this empirical Bayes approach has been used by Ferguson et al. (2013) to tackle the winner's curse in a setting where multiple genetic effects are dealt with jointly. When the original data are available, one could use the alternative bootstrap-based method reviewed by Bull, Graham, and Greenwood in Chapter 8.

Bayesian methods have also been successfully used in other genetic settings. For example, Lo and Gottardo (2007) coupled a hierarchical Bayesian model with an empirical Bayes specification of the prior to produce inference about differential expression in microarray studies; Gottardo et al. (2008) proposed a Bayesian analysis of Chromatin-immunoprecipitation microarrays in which the hierarchical structure of the model accounts for the spatial correlation present between neighboring probes; Wu et al. (2009) proposed a Bayesian segmentation approach that identifies copy number variations (DNA segments that exhibit duplications and deletions when compared to a reference genome); and Scott-Boyer et al. (2012) introduced a Bayesian hierarchical model that can combine genotypic and gene expression data to detect the so-called "expression quantitative trait loci (eQTL)."

Recent progress in genetic association studies has taught us some lessons. Although simple statistical techniques can identify many trait-associated genetic variants, these "low hanging fruits" are only a few small pieces of a much bigger puzzle. To solve the remaining puzzle, more sophisticated methodology is needed to mine the already available data, to analyze new kinds of data, and to combine different sources of data. We believe that the Bayesian methodology can play an important role in solving some of these issues. Dialog between statisticians and other scientists is becoming a critical component of the analytical process for these emerging studies.

Acknowledgments

The authors thank the editor, Jerald F. Lawless, and two external reviewers for helpful comments. The authors' research is supported by the Natural Sciences and Engineering Research Council of Canada and by the Canadian Institutes of Health Research.

About the Authors

Radu V. Craiu is a professor of statistics at the University of Toronto. He studied mathematics at the Universitatea din Bucureşti (MS, 1996) and statistics at the University of Chicago (PhD, 2001). His research interests include computational methods for Bayesian inference, especially Markov Chain Monte Carlo sampling algorithms, copula models, model selection and statistical genetics. He is an associate editor for *The Canadian Journal of Statistics*.

Lei Sun is an associate professor of biostatistics and statistics at the University of Toronto. She studied mathematics at Fudan University and obtained her PhD in statistics from the University of Chicago in 2001. Her primary research interest is statistical genetics, developing statistical methods and computational tools for high-dimensional studies of complex human traits.

Bibliography

1000GenomesProjectConsortium (2012). An integrated map of genetic variation from 1,092 human genomes. *Nature*, 491:56–65.

Agresti, A. (2002). *Categorical Data Analysis*. Wiley, New York.

Begum, F., Ghosh, D., Tseng, G. C., and Feingold, E. (2012). Comprehensive literature review and statistical considerations for GWAS meta-analysis. *Nucleic Acids Research*, 40:3777–3784.

Craiu, R. V. and Rosenthal, J. S. (2014). Bayesian computation via Markov Chain Monte Carlo. *Annual Review of Statistics and its Applications*, 1:7.1–7.23.

Derkach, A., Lawless, J. F., and Sun, L. (2014). Pooled association tests for rare genetic variants: A review and some new results. *Statistical Science*, 29:in press.

Dudbridge, F. and Gusnanto, A. (2008). Estimation of significance thresholds for genomewide association scans. *Genetic Epidemiology*, 32:227–234.

Efron, B. (2011). Tweedie's formula and selection bias. *Journal of the American Statistical Association*, 106:1602–1614.

Ferguson, J. P., Cho, J. H., Yang, C., and Zhao, H. (2013). Empirical Bayes correction for the Winner's Curse in genetic association studies. *Genetic Epidemiology*, 37:60–68.

Ghosh, A., Zou, F., and Wright, F. A. (2008). Estimating odds ratios in genome scans: An approximate conditional likelihood approach. *American Journal of Human Genetics*, 82:1064–1074.

Gottardo, R., Li, W., Johnson, W. E., and Liu, X. S. (2008). A flexible and powerful Bayesian hierarchical model for ChIP-Chip experiments. *Biometrics*, 64:468–478.

Hill, W. G., Goddard, M. E., and Visscher, P. M. (2008). Data and theory point to mainly additive genetic variance for complex traits. *PLoS Genet*, 4(2):e1000008.

Kass, R. E. and Raftery, A. E. (1995). Bayes factors. *Journal of the American Statistical Association*, 90:773–795.

Kuo, L. and Mallick, B. (1998). Variable selection for regression models. *Sankhyā*, *Series B*, 60:65–81.

Lee, S., Wu, M. C., and Lin, X. (2012). Optimal tests for rare variant effects in sequencing association studies. *Biostatistics*, 13:762–775.

Lo, K. and Gottardo, R. (2007). Flexible empirical Bayes models for differential gene expression. *Bioinformatics*, 23:328–335.

Marchini, J., Howie, B., Myers, S., McVean, G., and Donnelly, P. (2007). A new multipoint method for genome-wide association studies by imputation of genotypes. *Nature Genetics*, 39:906–913.

Piegorsch, W. W. (1990). Fisher's contributions to genetics and heredity, with special emphasis on the Gregor Mendel controversy. *Biometrics*, 46:915–924.

Row, J. R., Brooks, R. J., Mackinnon, C. A., Lawson, A., Crother, B. I., White, M., and Lougheed, S. C. (2011). Approximate Bayesian computation reveals the factors that influence genetic diversity and population structure of foxsnakes. *Journal of Evolutionary Biology*, 24:2364–2377.

Scott-Boyer, M. P., Imholte, G. C., Tayeb, A., Labbe, A., Deschepper, C. F., and Gottardo, R. (2012). An integrated hierarchical Bayesian model for multivariate eQTL mapping. *Statistical Applications in Genetics and Molecular Biology*, 11:1–30.

Stephens, M. and Balding, D. J. (2009). Bayesian statistical methods for genetic association studies. *Nature Review Genetics*, 10:681–690.

Strömberg, U., Björk, J., Vineis, P., Broberg, K., and Zeggini, E. (2009). Ranking of genome-wide association scan signals by different measures. *International Journal of Epidemiology*, 38:1364–1373.

Tavaré, S., Balding, D. J., Griffith, R., and Donnelly, P. (1997). Inferring coalescence times from DNA sequence data. *Genetics*, 145:505–518.

Thomas, D. C., Casey, G., Conti, D. V., Haile, R. W., Lewinger, J. P., and Stram, D. O. (2009). Methodological issues in multistage genome-wide association studies. *Statistical Science*, 24:414–429.

Wakefield, J. (2007). A Bayesian measure of the probability of false discovery in genetic epidemiology studies. *American Journal of Human Genetics*, 81:208–227.

Wakefield, J. (2009). Bayes factors for genome-wide association studies: Comparison with p-values. *Genetic Epidemiology*, 33:79–86.

Wegmann, D., Leuenberger, C., and Excoffier, L. (2009). Efficient approximate Bayesian computation coupled with Markov Chain Monte Carlo without likelihood. *Genetics*, 182:1207–1218.

WTCCC, W. (2007). Genome-wide association study of 14,000 cases of seven common diseases and 3,000 shared controls. *Nature*, 447:661–678.

Wu, L. Y., Chipman, H. A., Bull, S. B., and Briollais, L. (2009). A Bayesian segmentation approach to ascertain copy number variations at the population level. *Bioinformatics*, 25:1669–1679.

Xu, L., Craiu, R. V., and Sun, L. (2011). Bayesian methods to overcome the winner's curse in genetic studies. *The Annals of Applied Statistics*, 5:201–231.

Yi, N. and Zhi, D. (2011). Bayesian analysis of rare variants in genetic association studies. *Genetic Epidemiology*, 35:57–69.

10

Bayesian Statistical Methodology for Observational Health Sciences Data

Paul Gustafson

University of British Columbia, Vancouver, BC

10.1   Introduction

The data available to address important questions about harms and benefits to human health are often rather limited. Ethical concerns, logistical constraints, and resource limitations often mean that only observational studies can be conducted to address a given question about a possible determinant of health. And the data from observational studies are indeed prone to be limited in fundamental ways. Thus, while statistical methodologies play a central role across the health sciences, they become acutely critical in the observational study realm. As will be showcased here, Bayesian statistical methods can be applied to infer as much as possible about the health question at hand, while acknowledging the limitations of the available data.

As the name implies, observational studies involve measuring, but not otherwise intervening on, human subjects. It is well understood that intervention, when possible, leads to stronger evidence. Specifically, studies in which participants are randomly assigned to different levels of a potential health determinant (e.g., drug A versus drug B) lead to the firmest conclusions. Referring to a study participant's level of the potential determinant as the exposure variable, and the consequent health outcome as the disease variable, randomization yields well-supported inferences about the exposure-disease relationship. The extent to which the disease variable varies systematically across groups of participants defined by different exposure levels directly translates to a level of statistical confidence about the role of exposure in causing disease.

Unfortunately, in a wide array of settings where understanding the exposure-disease relationship is critical, randomization is impractical, unethical, or unaffordable. The next best course of action is typically an observational study, in which exposure level and health outcome are among a slew of variables measured on study participants. In such a study, however, participants are essentially self-selected into exposure groups. Thus there is far more trepidation in ascribing an across-group difference in health outcome as being a direct consequence of the differing exposure.

The observational studies we have in mind here fall under the rubric of "risk-factor epidemiology," addressing questions such as: Among children, does residential exposure to magnetic fields from power lines increase the risk of leukemia? Among post-menopausal women, does hormone replacement therapy (HRT) reduce the risk of osteoporosis? Among middle-aged women and men, does the "distribution" of abdominal fat, as reflected by the waist-to-hip ratio, relate to the risk of heart disease? Such questions can be subtle, and researchers strive to be exceedingly careful in how observational studies are conducted, in terms of both how the data are collected, and how statistical methods are brought to bear. However, intentions notwithstanding, messages from the research community to the public about what exposures are good and bad for health can be mixed at times, if not downright contradictory. The research community is well aware of this state of affairs; see, e.g., Taubes and Mann (1995) and Ioannidis (2005). At the end of the day, however, it can be plain old difficult to navigate from observational studies to conclusions that are sufficiently robust to convince a broad range of stakeholders and stand the scrutiny of time.

One of the biggest recent "flip-flops" in public messaging concerned HRT for post-menopausal women. Largely on the basis of evidence accrued from a series of observational studies conducted over many years, HRT became widely prescribed for relief of menopausal symptoms, and/or reduction in the risk of osteoporosis, and/or reduction in the risk of heart disease. In 2002, however, reports from a "bombshell" randomized study actually linked HRT to an elevated risk of heart disease. Prescribing practices changed abruptly, and many assigned some blame for the debacle to the limitations of observational studies.

The HRT situation is in fact quite complex. It turns out that careful consideration of the definition of exposure, with a distinction between initiation and ongoing use of HRT, goes a long way to reconciling the gap between observational study and randomized study results (Prentice et al., 2005; Hernán et al., 2008). However, some observational study challenges can be explained in a simple, pared-down context. To illustrate this, in Section 10.2 we describe a common observational study design, the case-control study. This is followed by a brief discussion of the Bayesian approach to statistical inference in Section 10.3. The Bayesian paradigm quite naturally handles situations involving multiple sources of uncertainty, as applies to many challenging observational data problems. Section 10.4 is then our "main feature." We show the Bayesian approach in action, dealing with the possibility that exposure status may not be recorded accurately for all study participants. We close in Section 10.5 with some remarks about broader applications of Bayesian methods to challenging observational data problems.

10.2   Case-Control Studies

The case-control study is a common type of observational study. A simple form arises when the exposure status and disease status of study participants are both binary. Say the exposure status of a participant is denoted as X, with $X = 0$ and $X = 1$ indicating "unexposed" and "exposed" respectively. Similarly, let $Y = 0$ and $Y = 1$ respectively indicate absence and presence of the disease in question. A case-control study proceeds by sampling n_0 members of the $Y = 0$ subpopulation to be the "controls," and n_1 members of the $Y = 1$ subpopulation to be the "cases." Note here that the study investigators set n_0 and n_1, often choosing the two sample sizes to be comparable to one another. In this way, a substantial number of the study participants can be individuals with the disease, even if the disease is actually very rare in the general population. Logistically, often this can be achieved via the use of a disease registry from which cases can be sampled. After any individual, control or case, is recruited into the study, his or her exposure status X is measured. Consequently, the fundamental summary of the study data is a "2 × 2 table" providing counts of study participants for each of the four combinations of exposure status X and disease status Y.

Before proceeding to an example of a 2 × 2 table, we must comment on a fundamental point concerning case-control studies. We seek to understand how Y depends on X. Translating to the language of conditional probabilities and conditional distributions then, we seek to learn about $\Pr(Y = y | X = x)$, where the common "bar notation" indicates conditioning, and the mathematical definition is

$$\Pr(Y = y | X = x) = \frac{\Pr(Y = y, X = x)}{\Pr(X = x)}.$$

For instance, $\Pr(Y = 1 | X = 0)$ is read as the probability that $Y = 1$ given $X = 0$, and is equal to the probability that both $Y = 1$ and $X = 0$ divided by the probability that $X = 0$. On face value it seems concerning that we seek to understand the conditional distribution of Y given X, yet the data arise from the conditional distribution of X given Y, e.g., the data on controls (cases) arise according to how X values are distributed among the subset of the population for whom $Y = 0$ ($Y = 1$).

Fortunately, we can resolve this issue, particularly if we are willing to think in terms of the odds of events rather than probabilities. We would like to know to what extent the odds of contracting the disease are different for the exposed than for the unexposed. Mathematically, again using the bar notation to indicate conditioning, we can write the odds of disease given exposure status $X = x$ as

$$\text{Odds}(Y = 1 | X = x) = \frac{\Pr(Y = 1 | X = x)}{\Pr(Y = 0 | X = x)}.$$

TABLE 10.1: Data from a fictitious case-control study with self-reported exposure status.

		Diseased $(Y = 1)$	Disease-Free $(Y = 0)$	Total
Reported	Exposed	150	120	270
Status	Unexposed	290	320	610
	Total	440	440	880

Then the Y given X odds-ratio,

$$OR_{Y|X} = \frac{\text{Odds}(Y = 1|X = 1)}{\text{Odds}(Y = 1|X = 0)},$$

summarizes how the disease rate varies between the unexposed and exposed. For instance, $OR_{Y|X} = 1.3$ would tell us that the odds of contracting the disease are 30% higher for the exposed than for the unexposed.

On first glance, since case-control data arise from the distribution of X given Y, it isn't clear that such data tell us anything about the Y given X odds-ratio. However, case-control data do directly speak to the "reverse" odds-ratio for X given Y, defined

$$OR_{X|Y} = \frac{\text{Odds}(X = 1|Y = 1)}{\text{Odds}(X = 1|Y = 0)}.$$

Then a very fortuitous bit of mathematics kicks in. In fact the two odds-ratios are identical, i.e., $OR_{Y|X} = OR_{X|Y}$, and we can unambiguously use OR to denote their common value. Consequently case-control data can indeed tell us about how disease rates, expressed on the odds scale, differ between the unexposed and the exposed.

A fictitious example of a 2×2 table for a study with $n_0 = 440$ controls and $n_1 = 440$ cases is given in Table 10.1. The labeling of the rows as giving "reported" exposure status foreshadows the problem we will dwell on presently. For now, however, we assume that the table counts really do cross-classify the study participants according to exposure status X and disease status Y. We then note a higher rate of exposure for cases than for controls. Specifically, the odds of exposure are estimated as $(150/440)/(290/440) = .517$ for cases, but only $(120/440)/(320/440) = .375$ for controls. We emphasize that while the earlier mathematical statements of odds describe the population as a whole, the data give us only estimates, since we only have information on the randomly sampled subsets on the entire control and case populations. Using the common statistical convention of the "hat-notation" to distinguish estimates from the population quantities they are estimating, $\widehat{OR} = .517/.375 = 1.38$ is then the best guess for the association between the exposure and the disease.

That is, from the available data, we have estimated the odds of acquiring the disease to be 38% higher for those population members who are exposed, relative to their unexposed compatriots.

10.3   Bayesian Analysis

In fact, we can use Bayesian statistics to formalize our inferential procedure above, and to put "error bars" around our estimated odds-ratio. The essence of the Bayesian method is to specify a prior distribution describing the state of belief about unknown quantities in advance of receiving of the data. The laws of probability then update this distribution on the basis of data received, giving the posterior distribution which summarizes the remaining uncertainty about the unknowns. In generic terms, say the unknown parameters θ are assigned a prior distribution having density function $\pi(\theta)$, and the statistical model for observable data Z given the parameters θ involves the conditional density function $\pi(z|\theta)$. Then the posterior distribution can be expressed via its density function

$$\pi(\theta|Z = z) \quad \propto \quad \pi(z|\theta)\pi(\theta),$$

with the proportionality understood to be as a function of θ. This expression is widely known as Bayes Theorem.

In the present situation, θ has two components, namely

$$\theta_0 = \Pr(X = 1|Y = 0) \quad \text{and} \quad \theta_1 = \Pr(X = 1|Y = 1),$$

the population exposure rates for controls and cases respectively. We can assign a uniform prior which doesn't favor any values for these rates over any others. The posterior distribution for θ given by Bayes' Theorem then induces the posterior distribution for the odds-ratio, since mathematically

$$OR = \frac{\theta_1/(1 - \theta_1)}{\theta_0/(1 - \theta_0)}.$$

Further, quantiles of the posterior distribution of OR can be used to report a plausible range of values in light of the data. For instance, the interval from the 5th to 95th percentile of the posterior distribution comprises a 90% credible interval for the population OR. Moreover, the theory allows a directly probabilistic interpretation of this interval: we think it nine times more likely that the true value is inside the interval rather than outside.

Based on the Table 10.1 data, and using some computational details described in the Appendix, the 90% credible interval for the population odds-ratio ranges from 1.03 to 1.90. In practical terms this is quite a lot of uncertainty: a 3% elevation in the odds of illness for the exposed is quite a different matter than a 90% elevation. Still, at least these data speak clearly about the exposed being at elevated risk. In many studies the credible interval for the odds-ratio will straddle 1. When this happens, the data are consistent with either an elevated or diminished disease risk for the exposed.

10.4   Exposure Misclassification

With the basics of case-control studies and Bayesian inference in hand, we now admit that ascertaining the exposure status for study participants may be easier said than done. A pervasive concern in risk-factor epidemiology is that exposure status may be misrepresented for a portion of study participants. For instance, say the nature of the exposure variable in Table 10.1, perhaps along with resource constraints, limits the exposure ascertainment to self-report, e.g., each participant checks a box on a questionnaire to record his/her exposure status. This raises the possibility that a portion of participants may check the wrong box, and the consequent need to deal with this circumstance when analyzing the study data.

To delve into this concern about exposure misclassification, we introduce the technical terms of sensitivity — the probability of correct exposure classification for a truly exposed person, and specificity — the probability of correct classification for a truly unexposed person. Thus exposure misclassification is manifested if one or both of sensitivity and specificity are below 1. To keep the present example as accessible as possible, say the sensitivity can be safely assumed to be 1, but specificity may fall below 1. This could happen, for instance, if being unexposed is socially undesirable. Consider exposures such as "brushing one's teeth twice daily," or "exercising vigorously three or more times a week." With a self-report questionnaire, or even more so with a face-to-face participant interview, it is easy to imagine that all truly exposed participants report correctly. It is equally easy, however, to imagine that a portion of truly unexposed participants report incorrectly. Also to keep things simple, we assume this proportion — 1 minus the specificity — is the same for cases as it is for controls.

Pursuing this line of thinking further, if the specificity of exposure assessment is known, then the data in Table 10.1 can be analyzed accordingly. Some mathematical details are given in the Appendix. The estimated odds-ratios and 90% credible intervals based on some selected values of specificity appear in Figure 10.1. (The result corresponding to 100% specificity is identical to the "face value" result given in the previous section.) The trends seen in Figure 10.1 are in fact rather general. As we assume lower specificity, we assume there is more random variation in our Table 10.1 data. Such variation will weaken, or attenuate, the relationship between reported exposure and disease. Thus as we assume more randomness is present, we should boost, or

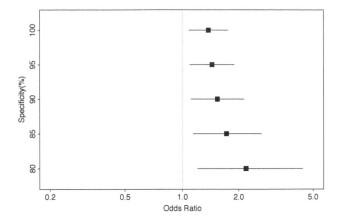

FIGURE 10.1: Inference for the exposure-disease odds-ratio based on Table 10.1 data, for selected values of specificity. The posterior estimate of the odds-ratio and the 90% equal-tailed credible interval are displayed. Technically, the posterior estimate is the exponentiated posterior mean of the log odds-ratio. Note the use of a logarithmic axis.

de-attenuate, the estimated odds-ratio for the actual exposure and disease. The upward march in the estimated odds-ratio as the assumed specificity is reduced is then intuitively anticipated. By the same token, admitting that the data are more afflicted by random error is tantamount to admitting that they contain less useful information. Therefore, we are not surprised to see the credible interval widen as the assumed specificity decreases. Finally, it is both interesting and typical that in some sense the two trends seen cancel each other out. As the assumed specificity decreases, the increase in estimated odds-ratio and the widening of the credible interval combine such that the lower endpoint of the credible interval is actually quite stable.

Reflecting upon Figure 10.1, we have five different statistical inferences that adjust for exposure misclassification, corresponding to five different assumptions about the magnitude of the misclassification. If the only overarching judgment we are comfortable with is that the specificity is 80% or more, then perhaps the ensemble of inferences is the take-away item. Such a "what-if" analysis is often referred to as a sensitivity analysis (as in sensitivity of the inference to the assumption, not the technical sense of sensitivity as a probability of correct classification), or a stress test. Of course the results of a sensitivity analysis are disquieting when, as at present, the inference changes considerably with the assumption.

Intuitively, at least, our Table 10.1 data do not appear to tell us anything about the extent of exposure misclassification. Since we don't observe the true exposure status for any individuals, we cannot infer how commonly the

TABLE 10.2: Data from a second fictitious case-control study with self-reported exposure status.

		Diseased ($Y = 1$)	Disease-free ($Y = 0$)	Total
Reported	Exposed	188	168	356
Status	Unexposed	692	712	1404
	Total	880	880	1760

true status and self-reported status are discordant. We don't have to rely on intuition here, however. Just as Bayesian inference yields the posterior distribution of the target parameter given the data and an assumed value of specificity, it can also give us the posterior distribution of the specificity itself. As always with Bayesian inference, such calculation of posterior probabilities is predicated on the assignment of prior probabilities. Say, for instance, we assign equal prior probabilities of .20 to each of the five specificity values (80%, 85%, ..., 100%) considered in Figure 10.1. The Table 10.1 data then induce corresponding posterior probabilities of .248, .220, .196, .176, .159 (again, some of the calculation details appear in the Appendix). Thus the data are very slightly more suggestive of a smaller value of specificity. This "tilt" is so small though, that we are really just left with the what-if message from Figure 10.1. This is a fairly common situation. We can posit different assumptions about the extent of a problem with observational data, while the data themselves have little to commend one assumption over the others.

Our plot thickens upon considering a second fictitious dataset, given in Table 10.2. The what-if analysis for these data appears in Figure 10.2. Again, as we lower the assumed specificity, the best-guess odds-ratio is boosted, and the associated credible interval widens. This time, however, the credible interval consistently crosses 1. Thus we are not completely convinced that exposure is associated with an elevation in disease risk, regardless of how much exposure classification error is assumed to be manifested in the data. Also, note that the change in inference upon reducing the assumed specificity from 85% to 80% is particularly marked.

Again we can start with equal prior probabilities for the five specificity values and then update these based on the data. This results in posterior probabilities of .068, .273, .243, .218, .197, for specificity ranging from 80% up to 100% by steps of 5%. A specificity of 80% is thus far less supported by the data than the four higher values. In one sense this is good news. Of the five analyses, the one that stands furthest apart from the others is also the one given by far the least credence by the data. Against this, however, the story has become more complex. For some datasets, such as that in Table 10.1, a sensitivity analysis is more or less just that: inferences under some different scenarios are reported, and the a priori uncertainty about which scenarios

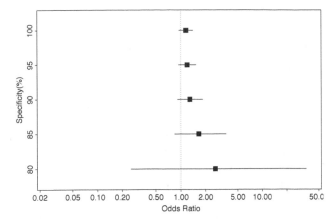

FIGURE 10.2: Inference for the exposure-disease odds-ratio based on Table 10.2 data, for selected values of specificity. The posterior estimate of the odds-ratio and the 90% equal-tailed credible interval are displayed. Technically, the posterior estimate is the exponentiated posterior mean of the log odds-ratio. Note the use of a logarithmic axis.

are most plausible remains largely unchanged upon receipt of the data. Other datasets, however, have considerably more to say, e.g., one of the five inferences in Figure 10.2 is considerably less plausible than the others. Also, note that for a new dataset coming onto the scene, we won't know which situation applies until we carry out the posterior analysis.

As a final thought about these examples, once we have the posterior distribution of the odds-ratio for each postulated value of specificity, as well as the posterior distribution of the specificity, the laws of probability dictate how the five what-if posterior analyses can be synthesized into an "overall" posterior analysis. Applied to the first dataset, this synthesis produces a best estimate of 1.67 for the odds-ratio, with a 90% credible interval running from 1.11 to 3.02. With reference to Figure 10.1, this looks about right for bringing the five inferences together. For the second dataset, we synthesize a best guess of 1.38 for the odds-ratio, with a 90% credible interval from .89 to 3.07. With reference to Figure 10.2, we see that the lower posterior probability attached to 80% specificity results in a synthesis relying less on this scenario than the others.

10.5   Going Further with Bayesian Analysis

We have seen through a simple, if somewhat contrived example of using Bayesian analysis to acknowledge a common limitation of observational study data. In the context of a simple exposure variable prone to a simple form of misclassification, Bayesian inference was used to estimate the exposure-disease association while acknowledging uncertainty about the extent of misclassification. We emphasize that in this sort of problem we are indeed mitigating, rather than repairing, the damage caused by the misclassification. No amount of statistical cleverness can recover the actual exposure status for each study participant and thereby reclaim the inference we would ideally have drawn. At best we can draw inferences with sufficiently wide credible intervals to acknowledge the damage done by misclassification.

In fact, issues of poor measurement in observational studies can arise in a much more complex manner, in the face of much more complex data structures. Moreover, measurement error is often mentioned as but one of three major challenges that make it hard to obtain broadly convincing evidence from an observational study. Measurement error, along with uncontrolled confounding and selection bias, have been referred to as a "holy trinity" of threats to validity for observational studies in risk-factor epidemiology (Greenland, 2009). Roughly, uncontrolled confounding refers to the complete absence of data on a crucial variable that is associated with exposure status and also associated with disease status. For instance, it may only make sense to look at the relationship between abdominal fat distribution and heart disease within subpopulations defined by a genotype, but if this genotype is not recorded in the available data, then we have a severe challenge on our hands. Selection bias, on the other hand, is somewhat self-explanatory. If the characteristics of study participants somehow deviate systematically from the characteristics of the population as a whole, then we need statistical analysis which acknowledges this lack of representation.

Statisticians are kept busy with devising methodology to deal with this trinity of threats, and others, in the observational study realm. The Bayesian approach provides a general framework for such methodologies. As in Section 10.4, but with more complex data structures, one can estimate a target quantity from less than ideal data, and report an honest amount of uncertainty surrounding the estimate. Arguably, the Bayesian paradigm is particularly good for this uncertainty management. For instance, the credible intervals reported at the very end of Section 10.4 seamlessly integrate three types of uncertainty. The first type is the usual statistical uncertainty arising from having measurements on only the study participants (a sample), not the whole population. The second type is the additional uncertainty arising from having less than ideal data on a per participant basis, e.g., having self-reported exposure status rather than actual exposure status. The third type is the yet further uncertainty arising because the extent to which the data are less than ideal is itself unknown. For instance, a priori we might plausibly think that the specificity of the exposure classification could be as low as 80% or as high as 100%, and, as we have seen, this prior belief may not be changed very much upon observing the data.

Also with regard to the trinity of threats to validity for observational studies, it is important to note that two of these threats, or even all three, can easily manifest themselves in the same study. Consequently, some of the most impressive work is that developing Bayesian (or nearly Bayesian) schemes is to tackle multiple failings of the data simultaneously. Great examples of this are the methodological developments of Greenland (2003, 2005), which are applied to observational studies concerning magnetic fields from power lines and childhood leukemia.

This volume celebrates the International Year of Statistics from a Canadian perspective. Happily, in the realm of Bayesian methods for observational study data, there is considerable Canadian activity to report. Researchers in Montréal and Vancouver are particularly active in this area. On unobserved confounding, see, e.g., McCandless et al. (2007, 2008), Gustafson et al. (2010), and McCandless (2012). Concerning measurement error, recent work includes Ladouceur et al. (2007), Hossain and Gustafson (2009), Liu et al. (2009), Dendukuri et al. (2010), Chu et al. (2010), Lu et al. (2010), and Espino-Hernandez et al. (2011), and Wang et al. (2012).

A considerable portion of the current excitement and challenge and societal impact in statistical science lies toward the "Big Data" frontier. While it may be a quieter cousin, the "Difficult Data" frontier also generates excitement and challenge and societal impact. Some of the most important decisions to be taken, such as those relating to what exposures help or harm our health, are informed by some of the most difficult data. As part of the modern statistical firmament, Bayesian inference applied to observational study data exemplifies the principled extraction of evidence from difficult data.

Appendix

Let C_0 and C_1 be the counts of apparent exposures among the n_0 controls and n_1 cases respectively. The analyses presented in Sections 10.3 and 10.4 arise from modeling these counts as binomially distributed. That is, C_i counts the number of "successes" among n_i independent trials, each of which has success probability θ_i^*. In formal notation, $C_i \sim \text{Bin}(n_i, \theta_i^*)$ independently for controls ($i = 0$) and cases ($i = 1$).

For given sensitivity SN and specificity SP of exposure classification, one has $\theta_i^* = \theta_i SN + (1 - \theta_i)(1 - SP)$, where $\theta_i = \Pr(X = 1 | Y = i)$ is the true proportion exposed. Assigning the prior distribution $\theta_i \sim \mathcal{U}(0, 1)$, independently for $i = 0, 1$ induces the prior $\theta_i^* \sim \mathcal{U}(1 - SP, SN)$. Consequently, the posterior distribution of θ_i^* is the $\mathcal{B}(c_i + 1, n_i - c_i + 1)$ distribution truncated to the interval $(1 - SP, SN)$, again independently for $i = 0$ or 1. A Monte Carlo sample can then be drawn directly from the posterior distribution of (θ_0^*, θ_1^*) and transformed to realizations from the (θ_0, θ_1) posterior.

To compare the posterior weight given to different values of (SN, SP), one must compute the marginal probability $\Pr(C_0 = c_0) \times \Pr(C_1 = c_1)$ for each pair of values considered. The desired quantity can be computed via a routine for the Beta distribution, given that $\Pr(C_i = c_i)$ can be written as

$$\frac{1}{SN + SP - 1} \binom{n_i}{c_i} \int_{1-SP}^{SN} \theta_i^{*(c_i+1)-1}(1 - \theta_i^*)^{(n_i-c_i+1)-1} \, d\theta_i^*.$$

About the Author

Paul Gustafson is a professor of statistics at the University of British Columbia. He received his undergraduate training at UBC and a PhD from Carnegie Mellon University, Pittsburgh, PA. His research interests include Bayesian inference, causal inference, epidemiology, measurement error, and the analysis of observational data. He has served as editor of *The Canadian Journal of Statistics* (2007–09) and was the 2008 winner of the CRM–SSC Award. He is a fellow of the American Statistical Association.

Bibliography

Chu, R., Gustafson, P., and Le, N. (2010). Bayesian adjustment for exposure misclassification in case-control studies. *Statistics in Medicine*, 29:994–1003.

Dendukuri, N., Bélisle, P., and Joseph, L. (2010). Bayesian sample size for diagnostic test studies in the absence of a gold standard: Comparing identifiable with non-identifiable models. *Statistics in Medicine*, 29:2688–2697.

Espino-Hernandez, G., Gustafson, P., and Burstyn, I. (2011). Bayesian adjustment for measurement error in continuous exposures in an individually matched case-control study. *BMC Medical Research Methodology*, 11:67.

Greenland, S. (2003). The impact of prior distributions for uncontrolled confounding and response bias: A case study of the relation of wire codes and magnetic fields to childhood leukemia. *Journal of the American Statistical Association*, 98:47–55.

Greenland, S. (2005). Multiple-bias modelling for analysis of observational data. *Journal of the Royal Statistical Society, Series A*, 168:267–306.

Greenland, S. (2009). Accounting for uncertainty about investigator bias: Disclosure is informative how could disclosure of interests work better in medicine, epidemiology and public health? *Journal of Epidemiology and Community Health*, 63:593–598.

Gustafson, P., McCandless, L. C., Levy, A. R., and Richardson, S. (2010). Simplified Bayesian sensitivity analysis for mismeasured and unobserved confounders. *Biometrics*, 66:1129–1137.

Hernán, M. A., Alonso, A., Logan, R., Grodstein, F., Michels, K. B., Willett, W. C., Manson, J. E., and Robins, J. M. (2008). Observational studies analyzed like randomized experiments: An application to postmenopausal hormone therapy and coronary heart disease. *Epidemiology*, 19:766–779.

Hossain, S. and Gustafson, P. (2009). Bayesian adjustment for covariate measurement errors: A flexible parametric approach. *Statistics in Medicine*, 28:1580–1600.

Ioannidis, J. P. A. (2005). Why most published research findings are false. *PLoS Medicine*, 2:e124.

Ladouceur, M., Rahme, E., Pineau, C. A., and Joseph, L. (2007). Robustness of prevalence estimates derived from misclassified data from administrative databases. *Biometrics*, 63:272–279.

Liu, J., Gustafson, P., Cherry, N., and Burstyn, I. (2009). Bayesian analysis of a matched case-control study with expert prior information on both the misclassification of exposure and the exposure-disease association. *Statistics in Medicine*, 28:3411–3423.

Lu, Y., Dendukuri, N., Schiller, I., and Joseph, L. (2010). A Bayesian approach to simultaneously adjusting for verification and reference standard bias in diagnostic test studies. *Statistics in Medicine*, 29:2532–2543.

McCandless, L. C. (2012). Meta-analysis of observational studies with unmeasured confounders. *International Journal of Biostatistics*, 8(2):art. 5.

McCandless, L. C., Gustafson, P., and Levy, A. R. (2007). Bayesian sensitivity analysis for unmeasured confounding in observational studies. *Statistics in Medicine*, 26:2331–2347.

McCandless, L. C., Gustafson, P., and Levy, A. R. (2008). A sensitivity analysis using information about measured confounders yielded improved assessments of uncertainty from unmeasured confounding. *Journal of Clinical Epidemiology*, 61:247–255.

Prentice, R. L., Langer, R., Stefanick, M. L., Howard, B. V., Pettinger, M., Anderson, G., Barad, D., Curb, J. D., Kotchen, J., Kuller, L., et al. (2005). Combined postmenopausal hormone therapy and cardiovascular disease: Toward resolving the discrepancy between observational studies and the women's health initiative clinical trial. *American Journal of Epidemiology*, 162:404–414.

Taubes, G. and Mann, C. C. (1995). Epidemiology faces its limits. *Science*, 269:164–169.

Wang, D., Shen, T., and Gustafson, P. (2012). Partial identification arising from non-differential exposure misclassification: How informative are data on the unlikely, maybe, and likely exposed? *International Journal of Biostatistics*, 8(1):Article 31.

11

Statistical Models for Disease Processes: Markers and Skeletal Complications in Cancer Metastatic to Bone

Richard J. Cook

University of Waterloo, Waterloo, ON

In many chronic diseases, the presence and concentration of molecular compounds circulating in the blood are associated with the stage of the disease and the risk of disease progression. In diabetes, for example, individuals with poor blood glucose control are at elevated risk of vascular complications involving the heart, eyes and kidneys. The level of glycosylated hemoglobin (HbA1c) in the blood reflects how well blood glucose levels are regulated, so measurements of HbA1c are used routinely in evaluating the effectiveness of medical or behavioral interventions in diabetes. In individuals with autoimmune diseases such as rheumatoid arthritis, levels of C-reactive protein (CRP) in the blood reflect the degree of inflammation which in turn predicts joint damage. In individuals with human immunodeficiency virus (HIV) infection, physicians monitor viral load, which is the concentration of the virus in the blood at any given time. In each of these settings, the measurements in the blood are termed markers which reflect disease activity or severity. They often play a useful role in predicting the course of disease and the occurrence of serious debilitating events. This chapter describes some ways to use marker data in patients with cancer metastatic to bone.

11.1 Skeletal Complications in Individuals Suffering from Cancer Metastatic to Bone

Bone resorption is the process by which osteoclast cells break down bone tissue and release minerals such as calcium into the bloodstream. In healthy individuals the process of bone resorption is usually well coordinated with that of bone formation, to ensure repair and maintenance of normal bone tissue. In many cancers, however, metastatic lesions develop in the skeleton, disrupt this equilibrium, and thereby compromise the structural integrity of the bone. This weakens the skeleton and puts individuals at increased risk of complications such as fractures, spinal cord compression, hypercalcemia, and bone pain requiring intervention. These complications significantly reduce a person's ability to carry out normal daily activities and decrease their quality of life. With the increased effectiveness of anti-tumor therapy in recent decades, patients are living longer following cancer diagnosis, so management of the skeletal metastases is an important aspect of patient care.

N-telopeptide of type I collagen (Ntx) is a marker of bone resorption that has been found to be associated with the presence and volume of skeletal metastases in breast cancer patients (Lipton et al., 2001), as well as the occurrence of skeletal complications (Coleman et al., 1997). In addition, the bone formation marker, bone-specific alkaline phosphatase (BALP) is often elevated in prostate cancer patients (Smith et al., 2011). Scientists are interested in relating the levels of bone formation and resorption markers to the occurrence of skeletal complications to better understand the disease process (Demers et al., 1995).

Bisphosphonates are a class of drugs which combat abnormal bone resorption to help maintain the integrity of the skeleton. These drugs are used extensively in the treatment of osteoporosis and substantial evidence has emerged that suggests treatment with bisphosphonates can reduce the incidence of skeletal complications from malignant bone disease. Three large randomized, multi-center, double-blind, phase III clinical trials evaluated the safety and efficacy of a bisphosphonate called zoledronic acid in more than 3,000 patients with malignant bone disease from a broad range of primary cancers (Rosen et al., 2003, 2004; Saad et al., 2004). Upon recruitment to these studies patients were randomly assigned to receive an intravenous infusion of zoledronic acid or a placebo control every three to four weeks for up to 24 months. Measurements of Ntx and BALP were taken at randomization, after one month of treatment, and quarterly thereafter during the course of the study.

Figure 11.1 shows the measurements of Ntx and BALP for four prostate cancer patients in the study by Saad et al. (2004). Also indicated on the horizontal axes are the times of skeletal events (E), death (D), and end of followup (C). It is apparent that there can be considerable variation in the bone marker values within patients over time, and that the two markers are not necessarily elevated at the same time. Also apparent is the variation in the period over which markers are measured due to incomplete assessments, study withdrawal, or death.

Here we highlight work of an international multidisciplinary collaborative team of investigators studying the utility of bone marker measurements collected in these trials. One goal of this team was to study the prognostic value of the marker measurements taken at study entry for skeletal events and death. Insight into their prognostic role could help researchers select high-risk patients for inclusion in experimental studies, or for more intensive monitor-

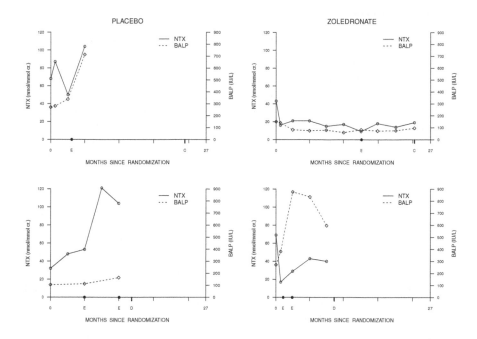

FIGURE 11.1: Sample profiles of an osteoclast (Ntx) and osteoblast (BALP) marker of bone disease along with indications of the times of skeletal events (E), death (D), and end of follow-up (C) for two prostate cancer patients receiving placebo (left panels) and two prostate cancer patients receiving zoledronate (right panels); data from Saad et al. (2004).

ing and treatment. Interest also lies in modeling how the markers vary over time and in exploration of the effect of treatment with zoledronic acid on the marker values. Several challenges arise when attempting to do this because (i) marker values are only available when blood samples are provided; (ii) measurement of markers ends upon withdrawal from the study; and (iii) the marker process naturally terminates upon death. Models which jointly examine marker values, skeletal events, and death are therefore needed to provide a complete description of the disease process and treatment effects. This research program addressed these challenges and led to considerable advances in the understanding of the pathophysiology and prognostic value of bone markers in patients with cancer metastatic to bone. Some statistical aspects of the works are described below in more detail.

11.2   Bone Markers and Prognosis for Adverse Events

Many studies examine the times to specific adverse events; such times are often termed lifetimes. In studies of lifetime data, events are often recorded as the time from study entry until their occurrence. For example, we may wish to study the times when skeletal events such as fractures occur. If T denotes the time of a disease-related event of interest, Kaplan–Meier estimates of $\Pr(T \leq t)$ reflect the cumulative risk of the event as a function of time (Kaplan and Meier, 1958). When interest lies in the effect of prognostic factors which change over time, however, it is more natural to model the risk of event occurrence at an instant in time. We describe next how this may be done.

Consider a set of factors measured at time t that may affect the risk of an event. These factors, denoted $x(t)$, are often called covariates (or covariables) since they arise along with ("co-") the variable (here, an event time) of primary interest. If there are p covariates then $x(t) = (x_1(t), \ldots, x_p(t))^\top$ is a $p \times 1$ covariate vector; some elements in $x(t)$ may be fixed but others may vary over time. In the present setting we may think of $x(t)$ as containing values for markers of bone metabolism and possibly other factors.

In time to event analyses, the intensity function is a mathematical function that specifies the instantaneous risk of an event among individuals who are event-free at a given time. It is denoted by

$$\lambda\{t|x(t)\} = \lim_{\Delta t \downarrow 0} \frac{\Pr\{t \leq T < t + \Delta t | t \leq T, x(t)\}}{\Delta t} ,$$

where the vertical line "|" demarcates the event that is being modeled ($t \leq T < t + \Delta t$) and the information that is being conditioned upon ($t \leq T, x(t)$). In other words, here we are modeling the risk of the event occurring in the interval $[t, t+\Delta t)$ given that it did not happen before t, and given the covariate values at time t.

The effects of the time-varying bone markers discussed earlier are naturally and most commonly studied through what are termed intensity-based regression models (Kalbfleisch and Prentice, 2002; Lawless, 2003). Such models require specification of a particular mathematical form for the intensity function to reflect the trends in risk over time, as well as how covariates alter this risk. The most widely used intensity-based model in this setting is called the Cox model (Cox, 1972). This model takes the form

$$\lambda\{t|x(t)\} = \lambda_0(t)e^{x(t)^\top \beta},$$

where $\lambda_0(t)$ is a reference intensity function characterizing risk for individuals with $x(t) = 0$, $\beta = (\beta_1, \ldots, \beta_p)^\top$ is a $p \times 1$ vector of unknown parameters called regression coefficients, and

$$x(t)^\top \beta = \sum_{j=1}^{p} x_j(t)\beta_j.$$

There is routinely a trade-off between the desire for simplicity and the desire for flexible models which describe a process of interest well. The appeal of the Cox regression model is that the form of the reference intensity function $\lambda_0(t)$ is left unspecified, thereby enabling the model to accommodate a wide range of shapes. This formulation also has the appealing property that the exponential of the jth element of β, e^{β_j}, is a relative risk (Prentice and Farewell, 1986). To see why, note that when $p = 1$ and x_0 is a specific value of $x(t)$,

$$\frac{\lambda\{t|x(t) = x_0 + 1\}}{\lambda\{t|x(t) = x_0\}} = e^{\beta_1},$$

is a ratio of instantaneous risks reflecting the effect of a one unit increase in the covariate (e.g., marker) value.

Despite the complexity of the disease process and the many types of events that patients with bone metastases may experience, it is common in clinical trials for treatments to be assessed based on the time of the first adverse event. These outcomes are called composite endpoints and here we consider one based on the time of the first skeletal event or death for patients in the metastatic prostate cancer trial of Saad et al. (2004). Marker data prior to this composite event are used in the analyses that follow.

The first analysis we consider involves examining the relationship of the bone marker values (Ntx and BALP) measured at the time of randomization on the event intensity. For this purpose, the osteoclast marker Ntx and the osteoblast marker BALP were used in the models as categorical variables with four categories based on the quartiles of the baseline distributions; these were 54.5, 89.0 and 180.5 nmol/mmol cr. for Ntx and 150.25, 267.50 and 529.75 IU/L for BALP. Because of this categorization of the markers, the relative risks convey the effect of an individual having a value in the second quartile versus the first quartile, the third quartile versus the first quartile, and the fourth quartile versus the first quartile, for each of the two markers.

Each factor is examined in a separate model in univariate (i.e., single marker variable) analyses, where the results reflect the association between the baseline marker value and the risk of the event in the absence of other information. Covariates often reflect similar information (and hence are correlated) and so it is customary to also examine the effect of covariates simultaneously by fitting multivariate (i.e., several variable) regression models.

These models are used to identify factors which convey prognostic information in the presence of the other covariates in the model. In addition to examining the joint effect of Ntx and BALP, the multivariate models adjusted for several non-marker baseline variables (see the footnote of Table 11.1). Finally, the argument t in the covariate $x(t)$ reflects the fact that one can use the serial (time-dependent) marker values and we also do this for both univariate and multivariate models. In this case marker measurements were carried forward between assessments for up to six months, at which point if no updated values were available individuals were dropped from the analysis.

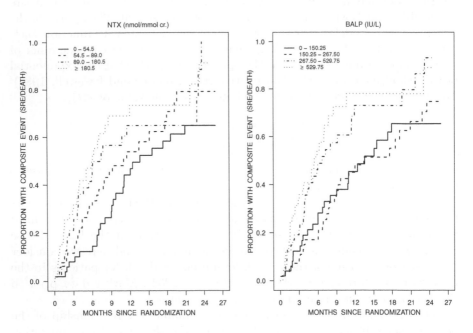

FIGURE 11.2: Kaplan–Meier estimates of the probability of an event by quartiles of urinary N-telopeptide (Ntx) (nmol/mmol cr.) and serum BALP (IU/L).

Further data from such individuals were used if subsequent measurements were made. The results of fitting univariate and multivariate models with baseline data are displayed in the top half of Table 11.1, with the corresponding results from time-dependent marker analyses in the bottom half of Table 11.1. These analyses are for simplicity restricted to placebo patients only. The table shows relative risks which are estimated by fitting Cox models to the data. It also shows 95% confidence intervals for each relative risk, which represent a standard margin of error for an estimate. Finally, p-values are shown for each relative risk; they are a measure of the evidence against the hypothesis that the relative risk is one, in which case the marker category is not associated with an increased risk of the event. Small p-values represent significant evidence against the hypothesis; values less than .05 are often considered to provide fairly strong evidence.

Table 11.1 indicates that when the baseline values are considered in univariate analyses, there is evidence of a strong relationship of both markers to event occurrence, with increasing risk associated with the higher marker quartiles. Figure 11.2 shows estimates of the probability that an event occurs as a function of time since entry to the study, estimated separately for the different quartiles of Ntx and BALP. The probability that an event occurs by a given time increases with the level of each marker, and the effects are remarkably similar for the osteoblast marker BALP (left panel) and the osteoclast marker Ntx (right panel). In the multivariate analyses of baseline marker values (top right in Table 11.1), covariates included age (years), cancer duration (years), presence of metastases in the lymph nodes (yes/no), an indicator of advanced disease (ECOG score ≥ 1 vs. 0), bone pain (present/absent), prostate specific antigen ($\log_{10}$PSA, where PSA is measured in ng/mL), hemoglobin (g/dL), creatinine (≥ 1.4 mg/dL), albumin (g/L), lactate dehydrogenase (LDH > 454 units/L), an indicator of analgesic use (yes/no), and an indicator of whether they had experience a prior skeletal event (yes/no). After controlling for all of these factors, there is less evidence of an association between the markers and event occurrence. We can infer from this that the baseline markers do not convey prognostic information about event occurrence beyond that contained in other available covariates (see the footnote of Table 11.1).

A different picture arises when we consider the serial (time-dependent) values of the two bone markers (bottom half of Table 11.1). In univariate analyses of these marker values there is stronger evidence of an association with event occurrence and the relative risks are considerably larger than those seen using only baseline values. Moreover, even in the multivariate analyses, the level of BALP is significantly associated with event occurrence (p-value = .0144); risk of the event is more than double when BALP is the upper quartile compared to the lowest quartile.

There are two points to be noted from these analyses. First, as time goes by, the prognostic relevance of a marker value measured at baseline might be expected to decrease relative to a more recent measurement of a marker. A comparison of the top and bottom portions of Table 11.1 suggests this is the case. Second, the prognostic utility of a marker depends on whether other potentially important variables are considered. That is, when other variables are available and in a model, one is implicitly examining the incremental prognostic utility of the marker in the presence of the other variables. This tends to reduce the effect (as measured by the relative risk) compared to the univariate model.

11.3 Joint Models for Dynamic Bone Markers and Adverse Events

Outcomes such as skeletal events and death are of clear importance and offer a natural basis for the comparison of treatments. Not all individuals experience these events during a study, however, and there is interest in modeling changes in marker values over time and studying how they may be affected by treatment. This can give insight into the effect of bisphosponate therapy on the

TABLE 11.1: Relative risks obtained from univariate and multivariate Cox regression analyses for intensity-based models of skeletal event-free survival in placebo patients of Saad et al. (2004); results based on baseline marker values are displayed in the top half of the table and serial marker values in the bottom half of the table.

Marker[1]	Univariate			Multivariate		
	RR	95% CI	p-value	RR	95% CI	p-value
Baseline Marker Values						
Ntx < 54.5	—	—	—	—	—	—
54.5 ≤ Ntx < 89.0	1.49	(.87, 2.55)	.151	1.52	(.81, 2.86)	.196
89.0 ≤ Ntx < 180.5	2.05	(1.15, 3.64)	.015	1.77	(.86, 3.63)	.120
180.5 ≤ Ntx	2.22	(1.28, 3.86)	.004	1.40	(.63, 3.13)	.411
Global Test, p-value			.021			.411
BALP < 150.25	—	—	—	—	—	—
150.25 ≤ BALP < 267.50	.91	(.53, 1.59)	.750	.67	(.36, 1.24)	.200
267.50 ≤ BALP < 529.75	1.60	(.91, 2.83)	.102	1.47	(.73, 2.96)	.284
529.75 ≤ BALP	2.24	(1.28, 3.94)	.005	1.39	(.67, 2.90)	.381
Global Test, p-value			.005			.102
Time-varying (Serial) Marker Values						
Ntx < 54.5	—	—	—	—	—	—
54.5 ≤ Ntx < 89.0	1.83	(.94, 3.56)	.075	1.64	(.80, 3.38)	.179
89.0 ≤ Ntx < 180.5	2.80	(1.53, 5.11)	.001	1.68	(.82, 3.46)	.158
180.5 ≤ Ntx	4.09	(2.23, 7.49)	< .001	1.77	(.78, 3.98)	.169
Global Test, p-value			< .001			.466
BALP < 150.25	—	—	—	—	—	—
150.25 ≤ BALP < 267.50	1.15	(.55, 2.40)	.703	.86	(.39, 1.90)	.711
267.50 ≤ BALP < 529.75	2.98	(1.59, 5.61)	.001	2.10	(1.02, 4.32)	.045
529.75 ≤ BALP	3.62	(2.00, 6.55)	< .001	2.33	(1.10, 4.91)	.027
Global Test, p-value			< .001			.014

[1] Ntx is measured in units of nmol/mmol cr. and the units for BALP are IU/L.

process of bone resorption and formation. The analyses in the previous section revealed that serial BALP measurements are significantly associated with the occurrence of skeletal events and death. Given the role of BALP in bone formation, it is natural to consider the effect of bisphosphonate treatment on this marker.

Levels of BALP change continuously over time but they are only measured when blood samples are taken at the periodic follow-up assessments. Serial marker values are often compared for different treatments by estimating average values at different times. When the assessment times are variable, due to random variation or possibly missed assessments, estimation and interpretation of mean values can be challenging since they correspond to individuals who survived to the respective time and who provided a blood sample. Often average values are computed at different assessment times and lines are drawn to connect these averages. There is a tendency to interpret these graphs as representing the average course of markers over time, but it is now well known that such profiles are not interpretable at the individual level (Wu, 2009). Another approach is to study the rate of change of marker values using individual-specific linear or non-linear models. When marker values, and trends in marker values, alter the risk of events such as death, models based on the rate of change are also problematic.

We address these challenges by considering joint models for the serial marker values and the event of interest by defining distinct marker states in terms of the quartiles of the marker values. Figure 11.3 contains a so-called multistate diagram with marker States 1 to 4 corresponding to ranges

$$0 \leq \text{BALP} < 150.25, \quad 150.25 \leq \text{BALP} < 267.50,$$
$$267.50 \leq \text{BALP} < 529.75, \quad 529.75 \leq \text{BALP},$$

and State 5 representing the clinical event of interest (death or a skeletal event). The idea behind this multi-state diagram is that at any given time, an individual's disease state can be represented by one of the states, and dynamic features of the marker and event process can be represented by stochastic

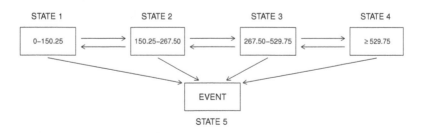

FIGURE 11.3: A multistate diagram for modeling changes in bone alkaline phosphotase prior to the occurrence of the clinical event of interest.

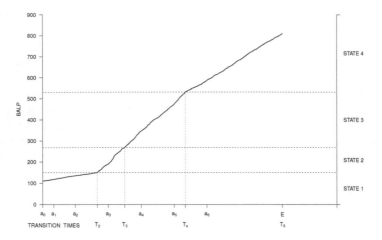

FIGURE 11.4: A plot of BALP values in continuous time for a hypothetical individual with the assessment times $(a_r, r \in \{0, \dots, 6\})$ and event time E indicated on the horizontal axis.

models for the transitions between the states. We consider a transition into State 5 the end of the process because one component of the composite event is death, and values will not be available after this. Moreover, the goal is to model marker dynamics before skeletal complications since the complications themselves may alter marker values.

We let $Z(t)$ represent the state occupied at time t so that $Z(t) = k$ if an individual is in state k at time t; for example, $Z(t) = 3$ if at time t an individual has a BALP value in the interval $[267.50, 529.75)$. Starting with $t = 0$ at the study entry, the path of the marker as depicted in Figure 11.3 is traced, and $\{Z(s), s > 0\}$ denotes the corresponding multi-state process. When the path is considered only up to a particular time t, we denote the history of the process up to that time as $\mathcal{H}(t) = \{Z(s), 0 < s < t\}$, which captures the nature and timing of all transitions over the interval $(0, t)$. The transitions between states are again governed by intensity functions, which in this context reflect the instantaneous risk of movement from one state to another given the history of the process. The transition intensity from state k to ℓ is denoted as $\lambda_{k\ell}\{t|\mathcal{H}(t)\}$ and defined formally as

$$\lambda_{k\ell}\{t|\mathcal{H}(t)\} = \lim_{\Delta t \downarrow 0} \frac{\Pr\{Z(t + \Delta t^-) = \ell | Z(t^-) = k, \mathcal{H}(t)\}}{\Delta t}, \quad k \neq \ell. \quad (11.1)$$

Figure 11.4 contains a plot of the continuous time course of BALP values for a hypothetical individual. The times at which measurements of BALP are available are indicated by a_r, $r \in \{0, \dots, 6\}$, along with the event time (T_5). The horizontal dashed lines in Figure 11.4 correspond to the quartiles of the baseline distribution of BALP which in turn define the states of Figure 11.3. The time that the BALP curve crosses the lowest dashed line (T_2) is therefore the time of the transition from State 1 to State 2 and subsequent transition times $(T_3$ and $T_4)$ are similarly defined. Because individuals are only seen periodically we do not observe the precise times of any of the transitions between the first four states; indeed, because an individual's BALP values can go either up or down over time, we do not even know the number of transitions between measurements. In contrast, if the event occurs during the course of followup, its time is known (see $T_5 = T$ in Figure 11.4), but in some cases all we know is that it did not happen while the individual was on study.

To formalize this, we let C denote the duration of time an individual was on study. If T denotes the time of the event, then we let $U = \min(T, C)$ and $\delta = \mathbf{1}(U = T)$ indicate whether U is the event time, or a lower bound for the event time (called a censoring time). Blood samples are drawn at times $a_0 < \cdots < a_R$, and so the data consist of the sequence of assessment times and states occupied at those times, along with the information on the event time: this is equivalent to $\{(a_r, Z(a_r)), r = 0, \ldots, R; (U, \delta)\}$. Under intermittent inspection we define the observed history as of the rth assessment as $H(a_r) = \{(a_s, Z(a_s)), s = 0, \ldots, r-1, a_r, a_r < U\}$; since data are not available between assessments, for $a_r \leq t < a_{r+1}$, we assume that $H(t) = H(a_r)$.

Statistical inference concerning the process in Figure 11.3 can be based on a likelihood function which is obtained from the probability of observing the realized data (Sprott, 2000). In this setting this is

$$\left[\prod_{r=1}^{R} \Pr\{Z(a_r) | H(a_r)\} \right]$$

$$\times \left[\sum_{k=1}^{4} \Pr\{Z(s^-) = k | H(s)\} \lambda_{k5}^{\delta}\{s | Z(s^-) = k, H(s)\} \right].$$

For many processes the probabilities above may be difficult to express in terms of the intensity functions used to specify the models. Researchers often model such data using Markov assumptions (Cook and Lawless, 2013) for which the transition intensities are of the form

$$\lambda_{k\ell}\{t | \mathcal{H}(t)\} = \lambda_{k\ell}\{t | Z(t^-) = k\},$$

i.e., the instantaneous risk of the event given the entire history, is governed by simply the current marker state. Jackson (2011) has written useful software which facilitates the sort of multi-state analyses which we employ here.

Figure 11.5 shows nonparametric Kaplan–Meier estimates of the cumulative probability $\Pr(T \leq t)$ of the composite event according to the baseline quartile of BALP. These estimates, provided separately for each treatment group, use only raw data and are not based on the model in Figure 11.3. Also displayed are the same probabilities estimated based on the model depicted

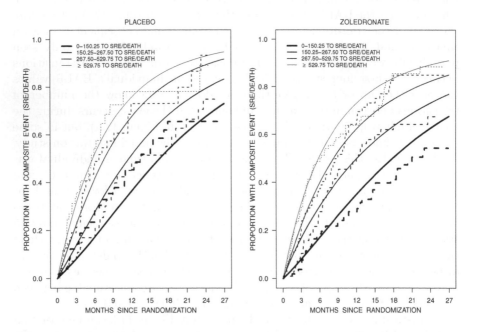

FIGURE 11.5: Kaplan–Meier estimates (dotted line) and $\widehat{P}\{Z(t) = 5 | Z(0) = k\}$ for $k = 1, 2, 3, 4$ based on a multistate model (solid line) for the probability of having a skeletal event or died by quartiles of serum BALP (IU/L).

in Figure 11.3. The two methods of analysis give estimates of cumulative risk which are in fairly good agreement.

An appeal of the multi-state analysis is that it can also be used to examine treatment effects on the transition intensities between the marker states. For example, if we let $x = 1$ for individuals receiving zoledronate and $x = 0$ for those receiving the placebo, we may fit proportional intensity models of the form

$$\lambda_{k\ell}\{t | H(t), x\} = \lambda_{k\ell}\{t | Z(t^-) = k\} e^{\beta_{k\ell} x}$$

for which the instantaneous risk of a $k \to \ell$ transition differs by a multiplicative factor $e^{\beta_{k\ell}}$ in the zoledronate versus placebo groups. Table 11.2 shows the results of fitting such a model and it can be seen that zoledronate significantly increases the transition rate to states of lower marker values; the rate of transition from State 2 ($150.25 \leq$ BALP < 267.50) to State 1 (BALP < 150.25) is 2.7 times higher in the treated group compared to the placebo control group, and similarly increased rates of transition are seen for the transitions from States 3 to 2 and 4 to 3. A key message is therefore that treatment with zoledronate can help reduce the value of BALP. Interestingly there appears to be

TABLE 11.2: Relative risks for the effect of zoledronate on the intensites of transitions between BALP marker states in the five-state Markov model depicted in Figure 11.3.

Transitions				
From	To	RR	95% CI	*p*-value
		Reduction in BALP (IU/L)		
0 − 150.25	—	—	—	—
150.25−267.50	0 − 150.25	2.72	(1.66, 4.47)	< .001
267.50−529.75	150.25−267.50	3.38	(1.84, 6.21)	< .001
≥ 529.75	267.50−529.75	1.84	(.92, 3.69)	.085
		Increase in BALP (IU/L)		
0 − 150.25	150.25−267.50	.96	(.63, 1.45)	.842
150.25−267.50	267.50−529.75	1.45	(.96, 2.18)	.076
267.50−529.75	≥ 529.75	1.14	(.76, 1.69)	.533
≥ 529.75	—	—	—	—

relatively little effect of zoledronate on the transition rates to higher BALP states based on this model.

11.4 Discussion

Marker data can play an important role in understanding disease processes, evaluating treatment effects and making predictions about the course of disease. We have focused here on the relationship between marker values and the primary response in the clinical trials, the time to the first skeletal event or death. We also considered the effect of treatment on the marker values through specification of a multi-state model. More elaborate multi-state models could of course be specified which may, for example, distinguish between the first skeletal event and death, accommodate the occurrence of multiple skeletal complications (Cook et al., 2009) or distinguish between the different types of skeletal complications. Such models would provide a better representation of the disease process but require strong assumptions warranting critical assessment.

Analyses must deal with the fact that marker values change in continuous time but are usually only available at periodic assessment times. In the bone marker studies, the assessments were scheduled according to a trial protocol and so measurement times are unrelated to the disease process. In observational cohort studies, however, marker measurements are often only made when individuals attend a clinic at which they receive medical care. In such circumstances, measurements of markers may be made more often when their values correspond to severe disease activity. This can make standard analyses invalid and may require more challenging analysis based on joint models of the assessment process and marker values (Chen et al., 2010; Sweeting et al., 2010).

There has been increased interest in the design and conduct of clinical trials aiming to evaluate use of markers to guide therapy, an objective that falls within the framework of personalized medicine. In asthma, inflammatory markers can be measured in sputum samples obtained from patients experiencing exacerbations and the cellular analysis can be used to guide therapy (Jayaram et al., 2006). A recent trial in the United Kingdom was designed to assess the utility of serial bone marker values in guiding the dose and frequency of bisphosphonate therapy in patients with cancer metastatic to bone. In such settings markers need to be highly predictive of events of importance to patients since otherwise marker driven dose reduction or other treatment modifications may increase the risk of adverse events.

Acknowledgments

The author thanks the editor for constructive comments and suggestions during the preparation of this chapter, Novartis for permission to use the data from the prostate cancer trial, Robert E. Coleman, Allan Lipton, Pierre Major, and Matthew Smith for collaboration on the bone marker research program, and Ker-Ai Lee for assistance with the statistical programming. This research was supported by grants from the Natural Sciences and Engineering Research Council of Canada, the Canadian Institutes for Health Research, and the Canada Research Chairs Program.

About the Author

Richard J. Cook is a professor of statistics and the holder of a Canada Research Chair in Statistical Methods for Health Research at the University of Waterloo. He also holds an adjunct appointment in the Department of Clinical Epidemiology and Biostatistics at McMaster University. He obtained a BSc in statistics from McMaster and his PhD from Waterloo. His research interests include the analysis of longitudinal and life history data, statistical methods for dealing with incomplete data, and the design and analysis of studies involving chronic disease. He was the recipient of the CRM–SSC Prize in 2007 and is a fellow of the American Statistical Association.

Bibliography

Chen, B., Yi, G. Y., and Cook, R. J. (2010). Analysis of interval-censored disease progression data via multi-state models under a nonignorable inspection process. *Statistics in Medicine*, 29:1175–1189.

Coleman, R. E., Purohit, O. P., Vinholes, J. J., and Zekri, J. (1997). High dose pamidronate: Clinical and biochemical effects in metastatic bone disease. *Cancer*, 80(Suppl):1686–1690.

Cook, R. J. and Lawless, J. F. (2013). Statistical issues in modeling chronic disease in cohort studies. *Statistics in Biosciences*, in press.

Cook, R. J., Lawless, J. F., Lakhal-Chaieb, L., and Lee, K.-A. (2009). Robust estimation of mean functions and treatment effects for recurrent events under event-dependent censoring and termination: Application to skeletal complications in cancer metastatic to bone. *Journal of the American Statistical Association*, 104:60–75.

Cox, D. R. (1972). Regression models and life tables (with discussion). *Journal of the Royal Statistical Society, Series B*, 34:187–220.

Demers, L. M., Costa, L., Chinchilli, V. M., Gaydos, L., Curley, E., and Lipton, A. (1995). Biochemical markers of bone turnover in patients with metastatic bone disease. *Clinical Chemistry*, 41:1489–1494.

Jackson, C. H. (2011). Multi-state models for panel data: The MSM package for R. *Journal of Statistical Software*, 38:1–28.

Jayaram, L., Pizzichini, M. M., Cook, R. J., Boulet, L.-P., Lemière, C., Pizzichini, E., Cartier, A., Hussack, P., Goldsmith, C. H., Laviolette, M., Parameswaran, K., and Hargreave, F. E. (2006). Determining asthma treatment by monitoring sputum cell counts: effect on exacerbations. *European Respiratory Journal*, 27:483–494.

Kalbfleisch, J. D. and Prentice, R. L. (2002). *The Statistical Analysis of Failure Time Data,* Second Edition. Wiley, New York.

Kaplan, E. L. and Meier, P. (1958). Nonparametric estimation from incomplete observations. *Journal of the American Statistical Association,* 53:457–481.

Lawless, J. F. (2003). *Statistical Models and Methods for Lifetime Data,* Second Edition. Wiley, Hoboken, NJ.

Lipton, A., Costa, L., Ali, S. M., and Demers, L. M. (2001). Bone markers in the management of metastatic bone disease. *Cancer Treatment Reviews,* 27:181–185.

Prentice, R. L. and Farewell, V. T. (1986). Relative risk and odds ratio regression. *Annual Review of Public Health,* 7:35–58.

Rosen, L. S., Gordon, D., Kaminski, M., Howell, A., Belch, A., Mackey, J., Apffelstaedt, J., Hussein, M. A., Coleman, R. E., Reitsma, D. J., Chen, B.-L., and Seaman, J. J. (2003). Long-term efficacy and safety of zoledronic acid compared with pamidronate disodium in the treatment of skeletal complications in patients with advanced multiple myeloma or breast carcinoma: A randomized, double-blind, multicenter, comparative trial. *Cancer,* 98:1735–1744.

Rosen, L. S., Gordon, D., Tchekmedyian, N. S., Yanagihara, R., Hirsh, V., Krzakowski, M., Pawlicki, M., de Souza, P., Zheng, M., Urbanowitz, G., Reitsma, D., and Seaman, J. (2004). Long-term efficacy and safety of zoledronic acid in the treatment of skeletal metastases in patients with nonsmall cell lung carcinoma and other solid tumors: A randomized, phase III, double-blind, placebo-controlled trial. *Cancer,* 100:2613–2621.

Saad, F., Gleason, D. M., Murray, R., Tchekmedyian, S., Venner, P., Lacombe, L., Chin, J. L., Vinholes, J. J., Goas, J. A., and Zheng, M., for the Zoledronic Acid Prostate Cancer Study Group (2004). Long-term efficacy of zoledronic acid for the prevention of skeletal complications in patients with metastatic hormone-refreactory prostate cancer. *Journal of the National Cancer Institute,* 96:879–882.

Smith, M. R., Cook, R. J., Lee, K.-A., and Nelson, J. B. (2011). Disease and host characteristics as predictors of time to first bone metastasis and death in men with progressive castration-resistant nonmetastatic prostate cancer. *Cancer,* 117:2077–2085.

Sprott, D. A. (2000). *Statistical Inference in Science.* Springer, New York.

Sweeting, M. J., Farewell, V. T., and De Angelis, D. (2010). Multi-state Markov models for disease progression in the presence of informative examination times: An application to hepatitis C. *Statistics in Medicine,* 29:1161–1174.

Wu, L. (2009). *Mixed Effects Models for Complex Data.* Chapman & Hall, London.

12

Analysis of Biased Survival Data: The Canadian Study of Health and Aging and beyond

Masoud Asgharian, Christina Wolfson, and David B. Wolfson
McGill University, Montréal, QC

Often researchers are faced with analyzing data from a biased sample that is unrepresentative of the population of interest. Studies that lead to samples that are biased-by-design are only useful if statisticians have a way to compensate for biased procedures that would result from naïve use of such non-representative data. This chapter tells the story of how we encountered biased survival data in a major Canadian study of dementia in the elderly. We relate how we overcame the problem of bias and developed methods to answer questions about dementia. Although the story is woven around the Canadian Study of Health and Aging, there are many other areas in which the type of bias that we encountered also arises.

12.1   Introduction

Epidemiologists define the prevalence of a disease as the number of individuals with the disease per 100,000 in the general population at a given time. They define the incidence rate to be the number of new cases of the disease per 100,000 individuals per unit of time (usually per year). A shocking statistic is that roughly 40% of Canadians over the age of 85 have some form of dementia. Consequently, as life expectancies of Canadians rise there will, in the foreseeable future, be a corresponding increase in both the incidence and prevalence of dementia. Not only will our health care system be hard-pressed to cope, there will also be a growing burden on caregiver families; most of us will be affected either directly or indirectly. It is vital, therefore, that we understand the natural history of dementia; that is, how it evolves in those stricken with it, and the factors that may hasten or slow its progression. At the population level, it is equally important to know how dementia incidence rates are changing and how this change, when combined with improving life expectancy, will affect the population burden of dementia. Further, we can only assess whether treatments or changes in lifestyle prevent or slow the course of a disease if we are able to measure changes in disease duration and incidence in populations.

Statisticians and epidemiologists collaborate to design studies and use the data collected from them to reveal patterns of changing incidence, and disease duration. With these goals in mind, in 1991 the Canadian Study of Health and Aging (CSHA) was launched (CSHA, 1994). A primary goal of the CSHA was to estimate the prevalence of dementia in elderly Canadians. Initially designed as a one time cross-sectional study, the investigator team obtained research funds to conduct two follow-ups on the participants over the subsequent decade.

Although there are many different types of dementia the two most common forms are those due to Alzheimer's disease and vascular dementia. Indeed, current research indicates that two thirds of all dementias that occur in the elderly are due to Alzheimer's disease. In this chapter, for simplicity, by the term "dementia" we shall mean either Alzheimer's disease or vascular dementia. Our story does not begin with incidence or prevalence but rather with the questions, "How long do people with dementia survive following onset of their disease?" and "What factors are associated with shorter (or longer) survival?"

These two questions were first posed by one of us, Christina Wolfson, Principal Investigator for the CSHA Progression of Dementia Study. Anticipating the use of data from the CSHA, she was concerned about the bias that invariably accompanies estimates of survival based on data collected from a study designed to follow-up a cohort of prevalent cases, i.e., cases that already have dementia at the time of their recruitment. Such studies are descriptively called prevalent cohort studies with follow-up. The CSHA was such a study.

In 1991 roughly 10,000 Canadians over the age of 65 were recruited and those living in the community were screened for dementia. All participants living in institutions, and those living in the community who screened positive at baseline were invited to undergo a thorough assessment for dementia that involved a battery of neurological and neuropsychological tests. Roughly 820 were diagnosed with prevalent (current) dementia in one of the two main categories above.

Although the full cohort of 10,000 was then followed forward, for the moment we focus on the 820 with prevalent dementia. By 1996, the end of the first phase of the CSHA, many of the 820 had died (and their dates of death recorded), and most of the rest were known to be still alive — said to have right censored survival times. (A survival time is said to be censored when only a lower bound of its value is known.) A small proportion of study subjects had been lost to follow-up in between 1991 and 1996 and the dates at which they were last known to be alive were recorded. Their survival times are also right censored.

In 1991 each caregiver provided an approximate date of dementia onset as well as covariate information, such as age at onset and number of years of education. That is, apart from their covariates, by 1996 each subject had contributed a survival time (possibly right censored) consisting of the time interval from their date of onset to the minimum of the date of death and their date of right censoring (see Figure 12.1).

To estimate how long people with dementia live from onset of their disease such data cannot be used without adjustment, because the sample is biased. The bias stems from the fact that in order to be among the 820 subjects identified with dementia in 1991 one would have to survive long enough to have a chance of being recruited into the study. For example, consider two subjects who had the same date of onset, say in 1988. If one of these subjects had died before 1991 they could not have become part of the CSHA, while the second subject would have been recruited into the CSHA if they had survived longer than three years. That is, the longer survivor would have been recruited.

This phenomenon occurs whenever prevalent cases are identified through a cross-sectional survey and then followed forward with a view to estimating survival from a meaningful origin such as date of onset, the origin that we shall use in this chapter. The observed survival times are said to be left-

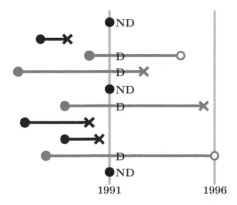

FIGURE 12.1: A schematic demonstrating the sampling mechanism and follow-up for the first phase of the Canadian Study of Health and Aging. The letters "D" and "ND" by the vertical line together depict some of the roughly 10,000 who were screened for dementia in 1991. Letters "D" depict those with prevalent dementia. Their ascertained onset times prior to 1991 are denoted by gray dots and their times of death/censoring, obtained from follow-up after 1991, are denoted by gray crosses/gray circles. The letters "ND" depict those without dementia in 1991. Also pictured are subjects with dementia who died prior to 1991; their unobserved onset times are denoted by black dots and their unobserved failure times are denoted by black crosses.

truncated, and if the incidence rate is constant (called the stationary case) the survival times are termed length-biased. It should be noted, however, that left-truncated survival times are, in general, length-biased and the two terminologies just provide a convenient way of differentiating between when the incidence process is stationary and when it is non-stationary (that is, the incidence rate changes over time).

Our first goal was to estimate the survivor function of those who develop dementia; we had carried out a literature search and found that based exclusively on prevalent cohort studies with follow-up, published estimated median survival times with dementia (for the two types combined) ranged from around 5 to 9 years; see Mölsä et al. (1986), Walsh et al. (1990) and Stern et al. (1997). Although the most recent of these studies mentioned survivor bias, none had included analyses that adjusted for it and we felt that proper adjustment would reduce the currently accepted range of median survival times.

12.2 Nonparametric Estimation of the Survivor Function

We begin with some basic terminology and notation. Let X_i be the survival time of subject i, measured from onset of dementia, had they been observed as an incident case (that is, a subject in an initially disease-free cohort, whose onset of disease is observed). Let $S_U(t) = \Pr(X_i > t)$ be the survivor function of X_i, the function that we wish to estimate. In a prevalent cohort study with follow-up there is an underlying incidence process that must be taken into account. Each point of incidence defines a random truncation time T_i, the time from the date of incidence to the date of recruitment. However, their onsets are not all observed and hence, neither are their truncation times (see Figure 12.2a).

Let T_i^0 be the observed (left) truncation time of a subject that is recruited. Other commonly used terminology for the observed time interval between disease onset and recruitment into the prevalent cohort (the observed truncation time), is the backward recurrence time. The forward recurrence time, R_i, is the time interval from recruitment until death (see Figure 12.2b). This might be censored by its associated censoring time C_i (see Figure 12.2c).

Let S_{LB} denote the survivor function of the observed length-biased survival times. It is not difficult to show (Asgharian et al., 2006) that there is a simple relationship between S_U and S_{LB}, viz.

$$S_{LB}(t) = \frac{\int_t^\infty \Pr(T \le x) \mathrm{d}S_U(x)}{\int_0^\infty \Pr(T \le x) \mathrm{d}S_U(x)}. \tag{12.1}$$

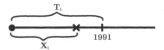

(a) An unobserved survival time X_i with its longer associated truncation time T_i, which is also not observed.

(b) An observed survival time $X_i^0 = T_i^0 + R_i$, which is the sum of the observed truncation time from onset to 1991 (backward recurrence time) and the time from 1991 to death, for a prevalent case (forward recurrence time).

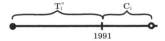

(c) An observed truncation time T_i^0 from onset to 1991 with its associated forward censoring time C_i. By definition, censoring can only occur during follow-up of prevalent cases.

FIGURE 12.2: Illustration of truncation, failure, and censoring times.

Let $X_i^0 = T_i^0 + R_i$ be the left-truncated survival time, with survivor function S_{LB}. Since X_i^0 is only observed if $X_i \geq T_i$, we have that

$$S_{LB}(t) = \Pr(X_i > t | X_i \geq T_i).$$

Under stationarity, Expression (12.1) becomes

$$S_{LB}(t) = \int_t^\infty x \, dS_U(x) \Big/ \int_0^\infty x \, dS_U(x), \qquad (12.2)$$

a relationship that is of central importance in this account.

Since the forward recurrence times are possibly right censored by their corresponding forward censoring times, the full data observed for n subjects are $\{T_i^0, \min(R_i, C_i), \delta_i\}$ for $i \in \{1, \ldots, n\}$, where $\delta_i = \mathbf{1}(R_i \leq C_i)$ is the censoring indicator. The censoring indicator takes the value 1 if $R_i \leq C_i$ and is 0 otherwise. Ignoring covariates for the moment, the CSHA survival data from the prevalent dementia cases were of this form, where T_i^0 corresponded to the interval between the date of onset of dementia and the date of recruitment into the CSHA, and the end point was death from any cause/censoring. In our initial analysis we exploited Expression (12.2) to find the nonparametric maximum likelihood estimator (MLE) of S_U by substituting the MLE of S_{LB}, obtained directly from the observed length-biased data. That is, we obtained an estimator of S_U that does not assume any specific parametric form for S_U.

We chose this route rather than use the more robust product limit estimator of Tsai et al. (1987) that does not depend on stationarity, because there was no reason to believe that the incidence rate of dementia had increased much, if at all, in the approximately twenty years prior to 1991. If the incidence rate of dementia could be regarded as roughly unchanged, we felt that with these stronger model assumptions we would obtain a more efficient estimator of the survivor function than the product limit estimator (that is, an estimator with a smaller variance). To our surprise, although other researchers such as Vardi (1982) had found the Nonparametric Maximum Likelihood Estimator (NPMLE) of S_U under stationarity, none had allowed for censoring. Alas, our first attempts ended in failure. For, we fell into the trap of using the Kaplan–Meier estimator as the NPMLE of the length-biased survivor function, S_{LB}; after all, did we not have randomly right censored survival data from S_{LB}?

Our mistake, pointed out by a referee, was that censoring for length-biased data obtained from a prevalent cohort study with follow-up is informative. Censoring is informative when censoring and survival times are not independent, as required by the methodology. Fortunately, we were able to repair the damage by using the correct NPMLE for S_{LB}. This had previously been obtained by Vardi (1989), although in a context entirely different from that of a prevalent cohort study with follow-up. Therefore, when we came to work out the asymptotic properties (that is, the properties in the ideal situation where samples are very large) of our NPMLE we ran into another problem. Clearly, from Expression (12.2) these properties would follow from the asymptotic properties of the NPMLE of S_{LB} and, even though Vardi's likelihood and our likelihood were proportional, the sampling properties of an MLE depend on how the data are obtained. We could not use the results of Vardi and Zhang (1992), who later derived the asymptotic properties of the NPMLE under multiplicative censoring. However, we were able to learn from their methods. The paper by Asgharian et al. (2002) in which we present our research also demonstrates the markedly smaller variance of the NPMLE that exploits the assumed stationarity of the incidence process over the product limit estimator, which does not require the assumption of stationarity. Our first article was followed by a detailed investigation (Asgharian and Wolfson, 2005) of various censoring mechanisms and a rigorous presentation of the arguments underpinning our paper.

While we had embarked on our methodological research in order to estimate the survival distribution from the onset of dementia until death, when we submitted our CSHA paper to the *New England Journal of Medicine* (NEJM) our methodology paper had not yet been published. The Editor of the NEJM therefore, required us to conduct our analyses using the product limit estimator, which was the standard method for analyzing right censored left-truncated survival data. We estimated median survival from onset of dementia to be 3.3 years. This was considerably less than had previously been thought and it caused quite a stir in the media.

Figure 12.3 is taken with permission from the NEJM (Wolfson et al., 2001), and demonstrates the difference between the unadjusted estimator of survival and the correct product limit estimator. A possible explanation, put forward by some, for the difficult-to-believe 3.3 years was that in the CSHA, those at least 85 years of age had been oversampled. However, without adjustment the estimated median survival of 6.6 years was consistent with the estimated median survival in other studies, obtained without adjustment. Also, we re-estimated the survivor function assuming a constant incidence rate for dementia, using our new approach, and obtained a similar estimated median of 3.75 years, although with a much narrower confidence interval (Asgharian et al., 2002). This was not surprising since the method we employed for the NEJM article and our newly developed approach yield consistent estimators, which have the property that they will be close to the target with high probability for large sample sizes. Following its publication, with an accompanying editorial, our NEJM paper has been cited over 400 times.

An incident cohort study, although logistically difficult and more expensive, is the gold standard for estimating disease incidence and the survival distribution. Under this design, a cohort of disease-free subjects is followed for the development of dementia. The dates of onset are recorded, and the cohort is followed over time until some end-of-study date. The dates of death or censoring are recorded for all who become incident cases while under follow-up. Of relevance are three recent studies, based on incident cohorts (Helzner et al., 2008; Xie et al., 2008; Matsui et al., 2009), which have confirmed our findings that survival with dementia is much shorter than previously thought. Further, stratification on type of dementia did not produce any anomalous results; the effect of length bias was strong, and survival from onset, short.

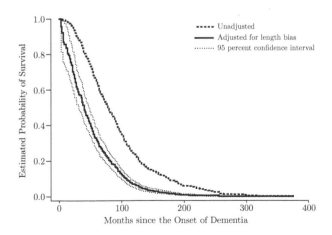

FIGURE 12.3: Estimated probability of survival, without adjustment (dotted line) and with adjustment (solid line, including 95% confidence bands).

Since in the NEJM article we had used the product limit estimator for left-truncated data we compared the point estimates given there, with our estimates based on the stationarity assumption; see Asgharian et al. (2002).

12.3   Checking for Stationarity of the Incidence Process

Carrying out research is akin to trying to slay the mythical Hydra; a problem solved creates two new problems. So it was with us. Our NPMLE and its asymptotic sampling properties had been developed under the assumption of a stationary incidence process. In order to trust our methods for the CSHA survival data we had to check our assumption that the incidence rate of dementia had remained fairly constant over the time period leading up to 1991. More generally, we wondered if one could test for stationarity using only data collected from a prevalent cohort study with follow-up. Wang (1991) had suggested how one might check for stationarity graphically using her estimator of the truncation time distribution; under stationarity this should be uniform. However, she proposed no formal test.

The problem is this: We must make inference about a stochastic process (the incidence process) based on incomplete information. Data from some individuals are missing because we only observe the onsets of those who survive to be recruited into the prevalent cohort. Equivalently, we need to make inference about the distributions of the T_i's based on an observed subset of them, the T_i^0's (see Figure 12.4). Our idea was this: From renewal theory (Karlin and Taylor, 1975) it is known that under stationarity of a renewal process the forward and backward recurrence times should have the same distribution. Even though our setting was not exactly that of renewal processes some of the results from renewal theory also hold in the setting of prevalent cohort studies with follow-up, albeit with different proofs. In particular we were able to show that the forward and backward recurrence time distributions are equal if and only if the full incidence process is stationary. Conversely, if the incidence rate is increasing one would expect an excess of short backward recurrence times and long forward recurrence times. This suggested a simple graphical check for stationarity: On the same axes plot the Kaplan–Meier estimates of the forward and backward recurrence times, respectively, and assess their similarity; see Asgharian et al. (2006). An application of our simple procedure to the CSHA data supported our speculation that the incidence rate of dementia did not appear to have changed over roughly a twenty year period prior to 1991.

Now, although like Wang, we had proposed a simple graphical check for stationarity, this would have limited value unless we could also put forward a formal test for stationarity. One of our PhD students, Vittorio Addona, addressed this problem as part of his doctoral dissertation. The difficulties that had to be overcome were that forward and backward recurrence times

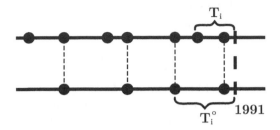

FIGURE 12.4: All onset times (top line) and observed onset times (bottom line).

are dependent. Further, forward recurrence times are often right censored. The dependence and censoring precluded a conventional test of equality of two distributions. A search of the literature, however, yielded a nonparametric Wilcoxon rank sum type test due to Wei (1980), that could be adapted to our situation. We were able to confirm what we had seen graphically: the CSHA data were not inconsistent with a roughly constant incidence rate of dementia (Addona and Wolfson, 2006).

12.4   Estimating the Incidence Rate

A new problem now emerged. One of the goals of the CSHA had been to estimate the incidence rate of dementia by following, for five years, the subjects that had been declared dementia-free in 1991. Ideally this cohort would have been monitored closely so that dates of dementia onset could be recorded near the time that they occurred. However, this cohort was only revisited in 1996, and information about those who had died between 1991 and 1996 was scant; in many cases there was considerable uncertainty whether they had died with dementia, and even if it was decided that they had (by means of an algorithm based on logistic regression), it was impossible to ascertain their dates of onset. By relying on subjects with prevalent disease only, we were able to avoid the ambiguity of whether or not subjects had dementia at the time of death. Although the date of onset for dementia is difficult to pin down, we at least had access to caregivers at the time that the subjects were diagnosed. By using a simple two-stage procedure that relied on the recollections of caregivers we obtained "best guess" dates of onset for each subject (Rouah and Wolfson, 2001).

Armed with this information we sought to exploit the well known epidemiological relationship between the prevalence odds, $P/(1-P)$, the (constant) incidence rate, λ_U, and the mean disease duration, μ_U (of the time from onset of dementia to death), to estimate the incidence rate, viz.

$$\frac{P}{1 - P} = \lambda_U \times \mu_U. \tag{12.3}$$

The proportion of the 10,000 sampled CSHA participants diagnosed with dementia in 1991 gave us an estimate of P, the prevalence of dementia. Then, using Expression (12.3) above, since we had estimated the survivor function, we were able to estimate μ_U, using $\hat{\mu}_U = \int_0^\infty \hat{S}_U(s)ds$, and hence the incidence rate as

$$\hat{\lambda}_U = \frac{1}{\hat{\mu}_U} \times \frac{\hat{P}}{1 - \hat{P}}. \tag{12.4}$$

Of course, overall incidence rates are of less interest than age-specific incidence rates and we next set about estimating these (Addona et al., 2009). This required a generalization of Expression (12.4) to allow for the changing age distribution in the general population, information that was obtained through Statistics Canada. The asymptotic properties of our estimators built on those of $\hat{S}_U$, which we had previously obtained in Asgharian and Wolfson (2005) and Asgharian et al. (2002). This was not the end of the incidence rate story, however, because Carone et al. (2012) have derived an estimator of the incidence rate without the restriction that it be constant.

Our research had been preceded by the work of Keiding (1991), who took a multi-state approach to the problem. He defined three states: disease-free, diseased, and death, and modeled the transitions between these using data that included no follow-up of the prevalent cases. In particular he showed how to estimate a constant incidence rate from such data. However, the avoidance of follow-up requires stronger model assumptions that may sometimes be difficult to justify.

12.5   Covariates

From the beginning, CSHA researchers had been interested in factors (covariates) associated with survival with dementia, such as sex, age at onset, years of education, and the presence of extrapyramidal signs and/or psychiatric symptoms. To investigate the association between selected factors and survival, in our NEJM article we chose to use a proportional hazards model with a Weibull baseline hazard (Lawless, 2002). That is, we assumed a model relating covariates to survival, as if our data had arisen from an incident cohort. Next, we corrected this miss-specified model by exploiting a well known relationship between a length-biased and unbiased model, to obtain a likelihood for our observed data. Finally, we maximized this likelihood with respect to the regression parameters inherited from the proportional hazards model. Having originated from a proportional hazards model, the estimated parameters were easily interpreted. Had we begun with a proportional hazards model for the length-biased data, we would not have been able to interpret the estimated regression coefficients as easily in the model of ultimate interest, the model that describes survival in incident cohorts.

We found, not surprisingly, that older age at onset is associated with shorter survival, that females with dementia tend to live longer than males and that there was a trend toward those with vascular dementia having shorter survival than those with Alzheimer's disease. We found no effect of level of education on survival.

Our PhD student Pierre-Jérôme Bergeron investigated the effect of length-bias on the sampling properties of the estimated regression parameters (Bergeron et al., 2008). At first we thought that the estimators would be asymptotically biased for the true parameters because covariates that accompany length-biased survival times would, themselves, not be representative of covariates in the general population. For example, one observes longer survivors in a prevalent cohort and it is known that females in general tend to live longer than males. Therefore, in a prevalent cohort one is more likely to find an excess of females among the prevalent cases, and might conclude that females with dementia survive longer than do males. However, our suspicions were unfounded, as it was found that bias is not the issue. Rather, length-biased sampling could lead to a loss in efficiency in the parameter estimators.

12.6   Concluding Remarks

An assumption almost always made when analyzing length-biased survival data is that survival is independent of the date of onset. If one assumes stationarity then it is possible to test the assumption that the survival distribution did not change over time. This problem can be regarded as a dual to the problem of testing for stationarity assuming unchanging survival, and also depends on a comparison between the forward and backward recurrence time distributions; see Addona et al. (2012).

The history of left-truncated survival data is long and the range of applications broad. We have listed many of the important historical articles in the references below which are by no means restricted to medical settings. Since we began working in the field of length-biased data there has been a surge of methodological papers on the subject; some of these are also listed.

Our saga began with a scientific research question to be addressed through data collected as part of the CSHA: "What can be said about survival with dementia?" Much substantive and methodological research has resulted from this modest beginning. The Canadian Longitudinal Study on Aging (CLSA) (Raina et al., 2009), the largest aging study undertaken to date in Canada and broader in scope than the CSHA, is now underway (www.clsa-elcv.ca). The CLSA will follow 50,000 Canadians aged 45 to 85 years for 20 years, to ascertain determinants not only of disease but also of healthy aging. Since the CLSA is a prospective study, and a large number of initially disease-free subjects will be followed over a long time period, it will permit the identification of incident cases of a number of late life diseases. This will facilitate the study of survival of these diseases through the gold standard of an incident cohort study. Indeed, the study design requires the implementation of several disease ascertainment algorithms, not only for dementia, but also for Parkinson's disease, epilepsy, diabetes, chronic airflow disruption and ischemic heart disease, among others. However, when recruited, not all participants will be disease-free, as some of them will have prevalent disease. Consequently, the CLSA is also a prevalent cohort with follow-up. For example, our first venture into the use of the CLSA is to investigate survival with Parkinson's disease from the follow-up of participants who have Parkinson's disease when they are recruited.

Will the CLSA provide ground as fertile as the CSHA for statistical research, and in particular for survival analysis? We are sure it will, through the necessity to extend survival models and methods for length-biased data to recurrent events, multivariate failure times, and joint inference for longitudinal and survival data.

Postscript

Studies on biased sampling can be traced as far back as Wicksell (1925) and his corpuscle problem which is now a classical example in stereology. The next important contribution can perhaps be attributed to Fisher (1934) on bias induced by the method of ascertainment. Neyman (1955) identified a type of bias that epidemiologists often encounter and coined the term incidence-prevalence bias. This was followed by the work of Cox (1969) in the quality control of fabrics. Zelen and Feinleib (1969) identified biased sampling in screening for chronic diseases. For more recent examples of biased sampling, see Morgenthaler and Vardi (1986) in econometrics, Drummer and McDonald (1987) in botany, Nowell et al. (1988) in land valuation, Nowell and Stanley (1991) in marketing, Terwilliger et al. (1997) in genetics and linkage mapping, Gilbert et al. (1999) in causal inference, Feuerverger and Hall (2000) in applied physics, De Uña Álvarez (2004) in labor force studies, Kvam (2008) in nano-physics, and Leiva et al. (2009) in water quality.

In recent years, there has been considerable interest in statistical methods for length-biased survival data, resulting in several important papers. We cite five of these: Andersen and Keiding (2002), Mandel and Fluss (2009), Qin and Shen (2010), Huang and Qin (2012), and Shen and Cook (2013).

Acknowledgments

The authors thank Ana Best for her invaluable technical assistance in preparing the manuscript. This work was supported in part by Discovery Grants from the Natural Sciences and Engineering Research Council to David Wolfson and to Masoud Asgharian, and by a grant from the Canadian Institutes of Health Research to Christina Wolfson and David Wolfson.

About the Authors

Masoud Asgharian is an associate professor of statistics at McGill University. He earned BSc and MSc degrees from Shahid Beheshti University, Tehran, and a PhD from McGill in 1998. His main areas of interest are survival analysis, causal inference, variable selection, clustering and classification, change-point problems, and longitudinal data analysis. He is an associate editor for *The Canadian Journal of Statistics*.

Christina Wolfson is a professor in the Departments of Epidemiology, Biostatistics and Occupational Health and of Medicine at McGill University. She holds degrees in mathematics, statistics, and epidemiology and biostatistics from McGill. Her research interests include the design and analysis of observational studies, as well as the epidemiology of neurodegenerative disorders. She is a fellow of the American College of Epidemiology.

David B. Wolfson is a professor in the Department of Mathematics and Statistics at McGill University. He completed undergraduate studies and a master's degree at the University of Natal in Durban, South Africa. He obtained his PhD in statistics from Purdue University in 1974 and has been a faculty member at McGill since then. His current research is focused on survival analysis, change-point problems and optimal design, along with statistical applications to medicine, in particular to dementia and multiple sclerosis.

Bibliography

Addona, V., Asgharian, M., and Wolfson, D. B. (2009). On the incidence–prevalence relation and length-biased sampling. *The Canadian Journal of Statistics*, 37:206–218.

Addona, V., Atherton, J., and Wolfson, D. B. (2012). Testing the assumptions for the analysis of survival data arising from a prevalent cohort study with follow-up. *International Journal of Biostatistics*, 8:Online Publication.

Addona, V. and Wolfson, D. B. (2006). A formal test for the stationarity of the incidence rate using data from a prevalent cohort study with follow-up. *Lifetime Data Analysis*, 12:267–284.

Andersen, P. K. and Keiding, N. (2002). Multi-state models for event history analysis. *Statistical Methods in Medical Research*, 11:91–115.

Asgharian, M., M'Lan, C.-É., and Wolfson, D. B. (2002). Length-biased sampling with right censoring: An unconditional approach. *Journal of the American Statistical Association*, 97:201–209.

Asgharian, M. and Wolfson, D. B. (2005). Asymptotic behavior of the unconditional NPMLE of the length-biased survivor function from right censored prevalent cohort data. *The Annals of Statistics*, 33:2109–2131.

Asgharian, M., Wolfson, D. B., and Zhang, X. (2006). Checking stationarity of the incidence rate using prevalent cohort survival data. *Statistics in Medicine*, 25:1751–1767.

Bergeron, P.-J., Asgharian, M., and Wolfson, D. B. (2008). Covariate bias induced by length-biased sampling of failure times. *Journal of the American Statistical Association*, 103:737–742.

Canadian Study of Health and Aging Working Group (CSHA) (1994). Canadian study of health and aging: Study methods and prevalence of dementia. *Canadian Medical Association Journal*, 150:899–913.

Carone, M., Asgharian, M., and Wang, M.-C. (2012). Nonparametric incidence estimation from prevalent cohort survival data. *Biometrika*, 99:599–613.

Cox, D. R. (1969). Some sampling problems in technology. In *New Developments in Survey Sampling*, pp. 506–527. Wiley, New York.

De Uña Álvarez, J. (2004). Nonparametric estimation under length-biased sampling and Type I censoring: A moment based approach. *Annals of the Institute of Statistical Mathematics*, 56:667–681.

Drummer, T. D. and McDonald, L. L. (1987). Size bias in line transect sampling. *Biometrics*, 28:13–21.

Feuerverger, A. and Hall, P. (2000). Methods for density estimation in thick-slice versions of Wicksell's problem. *Journal of the American Statistical Association*, 95:545–546.

Fisher, R. A. (1934). The effect of methods of ascertainment upon the estimation of frequencies. *Annals of Eugenics*, 6:13–25.

Gilbert, P. B., Lele, S. R., and Vardi, Y. (1999). Maximum likelihood estimation in semiparametric selection bias models with application to AIDS vaccine trials. *Biometrika*, 86:27–43.

Helzner, E., Scarmeas, N., Cosentino, S., Tang, M., Schupf, N., and Stern, Y. (2008). Survival in Alzheimer disease: A multiethnic, population-based study of incident cases. *Neurology*, 71:1489–1495.

Huang, C.-Y. and Qin, J. (2012). Composite partial likelihood estimation under length-biased sampling, with application to a prevalent cohort study of dementia. *Journal of the American Statistical Association*, 107:946–957.

Karlin, S. and Taylor, H. M. (1975). *A First Course in Stochastic Processes*, Second Edition. Academic Press, New York.

Keiding, N. (1991). Age-specific incidence and prevalence: A statistical perspective. *Journal of the Royal Statistical Society, Series A*, 154:371–412.

Kvam, P. (2008). Length bias in the measurements of carbon nanotubes. *Technometrics*, 50:462–467.

Lawless, J. F. (2002). *Statistical Models and Methods for Lifetime Data*, Second Edition. Wiley, Hoboken, NJ.

Leiva, V., Sanhueza, A., and Angulo, J. M. (2009). A length-biased version of the Birnbaum–Saunders distribution with application in water quality. *Stochastic Environmental Research and Risk Assessment*, 23:299–307.

Mandel, M. and Fluss, R. (2009). Nonparametric estimation of the probability of illness in the illness-death model under cross-sectional sampling. *Biometrika*, 96:861–872.

Matsui, Y., Tanizaki, Y., Arima, H., Yonemoto, K., Doi, Y., Ninomiya, T., Sasaki, K., Iida, M., Iwaki, T., Kanba, S., et al. (2009). Incidence and survival of dementia in a general population of Japanese elderly: The Hisayama study. *Journal of Neurology, Neurosurgery & Psychiatry*, 80:366–370.

Mölsä, P. K., Marttila, R., and Rinne, U. (1986). Survival and cause of death in Alzheimer's disease and multi-infarct dementia. *Acta Neurologica Scandinavica*, 74:103–107.

Morgenthaler, S. and Vardi, Y. (1986). Choice-based samples: A non-parametric approach. *Journal of Econometrics*, 32:109–125.

Neyman, J. (1955). Statistics — Servant of all science. *Science*, 122:401–406.

Nowell, C., Evans, M. A., and McDonald, L. (1988). Length-biased sampling in contingent valuation studies. *Land Economics*, 64:367–371.

Nowell, C. and Stanley, L. R. (1991). Length-biased sampling in mall intercept surveys. *Journal of Marketing Research*, 28:475–479.

Qin, J. and Shen, Y. (2010). Statistical methods for analyzing right-censored length-biased data under the Cox model. *Biometrics*, 66:382–392.

Raina, P., Wolfson, C., Kirkland, S. A., Griffith, L. E., Oremus, M., Patterson, C., Tuokko, H., Penning, M., Balion, C. M., Hogan, D., et al. (2009). The Canadian Longitudinal Study on Aging (CLSA). *Canadian Journal on Aging*, 28:221–229.

Rouah, F. and Wolfson, C. (2001). A recommended method for obtaining the age at onset of dementia from the CSHA database. *International Psychogeriatrics*, 13:57–70.

Shen, H. and Cook, R. J. (2013). Regression with incomplete covariates and left-truncated time-to-event data. *Statistics in Medicine*, 32:1004–1015.

Stern, Y., Tang, M.-X., Albert, M. S., Brandt, J., Jacobs, D. M., Bell, K., Marder, K., Sano, M., Devanand, D., Albert, S. M., et al. (1997). Predicting time to nursing home care and death in individuals with Alzheimer disease. *Journal of the American Medical Association*, 277:806–812.

Terwilliger, J. D., Shannon, W. D., Lathrop, G. M., Nolan, J. P., Goldin, L. R., Chase, G. A., and Weeks, D. E. (1997). True and false positive peaks in genomewide scans: Applications of length-biased sampling to linkage mapping. *The American Journal of Human Genetics*, 61:430–438.

Tsai, W.-Y., Jewell, N. P., and Wang, M.-C. (1987). A note on the product limit estimator under right censoring and left truncation. *Biometrika*, 74:883–886.

Vardi, Y. (1982). Nonparametric estimation in the presence of length bias. *The Annals of Statistics*, 10:616–620.

Vardi, Y. (1989). Multiplicative censoring, renewal processes, deconvolution and decreasing density: Nonparametric estimation. *Biometrika*, 76:751–761.

Vardi, Y. and Zhang, C.-H. (1992). Large sample study of empirical distributions in a random-multiplicative censoring model. *The Annals of Statistics*, 20:1022–1039.

Walsh, J. S., Welch, H. G., and Larson, E. B. (1990). Survival of outpatients with Alzheimer-type dementia. *Annals of Internal Medicine*, 113:429–434.

Wang, M.-C. (1991). Nonparametric estimation from cross-sectional survival data. *Journal of the American Statistical Association*, 86:130–143.

Wei, L. (1980). A generalized Gehan and Gilbert test for paired observations that are subject to arbitrary right censorship. *Journal of the American Statistical Association*, 75:634–637.

Wicksell, S. D. (1925). On the size distribution of sections of a mixture of spheres. *Biometrika*, 17:84–99.

Wolfson, C., Wolfson, D. B., Asgharian, M., M'Lan, C.-É., Østbye, T., Rockwood, K., and Hogan, D. B., for the Clinical Progression of Dementia Study Group (2001). A reevaluation of the duration of survival after the onset of dementia. *New England Journal of Medicine*, 344:1111–1116.

Xie, J., Brayne, C., and Matthews, F. E. (2008). Survival times in people with dementia: Analysis from population based cohort study with 14 year follow-up. *British Medical Journal*, 336:258–262.

Zelen, M. and Feinleib, M. (1969). On the theory of screening for chronic diseases. *Biometrika*, 56:601–614.

13

Assessing the Effect on Survival of Kidney Transplantation with Higher-Risk Donor Kidneys

Douglas E. Schaubel and John D. Kalbfleisch

University of Michigan, Ann Arbor, MI

13.1   Introduction

Patients with end-stage renal disease (ESRD; also known as kidney failure) must address the deficit in renal function through dialysis or kidney transplantation. Kidney transplantation has been repeatedly demonstrated to be superior to dialysis in terms of patient survival (Schaubel et al., 1995; Wolfe et al., 1999; Rabbat et al., 2000). However, there are tens of thousands more patients in need of a kidney transplant than there are available donor kidneys. As a result, patients typically begin renal replacement therapy with dialysis, and those deemed medically suitable are placed on a wait list for transplant. Once on the wait list, a patient may later be removed if his or her health condition declines to the point where transplant surgery is considered futile. In the United States, deceased-donor kidneys are allocated on a first-come first-served basis. Patients on the wait list move toward the top of the list when patients above them on the list die, receive a transplant, or are removed.

The continuing shortage of deceased-donor kidneys has prompted the increased frequency of transplantation with expanded criteria donor (ECD) kidneys (Port et al., 2002). These are kidneys obtained from deceased donors who were age ≥ 60; or of age 50–59 with one or more of the following characteristics: death due to stroke, history of hypertension, serum creatinine $\geq 1.5 \mu$mol/L. As established by Port et al. (2002), ECD kidneys are associated with somewhat poorer outcomes (e.g., a 70% relative increase in the rate of graft failure). On the other hand, ECD kidneys are more available than non-ECD kidneys since patients are generally more likely to decline ECD offers. The question we address is whether a patient should accept an ECD transplant (considered as an experimental treatment), or whether he/she should opt for "conventional therapy." The latter would entail refusing ECD transplantation, with the hope of later receiving a non-ECD transplant; this comes with the risk that they will have to wait so long that the patient dies on the wait list or is removed.

In clinical settings, patients often have to choose between different therapies. For chronic conditions (particularly those associated with high rates of adverse events), a patient may choose the treatment course that offers the longest survival time, although other criteria such as quality of life may also enter the decision. The randomized controlled trial, in which patients are randomly assigned to experimental treatment or control groups, is widely viewed as the gold standard for the evaluation of treatments. However, the randomization of treatments is not possible in many settings due to ethical and/or logistical considerations. In such settings, so-called observational data offer the primary basis of treatment evaluations. The lack of randomization requires that the statistical analyses accurately account for imbalances between treatments with respect to measured patient characteristics, such as age, BMI or co-morbid conditions. These are often referred to as covariates, and of particular concern are covariates that are strongly related to the risk of death. Failure to adjust for such risk factors may artificially make one treatment appear to be better than another when, in fact, treatment-specific differences in outcomes rates were due only to differences in treatment-specific covariate distributions.

In clinical studies concerning adverse events, the observation period typically concludes before some subjects have experienced the event of interest. In such cases, a subject's event time (often termed a "failure" time) is said to be right censored; it is known only to be greater than their follow-up time. Survival analysis methods are well-suited to handling right censored failure times. For example, the Kaplan–Meier estimator (Kaplan and Meier, 1958) gives a simple estimate of the probability of remaining alive at various follow-up times. In the case of observational studies of patient survival, it is often desirable to assume a particular statistical model that expresses the death rate (more formally known as the hazard function) as a function of patient characteristics or covariates. Such models are referred to as regression models, and for close to forty years, the Cox regression model (Cox, 1972) has been the method of choice for regression analysis of censored survival data. This model allows for the estimation of covariate effects and, as such, yields covariate-adjusted comparisons of treatment options. Covariates in a Cox model may be fixed at baseline (i.e., time 0, the start of follow-up), or may vary during follow-up. The description of such methods is available from many sources; see, e.g., Kalbfleisch and Prentice (2002), Klein and Moeschberger (2003), and Lawless (2003).

In this chapter, we describe the application of both traditional and more recently developed methods in survival analysis for the comparison of therapies. Here, the available data are complicated by several issues: treatment is not assigned at time 0, but rather at some point after initial eligibility; although experimental and conventional forms of treatment are available, treatment availability is limited due to a perpetual excess of demand relative to supply;

the treatment received by a patient is not randomized; and patients can be declared ineligible for treatment.

In the next section, we describe the kidney transplant registry data used for our analyses and formulate the problem more precisely. In Section 13.3, we describe a traditional attempt to address the issues described in the preceding paragraph. In Section 13.4, we describe a modification of the methods used in Section 13.3. A more recently developed and generally more satisfactory approach to the problem is described in Section 13.5. Some discussion is provided in Section 13.6, including a comparison of related methods not used in this chapter.

13.2   Study Population and Notation

Data were obtained from the Scientific Registry of Transplant Recipients (SRTR; www.srtr.org), a nationwide population-based organ failure registry. Patient-specific data are reported by the transplant centers to the Organ Procurement and Transplantation Network, which oversees solid organ transplantation in the United States.

For analyses presented in this chapter, the study population was comprised of adult (age ≥ 18) patients initially wait-listed for kidney transplantation in the US between January 1, 1998 and December 31, 2006. For each patient, follow-up began at the date of wait-listing and concluded at the earliest of death, receipt of living-donor transplant, loss to follow-up, or the end of the study's observation period: December 31, 2006.

Table 13.1 presents a summary of events for the study population. Patients entered the SRTR database at the time of being placed on the wait list, which serves as the natural time 0. Just under 30% of the study population received a deceased-donor kidney transplant, and less than one-fifth of these received an ECD kidney. More than 10% of patients were removed from the wait list prior to transplant or death. Since the death time is known for all patients

TABLE 13.1: Analysis of SRTR data: Event counts.

Event	Count	Percentage of Wait-Listed
Wait-Listed	170,415	100
Transplanted: ECD	9,423	5.5
Transplanted: non-ECD	49,382	29.0
Died	39,475	23.2

who died before December 31, 2006, it is not censored by removal. There were approximately 40,000 deaths observed either before or after transplant.

As described in Section 13.1, the objective is to compare the experimental treatment (ECD kidney transplantation) with conventional therapy. In the seminal paper by Wolfe et al. (1999), kidney transplantation was shown to be associated with a strong and significant decrease in mortality relative to dialysis among wait-listed patients. Rabbat et al. (2000) replicated this result through a registry linkage study using data from Ontario, Canada. The main contribution of these works was to restrict the study population to patients actually wait-listed for transplantation. Previous studies, e.g., Schaubel et al. (1995), had compared kidney transplant to dialysis; these studies suffer the limitation that some of the dialysis patients in the comparison group were not actually wait-listed and, hence, not really eligible to receive a kidney transplant. The survival benefit of ECD transplantation is not obvious, based on the above-listed studies. In aggregate, such studies reveal that ECD kidneys are significantly more likely to fail, and that kidney transplantation (averaging over ECD and non-ECD transplants) offers reduced mortality.

We now introduce some notation that will allow us to describe the data for specific individuals, and the analyses that we consider in subsequent sections. Let D_i denote death time for patient i, measured in days since wait-listing. Let C_i represent censoring time. Since patients either leave the study by dying or being lost to follow-up, only the minimum of D_i and C_i is observed, and we set $X_i = \min(D_i, C_i)$. The time of transplant (if it occurs) is given by T_i. As mentioned earlier, a patient may be removed from the wait list prior to transplantation and, if this occurs, we denote the removal time by R_i. We let $A_{1i} = 1$ if patient i receives an ECD kidney transplant, and 0 otherwise. Analogously, A_{2i} is a 0/1 indicator for receiving a non-ECD transplant. The covariate vector is represented by $\mathbf{Z}_i$ and, in the present context, is intended to contain information on characteristics on patient i associated with both patient survival and the probability of being transplanted. Note that, for purposes of this chapter, the covariate information is recorded at time $t = 0$ and not updated.

It is useful to think of the data structure in terms of a state diagram, as in Figure 13.1. From this perspective, all patients enter the wait list (WL) state at time $t = 0$. From there, the patient will either transit to one (and only one) of the ECD, non-ECD, or removal (R) states, or the patient may reach the death state (D) without experiencing any of these three events. The death state can be accessed from any other state. We treat transitions into each of the ECD, non-ECD and removal states as non-reversible in that a patient that receives an ECD transplant is considered to be an ECD patient thereafter; the same holds for non-ECD transplantation. Examples of elements of $\mathbf{Z}_i$ include age, gender (represented as an indicator covariate: 0 = male, 1 = female), race, blood group, and diagnosis category.

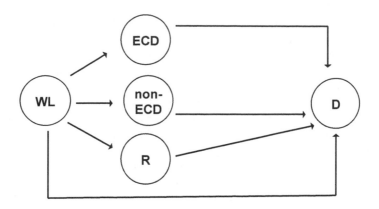

FIGURE 13.1: End-stage renal disease states. WL = wait list; ECD = expanded criteria donor kidney transplant; non-ECD= non-ECD kidney transplant; R = removed from wait list; D = death.

13.3 Analysis Based on Time-Dependent Treatment Indicator

In this section, we describe results from a traditional analysis. As a lead-in, we discuss some ideas related to the analytic approach we take in this and all subsequent sections of the chapter.

For a potentially censored failure time variate such as D_i, interest often lies in the hazard function, $\lambda_i(t)$, which represents the death rate for individual i at time t; this rate is conditional on survival up to time t. Thus,

$$\lambda_i(t) = \Pr(D_i = t | D_i \geq t).$$

Suppose that interest lies chiefly in comparing treated versus untreated subjects with respect to survival, and that treatment is randomly assigned at the beginning of follow-up ($t = 0$). We let $A_i = 1$ for treated subjects and $A_i = 0$ for untreated subjects. In this simple case, the following Cox regression model is often used:

$$\lambda_i(t) = \lambda_0(t) \exp(\beta A_i) = \begin{cases} \lambda_0(t) & \text{if } A_i = 0, \\ \lambda_0(t) \exp(\beta) & \text{if } A_i = 1. \end{cases}$$

The quantity $\lambda_0(t)$ is the baseline hazard function which, in this case, equals the hazard function for an untreated subject ($A_i = 0$). The hazard function for a treated subject is $\lambda_0(t) \exp(\beta)$; thus, the treatment multiplies the hazard function by a factor $\exp(\beta)$, which is commonly referred to as the relative risk due to treatment. If, for example, $\exp(\beta) = .6$, then treated subjects have a death rate that is 60% that of untreated subjects (i.e., there is a 40% reduction in mortality), which this model assumes to be constant throughout the follow-up period. In this model, no functional form is assumed for $\lambda_0(t)$ and, somewhat surprisingly, β can be estimated without simultaneously estimating $\lambda_0(t)$. Various extensions of this simple model are used in the analyses presented in this chapter.

Three features of the kidney transplantation study require us to extend the model described in the preceding paragraph. First, treatment (kidney transplantation) is not randomized, so a valid comparison of transplanted and wait list patients requires that patient characteristics be accounted for in the analysis. Second, treatment is inherently time-dependent; all patients begin follow-up on the wait list (i.e., untransplanted), with some eventually receiving a transplant. Moreover, there are are two forms of treatment: ECD kidney transplant and non-ECD transplant. Each of these aspects is accommodated by the first analysis we present, which is based on the following model,

$$\lambda_i(t) = \lambda_0(t)\exp\{\beta_1 A_{1i}(t) + \beta_2 A_{2i}(t) + \beta_3^\top \mathbf{Z}_i\}, \tag{13.1}$$

where $A_{1i}(t) = 1$ if patient i receives an ECD transplant before time t (and 0 otherwise); $A_{2i}(t) = 1$ if patient i receives a non-ECD transplant before time t (0 otherwise); and $\mathbf{Z}_i$ is a vector of covariates thought to be associated with both mortality and kidney transplantation.

For example, if the covariates considered were age (in years) and diabetes status (coded as 1 for diabetics and 0 for non-diabetics), then the covariate vector for a 60 year-old diabetic would be equal to $(60, 1)^\top$, with the $^\top$ denoting vector transpose. For this patient, the component in Model (13.1) corresponding to $\mathbf{Z}$ would be $\beta_3^\top \mathbf{Z} = 60\beta_{31} + \beta_{32}$, where β_{31} and β_{32} are the parameters measuring the effect of age and diabetic status on the death rate.

The analysis focuses on estimating the parameters β_1, β_2 and β_3. The inclusion of the covariate vector in Model (13.1) affects the interpretation of β_1 and β_2. Specifically, $\exp(\beta_1)$ is the relative risk (or hazard ratio) for an ECD-transplanted patient versus a patient on the wait list at time t, assuming the patients have identical covariate vectors. Analogously, $\exp(\beta_2)$ is the relative risk for a non-ECD patient versus wait-listed patient, with all covariates equal. The essence of the covariate adjustment is that between-treatment comparisons are interpreted as being among patients with identical covariate vectors. The distribution of patient characteristics may be quite different across treatment groups, but, provided Model (13.1) is correct, this does not introduce bias, since the model has explicitly captured the effect of such factors.

Since we seek to evaluate the merits of ECD kidney transplantation specifically, the parameter of chief interest is β_1. The key feature of Model (13.1) is the use of time-dependent treatment indicators, $A_{1i}(t)$ and $A_{2i}(t)$. At $t = 0$, both will be set to 0, since by definition patients begin follow-up at the time of wait-listing. At most one of $A_{1i}(t)$ and $A_{2i}(t)$ will become 1 during follow-up. The treatment indicators are non-reversible in that, once they jump from 0 to 1, they stay at 1 for the remainder of follow-up.

TABLE 13.2: Analysis of SRTR data: Results from time-dependent models (13.1) and (13.2).

Model	Contrast	Hazard Ratio	(95% CI)
(13.1)	ECD vs. WL	.72	(.69, .75)
	non-ECD vs. WL	.49	(.48, .50)
(13.2)	ECD vs. (WL + non − ECD + R)	.98	(.94, 1.02)

Results based on fitting Model (13.1) to the study data are given in Table 13.2. We find that the estimated hazard ratio (relative risk) is $\exp(\widehat{\beta}_1) = .72$, indicating that the death rate or hazard is reduced by 28% for a patient receiving ECD transplantation, compared to a patient not transplanted (the reference group). The confidence intervals in Table 13.2 represent a margin of error for the relative risk estimates. The estimated relative risk for a non-ECD transplant is $\exp(\widehat{\beta}_2) = .49$, indicating a 51% reduction in death rate for a patient who has received a non-ECD transplant, as compared to the same reference group. Notwithstanding the importance of this result, it does not completely answer our research question, due to the manner in which removals and non-ECD transplants are accommodated; issues that we now describe.

In Model (13.1), a patient who is removed from the wait list stays in the reference group; this is to avoid what is termed dependent censoring. To clarify, although censoring is an inherent feature of survival analysis, it is typically assumed that a subject being censored at time t carries no information about the future death time (other than $D_i > t$, of course) that would have been observed in the absence of censoring. Here, patients tend to get removed from the wait list when their condition has deteriorated to the point at which transplantation is considered futile. Therefore, censoring a patient upon removal would induce dependent censoring.

Although Model (13.1) avoids dependent censoring from removals, the inclusion of removals in the comparison group is also of concern. The motivation for covariate adjustment is to compare an ECD patient to an otherwise equivalent non-transplant patient. However, a patient who is removed is no longer eligible for an ECD (or a non-ECD) kidney transplant. Therefore, the interpretation of $\exp(\beta_1)$ as reflecting "otherwise equal" patients is certainly suspect.

Moreover, Model (13.1) does not handle non-ECD transplantation in a manner consistent with the primary research goal, which was to evaluate the benefit of receiving an ECD transplant. This we consider to be the comparison of the outcomes of an ECD kidney transplantation with the outcomes that would have been observed in the absence of an ECD transplant. A patient who foregoes ECD transplantation (i.e., rules it out as a therapeutic option) may have the opportunity to later accept a non-ECD transplant. In fact, patients who decline ECD transplantation are not doing so because they think ECD transplantation is no better than remaining perpetually on the wait list, but because they hope later to receive a non-ECD transplant. Since Model (13.1) includes the $A_{2i}(t)$ term, patients are essentially transferred out of the reference group (to which ECD is compared) upon receipt of a non-ECD transplant. One could argue that such patients should be left in the comparison group.

Due to the these limitations associated with the analysis of Model (13.1), our questions regarding the benefit of ECD kidney transplantation remain unanswered at this point. We investigate a simple alternative to this model in the next section.

13.4   Modification to Time-Dependent Analysis

In an attempt to remedy the handling of non-ECD transplants, an alternative time-dependent model is given by

$$\lambda_i(t) \quad = \quad \lambda_0(t) \exp\{\beta_E A_{1i}(t) + \boldsymbol{\beta}_0^\top \mathbf{Z}_i\}, \tag{13.2}$$

which results from deleting the $A_{2i}(t)$ component from Model (13.1). Under this approach, patients appropriately remain in the reference category both after being removed and after receiving a non-ECD transplant.

Results based on Model (13.2) are given in Table 13.2, where it now appears that there is essentially no difference between receiving ECD transplantation and not receiving an ECD transplant, with $\exp(\widehat{\beta}_1) = .98$ and $p = .25$. The reference category is fundamentally different from Model (13.1), in that survival of the ECD group is being compared to survival of all others (wait list, non-ECD, removed) combined.

However, Model (13.2) also has some important drawbacks. It treats removals exactly as in Model (13.1) and so is subject to the same criticism as above. Thus, a patient receiving an ECD transplant at time t is still compared to a reference group that includes patients who were removed prior to time t and so were not eligible for transplantation (ECD or non-ECD) at time t.

This is symptomatic of a more general problem with both Models (13.1) and (13.2). Essentially, the timing of events is not accounted for properly. Consider Figure 13.2, which shows event histories for six hypothetical patients. The patient $i = 1$ receives an ECD transplant (labeled in the figure by E); at the time of that ECD transplant, patient $i = 2$ had already received a non-ECD transplant (labeled by N). Under Model (13.2), patient $i = 2$ is included in the comparison group for patient $i = 1$. But this seems inappropriate since at the time patient $i = 1$ chose to undergo ECD transplantation, patient $i = 2$ had already received a non-ECD transplant and could not have made the

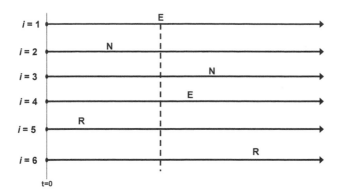

FIGURE 13.2: Event occurrences for $n = 6$ patients $(i = 1, \ldots, 6)$: E = expanded criteria donor (ECD) kidney transplant; N = non-ECD kidney transplant; R = removed from wait list.

same treatment decision. However, patient $i = 3$, who received a non-ECD transplant, would be appropriate to include in the comparison group, since that patient was transplant-eligible at the time patient $i = 1$ underwent ECD transplantation. Clearly, analogous comments apply to the two patients in Figure 13.2 that were removed from the wait list $(i = 5$ and $i = 6$, whose time of removal is identified by R). It would not be appropriate to include patient $i = 5$ in the comparison group, since this patient was removed before patient $i = 1$ received an ECD transplant. However, it would be appropriate to include patient $i = 6$ since the removal time follows the time of ECD transplantation for patient $i = 1$.

In considering the limitations associated with model (13.2), and Figure 13.2, it appears that blanket rules regarding non-ECD transplantation and removal are likely to result in an interpretation for β_1 that does not address the research question. However, the preceding paragraph suggests that definition of the appropriate comparison group is straightforward for any individual patient. This is implemented in the next section where we customize the comparison groups for each ECD recipient individually.

13.5 Sequential Stratification

Ideally, ECD kidney transplantation would be evaluated through a randomized controlled trial. In this trial, all patients would be placed on the wait list at the same time, and each time an ECD kidney was procured, a patient currently on the wait list (i.e., not already removed or transplanted) would be randomly selected to receive the donor kidney; this patient would then represent the "experimental treatment" group. A set of patients who were also eligible to (but did not) receive the ECD kidney would be used as a comparison group, perhaps through matching and/or random selection from those available. These patients represent the "conventional therapy" group, where conventional therapy allows all subsequent events except receipt of the experimental treatment. Note that not being already removed and not having already received a kidney transplant at the time of the index patient's ECD transplant are entry criteria to the comparison group. Note that, in the comparison group, a comparator patient is not censored at subsequent (post-matching) removal or non-ECD transplantation, since such events are standard sequelae of conventional therapy. Comparison patients would be censored when they receive an ECD transplant, since they have then essentially crossed over into the experimental treatment arm and, therefore, no longer contribute follow-up pertinent to the conventional therapy arm. The process of randomizing an incoming ECD kidney and selection of matched conventional therapy patients would be repeated many times, and survival outcomes could then be compared.

Such a trial can never occur, due to logistical and ethical considerations. However, since we take the time of wait list as the origin for all patients, the observed data are quite similar to this; of course the observed data are not randomized, but we could attempt to replace the randomization with careful covariate adjustment. In fact, the largest discrepancies in the setup described in the preceding paragraph and that presented in Sections 13.3–13.4 are related to the method of analysis, as opposed to the actual data structure. The implied analysis features ECD patient-specific comparison groups and, within a given matched set, the handling of removals and non-ECD transplants seems clear cut.

The sequential stratification method was motivated by such considerations, in the context of kidney (Schaubel et al., 2006) and liver transplantation (Schaubel et al., 2009). In essence, this method reorganizes the observed data, then replicates the analysis described above for the ideal experiment. To formalize the ideas, suppose that subject j receives an ECD transplant at time T_j. Patient i (where $i \neq j$) is eligible to be matched to patient j if i is alive and has not been transplanted prior to time T_j (i.e., $\min(T_i, C_i, D_i, R_i) > T_j$) and $s_i = s_j$, where s denotes the matching criteria, which we assume are known at $t = 0$. For example, in our application, age and organ donation service area (the geographic region to which a wait list is intended to apply) served as matching criteria. Thus, we find a set of patients "at risk," denoted by $\mathcal{A}_j$, which comprises patient j as well as those patients who are matched to patient j at time T_j. We refer to $\mathcal{A}_j$ as a matched set of patients, or stratum. The outcomes of patient j are compared to the outcomes of patients $i \neq j$ in this set. This could be done by assuming a Cox type model

$$\lambda_{ji}(t) = \lambda_{0j}(t) \exp(\theta_1 \mathrm{ECD}_{ji} + \boldsymbol{\theta}_0^\top \mathbf{Z}_i), \quad t > T_j \qquad (13.3)$$

TABLE 13.3: Analysis of SRTR data: Results from sequential stratification Model (13.3).

Model	Contrast	Hazard Ratio	(95% CI)
(13.3)	ECD vs. Conventional Therapy	.78	(.75, .81)

for each patient i in $\mathcal{A}_j$, where ECD_{ji} takes the value 1 if $i = j$, the ECD recipient, and takes the value 0 otherwise. Note that the covariate vector $\mathbf{Z}_i$ would typically not contain elements used as matching criteria. The sequential stratification method handles both removal and non-ECD transplants appropriately; each is an entry criterion but not a censoring criterion for a given matched set.

A pictorial representation of sequential stratification can be obtained by referring to Figure 13.2. Here, $i = 1$ receives an ECD kidney transplant. Patient $i = 2$ is excluded as a match due to prior (non-ECD) transplantation (i.e., $T_2 < T_1$), while patient $i = 5$ was already removed by the time that $i = 1$ received an ECD ($R_5 < T_1$). Thus $\mathcal{A}_1 = \{1, 3, 4, 6\}$ is the set of patients who were alive and had not been removed or transplanted as of T_1. In terms of post-matching follow-up, patient $i = 3$ is not censored at T_3, since receipt of a non-ECD transplant is a possible sequelae of foregoing ECD transplantation. Analogously, patient $i = 6$ is not censored from the matched set at R_6. Note that patient $i = 4$ would be censored from $\mathcal{A}_1$, at the time of receiving an ECD kidney, since such receipt does not fall under the rubric of conventional therapy, and would be deemed a cross-over to the experimental treatment. In fact, T_4 would generate another stratum, $\mathcal{A}_4 = \{3, 4, 6\}$.

Results from the sequential stratification analysis are presented in Table 13.3. This more complete analysis demonstrates that ECD kidney transplantation is associated with a significant 22% reduction in mortality relative to conventional therapy, with $\exp(\widehat{\theta}_1) = .78$ and $p < .0001$.

13.6 Discussion

In this chapter, we contrast different methods of quantifying the effect of expanded criteria donor (ECD) kidney transplantation with conventional therapy, for which a person stays on the wait list until they are offered a non-ECD kidney, are de-listed, or die. There are different ways of defining a treatment effect, and the one of interest in this chapter is the effect-of-treatment-on-the-treated (Pearl, 2000), for which one aims to compare the outcomes for treated subjects against those which would have been observed (for the treated subjects) in the absence of treatment. The scientific question is not well addressed by standard time-dependent survival analysis methods, which do not accommodate some important elements of the data structure. Sequential stratification, the method we advocate in this setting, involves reorganizing the observed data in a manner more in line with the research question. Aspects of the data, which are difficult to address appropriately through standard methods, can then be dealt with in a transparent and straightforward manner. The method has been used in several analyses targeting the benefit of liver or kidney transplantation; see, e.g., Miles et al. (2007), Schaubel et al. (2008), Lucey et al. (2009), Sharma et al. (2009), Englesbe et al. (2010), and Snoeijs et al. (2010).

A standard method of accommodating a time-dependent treatment in the survival analysis setting is through Cox regression with a time-dependent treatment indicator. Although this approach is suitable for some data structures, its use will generally be problematic when other time-dependent processes are at work. In the ECD setting, both the receipt of a non-ECD transplant and removal from the wait list are time-dependent processes that are important in comparing ECD transplantation to conventional treatment without the ECD option. However, such processes must be handled correctly for the model to be consistent with the particular research question. In other cases, the concomitant time-dependent process may be a covariate observed longitudinally which affects both treatment assignment and the death hazard. Sequential stratification can also be applied to this setting.

The evaluation of time-dependent treatments has received considerable attention in the survival analysis literature in the last 10–15 years. In particular, marginal structural models (MSMs) have gained much popularity in the past decade; see, e.g., Hernán et al. (2000), Robins et al. (2000), and Hernán et al. (2001). The treatment effects estimated through sequential stratification and MSMs are fundamentally different, as described by Schaubel et al. (2009) and Kennedy et al. (2010). The MSM leads to an average treatment effect, where the average is essentially taken across all patients and all possible treatment times. The MSM estimates what is known in the literature as the average causal effect, where this term has come to mean contrasting a scenario in which all patients are treated versus one in which none are treated. As noted in the previous paragraph, however, the treatment effect targeted by sequential stratification is similar to the effect-of-treatment-among-the-treated. Rather than averaging over all patients (treated or not) as would be the case in an MSM, sequential stratification implicitly averages over the patients observed to be treated with ECD and makes no effort toward inference on what would have been the treatment experience for untreated patients. The matching of similar subjects is an attempt to represent what would have been the outcomes of the treated subjects if (contrary to fact) they were untreated.

In the sequential stratification method, a patient selected as a match is censored upon receipt of an ECD transplant; that is, the time to death in the conventional therapy group is censored. Such cross-overs to the experimental group generally constitute dependent censoring, which necessitates a variant of Inverse Probability of Censoring Weighting; see, e.g., Robins and Rotnitzky (1992). This was explored in detail and implemented by Schaubel et al. (2009) in the context of liver transplantation, where such cross-overs were relatively common. Among end-stage renal disease patients, ECD kidney transplantation is sufficiently rare that bias due to cross-overs to the ECD group among matched patients would be expected to be small.

Since the model assumed in sequential stratification is a Cox-type model, its assumptions are analogous to those in the standard use of a Cox model. Of greatest concern would be the assumption of treatment and adjustment covariate effects that are constant throughout follow-up time. The evaluation of this assumption can be accomplished by extending the model to incorporate interactions between the treatments or covariates and time and appropriate tests carried out; such extensions also can be used to describe more complicated treatment effects than a constant relative risk as discussed in Kalbfleisch and Prentice (2002). When treatment is assigned at baseline, Wei and Schaubel (2008) proposed a method that (like the baseline hazard for the Cox model) assumes no particular functional form for the treatment effect; see also Schaubel and Wei (2011). More recently, Li et al. (2013) proposed matching methods to contrast survival functions when treatment assignment is time-dependent. On a related note, methods developed by Gong (2012) measure the treatment effect in terms of either the difference in survival function, or the area between the survival curves. The work of Gong (2012) requires more modeling assumptions than Li et al. (2013), but allows for more complicated data structures.

It is likely that the survival benefit of ECD transplantation depends on several patient characteristics. For instance, older patients, or patients with diabetes, may have more to gain (relative to conventional therapy) through ECD transplantation, since their prognosis on the wait list is less favorable. In addition, ECD transplantation may be a better option for patients in regions of the country that typically have longer wait lists. Each of these directions would be a useful topic of further investigation.

Acknowledgments

This work was supported in part by National Institutes of Health Grant 5R01 DK–70869. The authors are grateful to the editor and associate editor for their many constructive suggestions. They also thank the Minneapolis Medical Research Foundation for access to the Scientific Registry of Transplant Recipients database. The Scientific Registry of Transplant Recipients is funded by the Health Resources and Services Administration, US Department of Health and

Human Services. The views expressed herein are those of the authors and not necessarily those of the US Government.

About the Authors

Douglas E. Schaubel is a professor of biostatistics at the University of Michigan. His research interests include the evaluation of time-dependent treatments, multivariate survival analysis and collaborative work on liver transplantation and end-stage kidney disease. He works closely with clinicians and surgeons at Arbor Research Collaborative for Health, and the University of Michigan Kidney Epidemiology and Cost Center. He earned a BSc from the University of Waterloo, an MSc from McGill University, and a PhD in biostatistics from the University of North Carolina at Chapel Hill. He is a fellow of the American Statistical Association.

John D. Kalbfleisch is a professor emeritus of biostatistics and statistics at the University of Michigan where he has been on faculty since 2002. Previously, he was at the University of Waterloo, where he also served as chair of the Department of Statistics and Actuarial Science and dean of the Faculty of Mathematics. His research interests include survival and life history analysis, statistical inference and applications of statistics. He is a fellow of the Royal Society of Canada, the American Statistical Association, and the Institute of Mathematical Statistics. He was the recipient of the 1994 SSC Gold Medal for research, served as SSC president in 1999–2000, and was an R. A. Fisher Lecturer for the Committee of Presidents of Statistical Societies (COPSS).

Bibliography

Cox, D. R. (1972). Regression models and life-tables (with discussion). *Journal of the Royal Statistical Society, Series B*, 34:187–220.

Englesbe, M. J., Schaubel, D. E., Cai, S., Guidinger, M. K., and Merion, R. M. (2010). Portal vein thrombosis and liver transplant survival benefit. *Liver Transplantation*, 16:999–1005.

Gong, Q. (2012). *Semiparametric Methods for Estimating the Effect of a Longitudinal Covariate and Time-Dependent Treatment on Survival Using Observational Data with Dependent Censoring*. Doctoral dissertation, University of Michigan, Ann Arbor, MI.

Hernán, M. A., Brumback, B., and Robins, J. M. (2000). Marginal structural models to estimate the causal effect of zidovudine on the survival of HIV-positive men. *Epidemiology*, 11:561–570.

Hernán, M. A., Brumback, B., and Robins, J. M. (2001). Marginal structural models to estimate the joint causal effect of nonrandomized treatments. *Journal of the American Statistical Association*, 96:440–448.

Kalbfleisch, J. D. and Prentice, R. L. (2002). *The Statistical Analysis of Failure Time Data*, Second Edition. Wiley, Hoboken, NJ.

Kaplan, E. L. and Meier, P. (1958). Nonparametric estimation from incomplete observations. *Journal of the American Statistical Association*, 282:457–481.

Kennedy, E. H., Taylor, J. M. G., Schaubel, D. E., and Williams, S. (2010). The effect of salvage therapy on survival in a longitudinal study with treatment by indication. *Statistics in Medicine*, 29:2569–2580.

Klein, J. P. and Moeschberger, M. L. (2003). *Survival Analysis: Techniques for Censored and Truncated Data*, Second Edition. Springer, New York.

Lawless, J. F. (2003). *Statistical Models Methods for Lifetime Data*, Second Edition. Wiley, Hoboken, NL.

Li, Y., Schaubel, D. E., and He, K. (2013). Matching methods for obtaining survival functions to estimate the effect of a time-dependent treatment. *Statistics in Biosciences*, in press.

Lucey, M. R., Schaubel, D. E., Guidinger, M. K., Tome, S., and Merion, R. M. (2009). Effect of alcoholic liver disease and hepatitis C infection on waiting list and posttransplant mortality and transplant survival benefit. *Hepatology*, 50:400–406.

Miles, C. D., Schaubel, D. E., Jia, X., Ojo, A. O., Port, F. K., and Rao, P. S. (2007). Mortality experience in recipients undergoing repeat transplantation with expanded criteria donor and non-ECD deceased-donor kidneys. *American Journal of Transplantation*, 7:1140–1147.

Port, F. K., Bragg-Gresham, J. L., Metzger, R. A., Dykstra, D. M., Gillespie, B. W., Young, E. W., Delmonico, F. L., Wynn, J. J., Merion, R. M., Wolfe, R. A., and Held, P. J. (2002). Donor characteristics associated with reduced graft survival: An approach to expanding the pool of donor kidneys. *Transplantation*, 74:1281–1286.

Rabbat, C. G., Thorpe, K. E., Russell, J. D., and Churchill, D. N. (2000). Comparison of mortality risk for dialysis patients and cadaveric first renal transplant recipients in Ontario, Canada. *Journal of the American Society of Nephrology*, 11:917–922.

Robins, J. M., Hernán, M. A., and Brumback, B. (2000). Marginal structural models and causal inference in epidemiology. *Epidemiology*, 11:550–560.

Robins, J. M. and Rotnitzky, A. (1992). Recovery of information and adjustment for dependent censoring using surrogate markers. In *AIDS Epidemiology: Methodological Issues*, pp. 297–331. Birkhäuser, Boston, MA.

Schaubel, D. E., Desmeules, M., Mao, Y., Jeffery, J., and Fenton, S. (1995). Survival experience among elderly end-stage renal disease patients: A controlled comparison of transplantation and dialysis. *Transplantation*, 60:1389–1394.

Schaubel, D. E., Sima, C. S., Goodrich, N. P., Feng, S., and Merion, R. M. (2008). The survival benefit of deceased donor liver transplantation as a function of candidate disease severity and donor quality. *American Journal of Transplantation*, 8:419–425.

Schaubel, D. E. and Wei, G. (2011). Double inverse weighted estimation of cumulative treatment effects under non-proportional hazards and dependent censoring. *Biometrics*, 67:29–38.

Schaubel, D. E., Wolfe, R. A., and Port, F. K. (2006). A sequential stratification method for estimating the effect of an time-dependent experimental treatment in observational studies. *Biometrics*, 62:910–917.

Schaubel, D. E., Wolfe, R. A., Sima, C. S., and Merion, R. M. (2009). Estimating the effect of a time-dependent treatment in by levels of an internal time-dependent covariate. *Journal of the American Statistical Association*, 104:49–59.

Sharma, P., Schaubel, D. E., Guidinger, M. K., and Merion, R. M. (2009). Effect of pre-transplant serum creatinine on the survival benefit of liver transplantation. *Liver Transplantation*, 15:1808–1813.

Snoeijs, M. G. J., Schaubel, D. E., Hene, R., Hoitsma, A. J., Idu, M. M., Ijzermans, J. N., Ploeg, R. J., Ringers, J., Christiaans, M. H. L., Buurman, W. A., and van Heurn, L. W. E. (2010). Kidneys from donors after cardiac death provide survival benefit. *Journal of the American Society of Nephrology*, 21:1015–1021.

Wei, G. and Schaubel, D. E. (2008). Estimating cumulative treatment effects in the presence of nonproportional hazards. *Biometrics*, 64:724–732.

Wolfe, R. A., Ashby, V. B., Milford, E. L., Ojo, A. O., Ettenger, R. E., Agodoa, L. Y. C., Held, P. J., and Port, F. K. (1999). Comparison of mortality in all patients on dialysis, patients on dialysis awaiting transplantation, and recipients of a first cadaveric transplant. *New England Journal of Medicine*, 341:1725–1730.

14

Risk-Adjusted Monitoring of Outcomes in Health Care

Stefan H. Steiner

University of Waterloo, Waterloo, ON

14.1 Introduction

There is increasing interest among surgeons and health care administrators in monitoring post treatment outcomes, such as hospital length of stay or 30-day mortality, defined as death within 30 days of surgery. Monitoring such outcomes over time allows better oversight of the health care process. In this way serious problems can be rapidly detected and performance can be compared across care providers. In the long run, more quickly eliminating causes of problems and adopting best practices will result in process improvement and better care for patients.

Effective monitoring of health care processes requires a statistical method for two main reasons. First, decisions should incorporate a measure of uncertainty to avoid overreacting to expected outcome variability while still promptly detecting important problems. Here we can borrow ideas developed over the last 80 years for monitoring industrial process outputs. Second, patients (unlike manufactured parts) are not expected to be homogenous and importantly can have dramatically different risks prior to treatment, due for instance to underlying health. Since we want to compare care providers fairly and, for example, do not want to penalize a surgeon who takes on difficult cases, the monitoring should be based on risk-adjusted outcomes.

The goal of this chapter is to outline the main ideas and challenges in risk-adjusted monitoring of health care outcomes. I begin by motivating the need for improvement in health care processes by briefly discussing three recent examples where a lack of proper oversight led to unacceptably high death rates continuing for an extended period of time. Then, in the next section, I introduce the basic concepts behind process monitoring with an industrial example where risk adjustment is not needed. Next, I highlight some of the challenges and unresolved issues in monitoring health care outcomes using a cardiac surgery example to explain the ideas, concerns and methods. The chapter concludes with a summary and a brief discussion of application areas and possible future extensions of the presented methods.

14.2   Motivation and Background

Medical errors and inefficiencies are a major concern and there is a great need to improve the care of patients. In the landmark report *To Err Is Human: Building a Safer Health System* (Kohn et al., 2000), it was estimated that up to 98,000 preventable deaths are caused by errors in the health system in the United States each year. More recently, Baker et al. (2004) estimated that each year between 9000 and 24,000 deaths in Canadian hospitals are due to mistakes that could have been prevented. Rothschild et al. (2005) concluded that in the intensive care unit, the rates for preventable adverse events and serious errors were 36.2 and 149.7 per 1000 patient-days, respectively.

In light of these problems, it is important to monitor health care outcomes following the provision of medical care or surgical intervention. Such monitoring is also motivated by an increased emphasis on public accountability. In addition, careful monitoring of health care outcomes was recommended by a number of public inquiries conducted after serious problems were identified. The following examples provide illustration.

During 1994, 12 children died as a result of surgery at the paediatric cardiac center at the Winnipeg Health Sciences Centre (Waldie, 1998; Davies, 2001). A subsequent government inquest (Sinclair, 2000) found that many of the deaths were potentially avoidable and that there were systemic problems at the center that could and should have been acted on sooner.

A similar problem occurred at a children's cardiac center in Bristol (England) between 1991 and 1995 (Treasure et al., 1997). The Bristol Royal Infirmary Inquiry (2001) concluded that 30 to 35 children undergoing heart surgery died who probably would have survived if treated elsewhere. The mortality rate at Bristol for cardiac surgery on infants was estimated to be double the rate for the rest of England. The Royal Inquiry made many recommendations for improvement, several of which dealt specifically with "monitoring standards and performance."

In a different context, between 1971 and 1998 the family doctor Harold Shipman killed over 250 mostly elderly patients under his care in England. The inquiry into these events (Smith, 2005) concluded that, among other things, there are "major flaws in the systems that govern death registration, the prescription of drugs and the monitoring of doctors."

These examples portray situations where undesirably high rates of deaths remained undetected, or at least were not acted on, for an undue length of time. This provides motivation for better oversight and the development of methods for monitoring health care outcomes. In such cases, the rapid detection of poor surgical performance or excess deaths is critical. Early detection of problems would have resulted in prompt investigation of the cause and possibly a criminal investigation in cases of misconduct. This would likely have prevented unnecessary deaths and led to improvement through responses such as procedural changes and retraining.

Statistical methods for monitoring health care outcomes have developed rapidly in recent decades, following earlier work on monitoring industrial processes. I will next briefly describe an example of monitoring an industrial process to highlight the basic ideas of statistical monitoring and then go on to consider some ways that health care outcomes are monitored.

14.3   Monitoring Industrial Processes

To develop methods for monitoring health care outcomes it is natural to build upon the statistical methods for monitoring industrial processes. Beginning with the pioneering work by Shewhart (Shewhart, 1931), over the last 80 years monitoring methods have been further developed and are currently widely used in industry. For an overview of industrial monitoring methods see the Automotive Industry Action Group Reference Manual (AIAG, 1992) and Montgomery (2005).

To illustrate the basic ideas behind statistical monitoring, consider an industrial process that makes oil pans. The oil pan, while not a glamorous product, is critical to the proper functioning of an automobile with an internal combustion engine. Oil pan leaks can lead to catastrophic engine failure if not detected and repaired promptly. Oil pans are produced in a stamping plant by bending and punching pre-produced sheets of metal called blanks. All finished oil pans are inspected and scrapped if any defects such as splits, wrinkles or excessive material thinning are found. To illustrate, a plot is shown in Figure 14.1 of the proportion of oil pans scrapped per day (out of roughly 1125 oil pans) for a 20 day period. The scrap rate varies from day to day due to changes in important process conditions, such as amount of lubricant used, material and geometric properties of the blanks, sharpness of the forming die used to cut and bend the blanks, location of the blank relative to the die in the press, the amount of force used to bend the blank, etc.

To statistically monitor this process, we use the observed variation in the scrap rate during the 20-day period to determine the heights of the three horizontal lines added to Figure 14.1. These lines represent decision limits, derived using statistical models and assumptions, described in Montgomery (2005). The center line represents the average daily scrap rate (.083 over this period). More important are the lower and upper decision limits at heights .058 and .108 respectively, which delineate the expected maximum range of the daily proportion of oil pans scrapped when the process is operating normally.

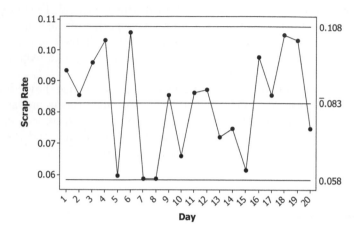

FIGURE 14.1: Chart for monitoring the daily oil pan scrap rate.

If, on the other hand, an unexpected large change in the oil pan production system occurs the scrap rate will fall outside the decision limits and trigger a "signal." For instance, suppose on day 21 the scrap rate is .12 (not shown in Figure 14.1) and thus the chart signals. In this case, based on the decision limits, we believe it is very unlikely that we would see such a large daily scrap rate if the process were operating normally. We therefore would conclude something unexpected happened on day 21 to increase the scrap rate, and would investigate what might have caused this. For instance, suppose we determine that on day 21 we started to use a new batch of blanks and that the specific cause of the signal in this case is that the new batch of blanks had low pliability. Then, to improve the process, one option would be to implement an inspection process to check each new batch of blanks before use, rejecting any with low pliability.

The oil pan example illustrates monitoring pass/fail outputs such as are typical in the health care context. However, many other types of monitoring charts have been developed for different types of outputs and goals (Montgomery, 2005). For instance, when monitoring a continuous output, e.g., the thickness of a part, two charts are often used to monitor the process average and variability separately. Charts for monitoring continuous features are common in industry, since in most cases monitoring a continuous rather than a pass/fail output is preferred due to the more detailed information provided by each observation.

Monitoring charts such as the one displayed in Figure 14.1, which only use data from a single day at a time to make decisions, are good at identifying relatively large process changes, as illustrated by the pliability of the batch of blanks in the example. However, with such charts, smaller, persistent process changes, such as the slow wearing of a die, will be hard to detect quickly. For smaller persistent changes, cumulative sum (CUSUM) or exponentially weighted moving average (EWMA) charts that accumulate information across multiple time periods are preferred. CUSUM and EWMA charts are commonly used in health care monitoring applications and are described in more detail in the next two sections.

14.4 Monitoring Outcomes in Health Care: Issues and Challenges

Monitoring health care outcomes has some of the same challenges as monitoring industrial processes. We still want rapid detection of problems without overreacting to process variation within acceptable bounds. Quick detection of comparatively poor, or exceptionally good, performance will lead to better public accountability and facilitate improvement by eliminating problems and adopting best practices. However, there are also some important differences between the industrial and health care contexts that we need to consider to successfully monitor health care outcomes.

First of all, there are potential privacy and ethical concerns with the use of health care outcomes data. For instance we need to ensure that any publicly available data cannot be linked to individual people. There is also a human element in responding to signals. If the chart suggests concerns about the performance of a particular surgeon or surgical team we must tread carefully. With any monitoring approach, false alarms can occur due to recent adverse outcomes attributable to bad luck rather than a real change in performance. Also, our monitoring method may be flawed. For example, we could be using an out-dated model for adjusting risk according to the patient mix (see below).

With important health care outcomes we prefer to update the monitoring chart after obtaining results for each individual patient rather than waiting until a fixed time period has passed, e.g. a day, as is typical in industry. In this way we will always be considering the most up-to-date information and have a better chance to quickly detect process changes. However, often health care outcomes are pass/fail, such as 30-day mortality, i.e., whether the patient dies within 30 days of surgery, or the presence or absence of complications. With only a single pass/fail outcome it is not possible to reliably detect process performance changes. As a result, in most health care applications we monitor with CUSUM or EWMA charts that accumulate information over time.

Health care outcomes are also not always available quickly or easy to determine. As an example, the commonly used outcome of 30-day mortality obviously requires 30 days to elapse before it can be observed. In addition, while determining whether a patient has died is clinically straightforward, some patients may be lost to follow-up. For example, imagine a patient survives the initial surgery and is discharged but then subsequently dies (perhaps after going back to a different hospital). We need to be careful, since without proper follow-up and record keeping we may make mistakes that will negatively affect the performance of the monitoring procedure.

For common health care procedures external standards of performance may be available from clinical research or historical data from other health care providers. Such external standards allow us to compare performance with the standard rather than look for changes in performance in the existing process. In industry external standards are typically not available due to differences in products and competitive pressures. To monitor a health care process by comparing results to an external standard, estimation of the historical process variation is no longer required. Thus, the use of external standards allows rapid development of an appropriate chart. This is a big advantage; however, it alters the nature of the chart and changes our interpretation of what a "signal," i.e., an observation falling outside the decision limits, means. Signals now no longer suggest the process has changed, but rather that we have accumulated enough evidence to conclude that the current process performance is substantially different than the external standard. This has important consequences. It implies that, given a signal, there is no sense in looking for a cause that acted recently since the process may well not have changed but rather that its average performance differs from the external standard. Thus, when comparing to an external standard, process improvement comes from reacting (appropriately) to evidence of significantly better or worse performance than the standard.

Another complication arises because in some health care contexts performance may be continually changing. For example, changes could be due to a learning curve where a novice surgeon improves with practice, or due to rapid innovations in surgical technique. In such cases the only option is to monitor the process by comparing performance to an external standard.

In a health care context it is also common to stratify (group) the outcomes in some way, for example by surgeon or type of surgery, and to compare performance across the strata. Here, the performance of each stratum must be estimated repeatedly as time goes on. Comparisons of this sort are also common in industry, i.e., we might compare machine, operators, production lines, etc. However, few monitoring methods have been developed to meet this goal. Liu et al. (2008) provides an exception.

Finally, and perhaps most importantly, patients do not all have the same risk of an adverse outcome. In industry we assume products are homogeneous and that prior to processing each is equally likely to yield poor results. In health care contexts we must take into account patient characteristics and the possible changing nature of the patient population, since the risk of an adverse outcome prior to treatment depends on numerous patient factors (termed covariates) such as age, gender, underlying health status, etc. Patient mix is the term used to describe the composition of the patient population which contributes to the natural variation of the process. We can not reduce the variation caused by patient mix in the actual health care process, but we can model the effect of the patient mix prior to medical treatment or surgery and use that model to "risk adjust" the individual observed outcomes. Using risk adjustment to remove the effect of patient mix makes the monitoring method more sensitive to important process changes that we can impact. Risk adjustment is also necessary to allow fair comparisons among surgical or medical care providers or institutions with different patient mixes. Intuitively, risk adjustment is needed because the death of the low risk patient is more indicative of poor performance than the death of a high risk patient, and similarly the survival of a high risk patient is more indicative of good performance than the survival of a low risk patient.

14.5   Monitoring Outcomes in Health Care: Methods

In this section, I briefly discuss health care outcome monitoring that involves some sort of risk adjustment. To illustrate the methods I use a cardiac surgery example presented previously in Steiner et al. (2000) consisting of 6,994 operations from a single UK surgical center over the seven year period 1992–98. The data on each patient includes surgery date, surgeon, type of procedure, surgical outcome and the pre-operative variables which comprise the Parsonnet score (Parsonnet et al., 1989) such as age, gender, hypertension, diabetic status, renal function and left ventricular mass. The Parsonnet score is an established overall measure of risk for adult cardiac surgery. To illustrate the risk-adjusted monitoring methodology we focus on the 30-day post-operative mortality rate. We define

$$y_t = \begin{cases} 1 & \text{if patient } t \text{ dies within 30 days,} \\ 0 & \text{otherwise.} \end{cases}$$

For all risk-adjusted monitoring methods we need to predict the risk of an adverse outcome of each patient prior to surgery. With 461 deaths within 30 days of surgery in these 6994 patients the overall mortality rate was .066. In this example, we could use a mapping of the Parsonnet score to a probability of death within 30 days of the surgery as an external standard. However, due to the large volume of data, I instead build a customized risk model. Following Steiner et al. (2000) and using the data from 1992–93, gives the risk prediction model

$$\log\left(\frac{p_t}{1 - p_t}\right) = -3.68 + .077\, x_t, \tag{14.1}$$

where x_t is the Parsonnet score and p_t is the predicted 30-day mortality probability for patient t. This model contains only a single covariate x_t but more complicated models are possible. For patients between 1992 and 1993 the Parsonnet scores ranged from 0 to 69 and thus the predicted risks of death prior to surgery from (14.1) ranged between .025 and .84.

A number of graphical methods for health care outcome monitoring have been proposed. They all plot some performance summary against time (represented by the patients ordered by surgery date). Here I consider three different approaches. For each approach I use the risk model (14.1) derived from the 1992–93 data and illustrate the proposed prospective monitoring with the patient outcomes for Surgeon 2 starting in 1994. Surgeon 2 left the surgical center in August 1996 and in the period 1994–96 operated on 330 patients. Here we use the internal standard from all surgeons at the center from 1992–93, i.e., the risk model (14.1), as an external standard for Surgeon 2 over the period 1994–96.

14.5.1 Variable Life-Adjusted Display Chart

Lovegrove et al. (1997) and Poloniecki et al. (1998) suggested monitoring using a plot of the difference between the cumulative observed deaths and cumulative predicted deaths given by

$$S_t = \sum_{i=1}^{t} (y_i - p_i), \quad t = 1, 2, 3, \ldots \tag{14.2}$$

That is, S_t is recomputed each time there is a new patient (and their 30-day mortality outcome becomes known). Plots of S_t versus t, known as variable life-adjusted display (VLAD) charts, provide a valuable visual aid where positive values of S_t suggest worse performance than expected. VLAD charts show changes in performance as changes in the slope. However, VLAD charts do not specify how much variation in the plot is expected due to chance under acceptable surgical performance, and hence it is not clear how large a deviation from the horizontal line at zero should raise concern. An example VLAD chart for Surgeon 2 is shown in the top panel of Figure 14.2. We see an increase in the VLAD chart showing that for Surgeon 2 at the end of the series there were more deaths than expected by the risk model (14.1) and the observed patient mix.

14.5.2 Exponentially Weighted Moving Average Chart

An alternative risk-adjusted performance summary for Surgeon 2 is given in the bottom panel of Figure 14.2. The exponentially weighted moving average (EWMA) chart plots M_t versus t, the ordered patient numbers, where we define

$$
\begin{aligned}
M_t &= \lambda w_t + (1 - \lambda) M_{t-1} \tag{14.3}\\
&= \lambda w_t + (1 - \lambda) \lambda w_{t-1} + (1 - \lambda)^2 \lambda w_{t-2} + \cdots
\end{aligned}
$$

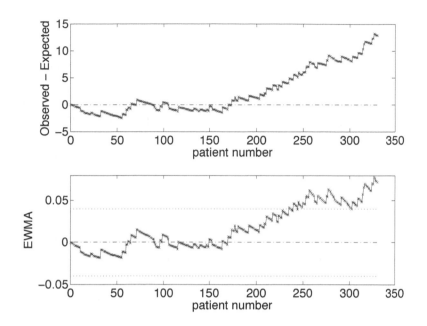

FIGURE 14.2: Risk-adjusted performance summary for Surgeon 2. Top panel: VLAD chart given by (14.2); bottom panel: EWMA chart given by (14.3) with $w_t = y_t - p_t$, $\lambda = .01$ and decision limits of $\pm.04$.

with $0 < \lambda \leq 1$, $M_0 = 0$ and w_t a score given to the tth patient. We can choose appropriate patient scores (w_t) and EWMA smoothing constant (λ) based on the monitoring goals. See Grigg and Spiegelhalter (2007) and Cook et al. (2011) for more details on risk-adjusted EWMA charts.

Here, similar to the VLAD charts in Section 14.5.1, I selected observed minus expected (O–E) patient scores; that is, for the tth patient

$$w_t = y_t - p_t. \tag{14.4}$$

The fifth column of Table 14.1 illustrates how the O–E scores (14.4) reflect surgical performance and provide the risk adjustment. Negative scores reflect good performance with a large negative score $(-.542)$ arising from successful surgery on a high risk patient (Parsonnet score = 50). Deaths, on the other hand, contribute a positive score with the death of a high risk patient resulting in a smaller positive score $(.458)$ than the death of a low risk patient $(.975)$.

Since pass/fail outcomes accumulate information slowly, small values of λ are desirable in Equation (14.3), because they allow many patient outcomes to contribute in a substantial way to the EWMA M_t. In Equation (14.3), λ determines how fast the relative weights drop off as patient outcomes are further in the past. For example, with $\lambda = .05$ the score of the most recent

TABLE 14.1: Example patient scores.

Description	Out-come y	Par-sonnet x	Prior Risk p	O–E Scores $y - p$	Likelihood Ratio Scores	
					$R = 2$	$R = .5$
Low Risk Success	0	0	.025	−.025	−.024	.012
High Risk Success	0	50	.542	−.542	−.433	.316
Low Risk Death	1	0	.025	.975	.669	−.681
High Risk Death	1	50	.542	.458	.260	−.377

patient has weight .05, the next most recent patient's weight is .05 × .95 = .0475, the patient weight just before that .05 × .95 × .95 = .0451, etc. In the bottom panel of Figure 14.2 we illustrate the O–E EWMA for Surgeon 2 with $\lambda = .01$, as recommended by Cook (2003). By contrast the VLAD chart in the top panel of Figure 14.2 uses the same patient scores but gives equal weight to all patients. As a result, the EWMA is better than the VLAD chart at detecting recent surgical performance changes. The EWMA also has the advantage that other patient scores, such as (14.6), can be used.

Run length, defined as the time (or number of observations) until a chart signals, can be used to compare the performance of different monitoring charts. When the process is operating as expected we want long run lengths, while if an important change has occurred we want short run lengths to ensure the change is detected quickly. Lucas and Saccucci (1990) provide a method to approximate the EWMA's average run length under different assumptions. Using this approximation I selected decision limits for the EWMA at ±.04 to give an approximate average run length of around 1500 patients when performance matches the predictions from the risk model (14.1). With these decision limits, shown in the bottom panel of Figure 14.2 as horizontal lines, the EWMA first signals evidence that the observed mortality rate for Surgeon 2 is larger than that predicted by the risk model at patient 235. Had this EWMA chart been used prospectively, i.e., as the patient outcomes arose, this signal would have triggered an investigation into the cause and might have resulted in a reaction such as retraining Surgeon 2.

14.5.3 Risk-Adjusted Cumulative Sum Chart

Another alternative monitoring approach, first proposed by Steiner et al. (2000), uses a risk-adjusted cumulative sum (RA-CUSUM) approach based on

$$X_t = \max(0, X_{t-1} + m_t), \qquad (14.5)$$

where

$$
m_t = \begin{cases} \log\left(\dfrac{1}{1 - p_t + Rp_t}\right) & \text{if } y_t = 0, \\[2ex] \log\left(\dfrac{R}{1 - p_t + Rp_t}\right) & \text{if } y_t = 1, \end{cases} \tag{14.6}
$$

$X_0 = 0$ and $R > 1$ is a chosen constant. Similar to the O–E scores (14.4), the patient scores given by (14.6) are positive when a death occurs and negative for a success, as shown in Table 14.1. The cumulative sum given by (14.5) then accumulates evidence of poor performance over time. The cumulative sum X_t of the scores is never allowed to be negative so that a deterioration of performance at any time (even after a series of favorable results immediately before the deterioration) will be quickly detected. The patient scores given by (14.6) use a different scaling and are a good alternative to the O–E scores given in (14.4) since they are optimal (Moustakides, 1986), in terms of average run length, to compare the hypotheses

$$
\begin{aligned}
\mathcal{H}_0 &: \text{odds of death for patient } t = p_t/(1 - p_t), \\
\mathcal{H}_A &: \text{odds of death for patient } t = Rp_t/(1 - p_t)
\end{aligned}
$$

repeatedly over time. Note that under $\mathcal{H}_0$ the odds of death equals what is expected by the risk model (14.1), while $\mathcal{H}_A$ corresponds to a worsening of performance. We choose R based on the size of the process change we are interested in quickly identifying. The change is measured in terms of the odds of death rather than the probability of death for mathematical reasons.

As defined in (14.5) and (14.6), the RA-CUSUM chart tends to increase if performance deteriorates, i.e., the mortality rate increases. To detect mortality rate decreases, i.e., performance improvements, we can also chart

$$
Z_t = \min(0, Z_{t-1} - m_t) \tag{14.7}
$$

with patient scores (m_t) as in (14.6) but with $R < 1$. Example patient scores based on (14.6) with $R = 2$ (double the odds of death) and $R = .5$ (half of the odds of death) are given in Table 14.1. We see that relative to the O–E scores (14.4) the scores based on (14.6) with $R = 2$ give a smaller positive penalty for low risk or high risk deaths while giving roughly the same (negative) credit for a low risk success. Since the RA-CUSUM scores are based on the optimal likelihood ratio test the RA-CUSUM with $R = 2$ is better, on average, than the O–E EWMA at detecting changes that result in a doubling of the mortality rate.

By using both X_t, given by (14.5), and Z_t, given by (14.7), the RA-CUSUM is sensitive to increases and decreases in the mortality rate just like the VLAD and O–E EWMA charts. Spontaneous process improvements are less likely than performance deterioration but if they occur we want to know as soon as possible, to learn from them and ensure the better performance will continue in the future.

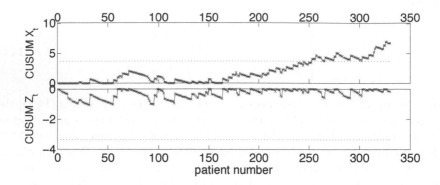

FIGURE 14.3: RA-CUSUM for Surgeon 2. X_t defined by (14.5) with $R = 2$ in top panel with decision limit at 3.65; Z_t defined by (14.7) with $R = .5$ in the bottom panel with decision limit at -3.35.

Since X_t is always non-negative and Z_t is always non-positive we can plot the two RA-CUSUM charts on the same horizontal axis. This is illustrated in Figure 14.3 that shows the RA-CUSUM chart for Surgeon 2 with $R = 2$ in the top half for X_t and with $R = .5$ in the bottom half of Z_t. The RA-CUSUM signals a change in performance if either X_t and Z_t fall outside the chosen decision limits. Setting appropriate decision limits for a RA-CUSUM is discussed in Steiner et al. (2000). In Figure 14.3 the decision limits are shown as horizontal lines at 3.65 and -3.35 for X_t and Z_t respectively and give a combined average run length of roughly 1500 patients (to match the O–E EWMA chart in Figure 14.2) when performance matches the predictions from the risk model (14.1). The RA-CUSUM in Figure 14.3 also shows the increased (risk-adjusted) mortality rate for Surgeon 2 at the end of the series. In this example the RA-CUSUM signals at patient 253; a little later than the O–E EWMA chart given in Figure 14.2.

The performance of the RA-CUSUM chart is explored in Hussein et al. (2011). Grigg and Farewell (2004) provide an overview of risk-adjusted monitoring that includes the RA-CUSUM and some other methods and conclude that in most circumstances the RA-CUSUM is preferred.

14.6 Monitoring Outcomes in Health Care: Uses and Future

I have highlighted here the need for, and uses of, risk-adjusted monitoring of health care outcomes. The three methods presented for process monitoring have proven helpful for controlling and improving health care by providing management oversight of critically important processes. Canadian contributions in this area have been substantial and are ongoing.

Risk-adjusted monitoring has been employed in many clinical settings and different contexts. The popularity of risk-adjusted monitoring is perhaps best exemplified by the publication "Variable Life-Adjusted Displays (VLAD) for Dummies" (Queensland Health's Clinical Practice Improvement Centre, 2008). The VLAD chart has been used extensively. Some examples include monitoring lung surgery outcomes in the Netherlands (Damhuis et al., 2006) and detecting deficiencies in trauma care in England (Tan et al., 2005). The EWMA method has been employed to look for excess deaths in an Australian intensive care unit (Pilcher et al., 2010) among other applications. Specific applications of the RA-CUSUM method include monitoring cardiac surgery outcomes in Canadian hospitals (Harris et al., 2005; Forbes et al., 2005; Novick et al., 2006) and monitoring the length of stay in an intensive care unit in Australia (Cook et al., 2003). In addition, Bottle and Aylin (2008) report the use of RA-CUSUM charts as a management tool to help drive improvement in a number of patient centered outcomes at nearly 100 English hospitals. As well, the RA-CUSUM method was employed to quickly detect problems related to cataract surgery in Western Australia (Ng et al., 2008), excessive radiation doses in Brisbane Australia (Smith et al., 2011) and infectious diseases in foxes in Germany (Höhle et al., 1991).

Research in monitoring health care outcomes is ongoing and appears to be growing. There are a number of important issues and application areas that need further study. In the next few paragraphs I give some examples.

In risk adjustment monitoring not much attention has been given to the goal of comparing performance across surgeons or centers. In the current practice, comparisons are typically only conducted at fixed time points, say every six months (Spiegelhalter, 2005). Methods that would allow comparisons on an ongoing basis as time passes could be valuable since they would allow more timely conclusions and reactions. Here we could use the existing methods to produce separate charts for each surgeon but this would only allow an informal comparison and become unwieldy if there are many surgeons.

An emerging related research area is monitoring public health data where the goal is to quickly detect outbreaks of disease such as influenza. Woodall (2006) and Shmueli and Burkom (2010) provide an overview of the methods and issues. A specific challenge here is the use of nonstandard data sources, such as emergency room visits, non prescription drug purchases, and even the volume of Internet searches for target key words. Also, the data usually come aggregated in geographical regions and so the best monitoring approaches incorporate the available spatial information. In addition, in most applications there are large numbers of outcomes to monitor. This means we need to be careful to control false alarm rates because too many signals may result in the monitoring charts being ignored. A specific application is given by Google Flu Trends (www.google.org/flutrends), which maps global influenza activity by country using aggregated data on Google Internet searches for key words that have been found to correlate with influenza activity.

All risk-adjusted methods for monitoring require specification and fitting of a risk adjustment model. Jones and Steiner (2012) found that the effect of estimation error and model specification error on the performance on the RA-CUSUM chart can be substantial. While their general conclusions are also applicable to other risk-adjusted monitoring approaches, more work is needed. Another related issue is how to best handle monitoring in contexts where we start with little data. Here an important question is when to re-estimate or update the risk model.

This chapter gave an example in which 30-day mortality, a pass/fail health care outcome, was monitored. If it is feasible, working with a continuous outcome is preferred since problems or process changes will then usually be more readily detected. There are many examples in medicine where continuous outcome data are already collected, e.g., hospital lengths of stay, post-surgery survival times, etc. For instance, Biswas and Kalbfleisch (2008), Sego et al. (2009) and Steiner and Jones (2010) all propose risk-adjusted EWMA based methods for monitoring time to death after surgery.

As discussed, monitoring health care processes can lead to improvement if signals are promptly addressed. However, in many applications we should also use more proactive methods that do not wait until we identify trouble before taking action. Lean Six Sigma (De Koning et al., 2006), a general quality improvement methodology that is also borrowed from industry, is at the forefront of such efforts. For more information on improving health care processes I suggest consulting *BMJ Quality and Safety* and the *Journal of Healthcare Quality*. In addition, there are professional organizations devoted to improving healthcare, including the Institute for Healthcare Improvement and the American Society for Quality Healthcare division.

About the Author

Stefan H. Steiner is a professor in the Department of Statistics and Actuarial Science as well as the director of the Business and Industrial Statistics Research Group at the University of Waterloo. He holds a PhD in business administration (management science/systems) from McMaster University. His research interests include quality improvement, statistical process control, experimental design and measurement system assessment. He is a fellow of the American Society for Quality.

Bibliography

AIAG (1992). *Statistical Process Control Reference Manual.* Chrysler Corporation, Ford Motor Company, and General Motors Corporation.

Baker, G. R., Norton, P. G., Flintoft, V., Blais, R., Brown, A., Cox, J., Etchells, E., Ghali, W. A., Hébert, P., Majumdar, S. R., O'Beirne, M., Palacios-Derflingher, L., Reid, R. J., Sheps, S., and Tamblyn, R. (2004). The Canadian adverse events study: The incidence of adverse events among hospital patients in Canada. *Canadian Medical Association Journal*, 170:1678–1686.

Biswas, P. and Kalbfleisch, J. D. (2008). A risk-adjusted CUSUM in continuous time based on the Cox model. *Statistics in Medicine*, 27:3382–3406.

Bottle, A. and Aylin, P. (2008). Intelligent information: A national system for monitoring clinical performance. *Health Services Research*, 43:1–31.

Cook, D. A. (2003). *The Development of Risk Adjusted Control Charts and Machine Learning Models to Monitor the Mortality Rate of Intensive Care Unit Patients.* Doctoral dissertation, The University of Queensland, Brisbane, Australia.

Cook, D. A., Coory, M., and Webster, R. A. (2011). Exponentially weighted moving average charts to compare observed and expected values for monitoring risk-adjusted hospital indicators. *BMJ Quality and Safety*, 20:469–474.

Cook, D. A., Steiner, S. H., Cook, R. J., and Farewell, V. T. (2003). Monitoring the evolutionary process of quality: Tracking outcomes in intensive care with the risk-adjusted CUSUM. *Critical Care Medicine*, 6:1676–1682.

Damhuis, R., Coonar, A., Plaisier, P., Dankers, M., Bekkers, J., Linklater, K., and Møller, H. (2006). A case-mix model for monitoring of postoperative mortality after surgery for lung cancer. *Lung Cancer*, 51:123–129.

Davies, J. M. (2001). Painful inquiries: Lessons from Winnipeg. *Canadian Medical Association Journal*, 165:1503–1504.

De Koning, H., Verver, J. P. S., van der Heuvel, J., Bisgaard, S., and Does, R. J. M. M. (2006). Lean Six Sigma in healthcare. *Journal for Healthcare Quality*, 28:4–11.

Forbes, T. L., Steiner, S. H., Lawlor, D. K., DeRose, G., and Harris, K. A. (2005). Risk-adjusted analysis of outcomes following elective open abdominal aortic aneurysm repair. *Annals of Vascular Surgery*, 19:142–148.

Grigg, O. and Farewell, V. T. (2004). An overview of risk-adjusted charts. *Journal of the Royal Statistical Society, Series A*, 167:523–539.

Grigg, O. and Spiegelhalter, D. A. (2007). Simple risk-adjusted exponentially weighted moving average. *Journal of the American Statistical Association*, 102:140–152.

Harris, J. R., Forbes, T. L., Steiner, S. H., Lawlor, D. K., DeRose, G., and Harris, K. A. (2005). Risk-adjusted analysis of early mortality following ruptured abdominal aortic aneurysm repair. *Journal of Vascular Surgery*, 42:387–39.

Höhle, M., Paul, M., and Held, L. (1991). Statistical approaches to the monitoring and surveillance of infectious diseases for veterinary public health. *Preventive Veterinary Medicine*, 2009:2–10.

Hussein, A. A., Steiner, S. H., and Gombay, E. (2011). Monitoring binary outcomes using risk-adjusted charts: A comparative study. *Statistics in Medicine*, 30:2815–2826.

Inquiry, B. R. I. (2001). *The Inquiry into the Management of Care of Children Receiving Complex Heart Surgery at the Bristol Royal Infirmary*. Stationery Office, London.

Jones, M. and Steiner, S. H. (2012). Assessing the effect of estimation error on risk-adjusted CUSUM chart performance. *International Journal for Quality in Health Care*, 24:176–181.

Kohn, L. T., Corrigan, J., and Donaldson, M. S. (2000). *To Err Is Human: Building a Safer Health System*. National Academy Press, Washington, DC.

Liu, X., MacKay, R. J., and Steiner, S. H. (2008). Monitoring multiple stream processes. *Quality Engineering*, 20:296–308.

Lovegrove, J., Valencia, O., Treasure, T., Sherlaw-Johnson, C., and Gallivan, S. (1997). Monitoring the results of cardiac surgery by variable life-adjusted display. *Lancet*, 18:1128–1130.

Lucas, J. M. and Saccucci, M. S. (1990). Exponentially weighted moving average control schemes: Properties and enhancements. *Technometrics*, 32:1–12.

Montgomery, D. C. (2005). *Introduction to Statistical Quality Control*, Fifth Edition. Wiley, New York.

Moustakides, G. V. (1986). Optimal stopping times for detecting changes in distributions. *The Annals of Statistics*, 14:1379–1387.

Ng, J. Q., Morlet, N., Franzco, F., Bremner, A. P., Bulsara, M. K., Morton, A. P., and Semmens, J. B. (2008). Techniques to monitor for endophthalmitis and other cataract surgery complications. *Ophthalmology*, 115:3–10.

Novick, R. J., Fox, S. A., Stitt, L. W., Forbes, T. L., and Steiner, S. H. (2006). Direct comparison of risk-adjusted CUSUM and non-risk-adjusted analyses of coronary artery bypass surgery outcomes. *Journal of Thoracic and Cardiovascular Surgery*, 132:386–391.

Parsonnet, V., Dean, D., and Bernstein, A. D. (1989). A method of uniform stratification of risks for evaluating the results of surgery in acquired adult heart disease. *Circulation*, 779:1–12.

Pilcher, D. V., Hoffman, T., Thomas, C., Ernest, D., and Hart, G. K. (2010). Risk-adjusted continuous outcome monitoring with an EWMA chart: Could it have detected excess mortality among intensive care patients at Bundaberg Base Hospital? *Critical Care and Resuscitation*, 12:36–41.

Poloniecki, J., Valencia, O., and Littlejohns, P. (1998). Cumulative risk-adjusted mortality chart for detecting changes in death rate: Observational study of heart surgery. *British Medical Journal*, 316:1697–1700.

Queensland Health's Clinical Practice Improvement Centre (2008). *VLADs for Dummies*. Wiley, Brisbane, Australia.

Rothschild, J. M., Landrigan, C. P., Cronin, J. W., Kaushal, R., Lockley, S. W., Burdick, E., Stone, P., Lilly, C., Katz, J., Czeisler, C. A., and Bates, D. (2005). The critical care safety study: The incidence and nature of adverse events and serious medical errors in intensive care. *Critical Care Medicine*, 33:1694–1700.

Sego, L. H., Reynolds Jr, M. R., and Woodall, W. H. (2009). Risk adjusted monitoring of survival times. *Statistics in Medicine*, 28:1386–1401.

Shewhart, W. A. (1931). *Economic Control of Quality of Manufactured Product*. Van Nostrand, New York.

Shmueli, G. and Burkom, H. (2010). Statistical challenges facing early outbreak detection in biosurveillance. *Technometrics*, 52:39–51.

Sinclair, C. M. (2000). *The Report of the Manitoba Pediatric Cardiac Surgery Inquest: An Inquiry into Twelve Deaths at the Winnipeg Health Sciences Centre in 1994*. Provincial Court of Manitoba, Winnipeg, MB.

Smith, I. R., Foster, K. A., Brighouse, R. D., Cameron, J., and Rivers, J. T. (2011). The role of quantitative feedback in coronary angiography radiation reduction. *International Journal of Quality in Health Care*, 23:342–348.

Smith, J. (2005). *The Shipman Inquiry: The Final Report.* http://www.shipman-inquiry.org.uk/.

Spiegelhalter, D. J. (2005). Funnel plots for comparing institutional performance. *Statistics in Medicine*, 24:1185–1202.

Steiner, S. H., Cook, R., Farewell, V. T., and Treasure, T. (2000). Monitoring surgical performance using risk-adjusted cumulative sum charts. *Biostatistics*, 1:441–452.

Steiner, S. H. and Jones, M. (2010). Risk-adjusted survival time monitoring with an updating exponentially weighted moving average (EWMA) control chart. *Statistics in Medicine*, 29:444–454.

Tan, H. B., Cross, S. F., and Goodacre, S. W. (2005). Application of variable adjusted display (VLAD) in early detection of deficiency in trauma care. *Emergency Medicine*, 22:726–728.

Treasure, T., Taylor, K., and Black, N. (1997). Independent review of adult cardiac surgery. Health Care Trust, Unite Bristol, Bristol.

Waldie, P. (1998). Crisis in the cardiac unit. *The Globe and Mail*, October 27 Edition, Section A:3 (column 1).

Woodall, W. (2006). The use of control charts in health-care and public-health surveillance. *Journal of Quality Technology*, 38:89–104.

15

Statistics in Financial Engineering

Bruno Rémillard

HEC Montréal, Montréal, QC

15.1 Introduction

As scientific disciplines, statistics and finance had their first date in 1900. They met through the work of French mathematician Louis Bachelier, whose thesis (Bachelier, 1900) used the familiar bell-shaped curve or Normal distribution to model differences between prices over time. He was also the first to use another well-known probabilistic object, Brownian motion, to evaluate stock options and model asset returns. Five years later, Albert Einstein used the same process to model the displacement of molecules in a liquid.

Statistics and finance have been together ever since. One of their offspring is financial engineering, a multidisciplinary field that combines financial theory with statistical techniques and computational tools. Financial engineers evaluate options, assess risks, manage portfolios and contribute to the development of financial products. Fundamentally, these various tasks all rely on the crucial ability to construct accurate statistical models for predicting the behavior of assets and other financial products. Bad models may lead to wrong decisions and considerable financial losses.

Selecting, fitting and validating models for financial data is key to all research and development activities in financial engineering. This is the hallmark of statistics. But as we shall see, statisticians also make significant contributions to more specialized areas of financial engineering in which estimation and prediction play a critical role, together with the numerical implementation of options and hedging evaluation techniques.

15.2   Modeling

To illustrate the challenges involved in financial data modeling, consider the following simple problem, which is nevertheless typical of the issues faced daily by traders on the stock market. Imagine that the current market value of one share from Apple is $450 and that for $18, you can purchase a call option on this asset. If you do, you earn the right to buy one share from Apple for $460 (the strike price) at any moment of your choice within the next two months (the maturity period). If the price of this share stays below $460 in that period, you will never exercise the option and you will have wasted $18. But if the price goes up and you exercise your option when the share is at $520, say, you will have saved $520 − $460 − $18 = $42. Is this an interesting business opportunity?

To answer this question, we must first choose a plausible model for the way in which the underlying asset varies over time. Then we must use previous data to check whether the assumptions of this model hold, at least approximately. If the model is satisfactory, we can then assess whether the price of the asset and the option are "fair" in some way or other. A rational decision to buy the option or not can then be made. But whether the decision is right depends critically on the choice of model and, more specifically, on its appropriateness for the data at hand.

The financial literature abounds with statistical models for option pricing. The most commonly used model was articulated by the American economist Fischer Black and his co-author, Canadian-born financial economist Myron Scholes, in their 1973 paper, "The Pricing of Options and Corporate Liabilities," published in the *Journal of Political Economy* (Black and Scholes, 1973). American economist Robert Merton later contributed to the refinement of the model.

15.2.1   Black–Scholes–Merton Model

The BSM model due to Black, Scholes, and Merton makes assumptions on the probability distribution of the return of an asset as it varies in time. Suppose for instance that the price of an asset is observed a times $1, 2, \ldots$ Denote these prices by $S_1, S_2, \ldots$ The return of the asset at time $k > 1$ is then defined by

$$R_k = \log{(S_k/S_{k-1})}.$$

In its simplest form, the BSM model then states that R_k has the same bell-shaped or Normal distribution at every time point k in time, and that this distribution depends in no way on past observed values $R_1, \ldots, R_{k-1}$. The latter hypothesis is called serial independence.

The hypotheses of the BSM model are quite strong. In particular, the choice of the bell-shaped curve for the returns means that wild variations in

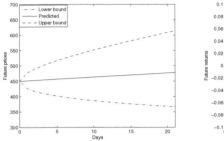 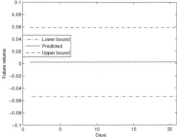

FIGURE 15.1: Prediction intervals for future prices (left) and returns (right) for the next 21 trading days predicted by the Black–Scholes model when the current price is $450.

prices are unlikely to occur. The standard deviation is a measure of the spread of a distribution. For a Normal distribution, there is only a .3% probability that a future return will lie more than three standard deviations away from the mean. For other distributions, this probability can be much greater, up to about 10 or 11%.

For example, based on recent observations of Apple shares (2009–11), typical values for the mean of the daily returns is .003, while the standard deviation is .02. Under the hypotheses of the BSM model, 99.7% of the future returns should thus lie in the interval $.003 \pm .06$, that is, between $-.057$ and $.063$. Prediction intervals of the prices and returns for the next 21 days are given in Figure 15.1. The interpretation of these intervals is that the probability the stock price, or return, on any given day lies between the limits in the figure is 99.7%. Notice that the prediction intervals for returns all have the same width; this is a consequence of the model assumptions.

Merton and Scholes won the Nobel prize in Economics in 1997 for their work on this subject (unfortunately, Black died in 1995). They were aware that both assumptions in their basic model were debatable; students and colleagues helped them to refine it in later work. The most glaringly simplistic assumption is the hypothesis of serial independence: in practice, financial returns are almost always dependent on previous values, and hence any decision based on a model that ignores this fact is bound to spell trouble eventually.

There is a whole slew of statistical procedures for testing whether a financial series is serially dependent or not. For recent contributions by Canadian statisticians, see, e.g., Genest and Rémillard (2004) and Genest et al. (2007). Having rejected the hypothesis of serial independence, one must turn to time series models. Such models are far better at reflecting the variation of returns and other financial variables in time, but they are also more complex. Their use requires much more sophisticated statistical tools, many of which are still under development.

15.2.2   Extensions of the BSM Model

In most financial time series models, the current value of returns is not only a function of previous values but also of exogenous variables such as interest rates or tax structures. Good models involve parameters that can be adjusted to fit the data at hand, and the residual (hopefully small) fluctuations that remain unexplained are accounted for by an unobservable error term called an innovation. For example, it is typical to assume that the return R_k at time k is of the form

$$R_k = \mu_k + \sigma_k \varepsilon_k, \tag{15.1}$$

where μ_k and σ_k represent the mean and standard deviation of the distribution of R_k. The innovations $\varepsilon_1, \varepsilon_2, \ldots$ are generally assumed to be independent and to have the same distribution with mean 0 and standard deviation 1. The formulas for μ_k and σ_k typically involve previous returns and other unknowns.

As an illustration, suppose that we take

$$\mu_k = \mu + \phi R_{k-1}, \quad \sigma_k = \sigma > 0.$$

The first equation amounts to saying that the average return value at time k is a linear function of R_{k-1} with slope ϕ and y-intercept μ. The second equation states that all returns $R_1, R_2, \ldots$ have the same standard deviation σ. This model is called the "autoregressive model of order 1" or AR(1) for short. The original Black–Scholes model is an AR(1) model with $\phi = 0$.

When the variability or volatility in the returns is thought to change with time, a popular option is to take

$$\sigma_k^2 = \omega + \alpha \sigma_{k-1}^2 \varepsilon_{k-1}^2 + \beta \sigma_{k-1}^2,$$

in which case the model is called GARCH, as in "generalized autoregressive conditionally heteroscedastic." American economist Robert Engle won the 2003 Nobel Prize in Economics, sharing the award with British economist Clive Granger, for methods of analyzing economic time series with time-varying volatility.

Figure 15.2 shows 99.7% prediction intervals for the prices and returns for 1 Apple share based on a GARCH model with Normal innovations and parameters $\mu = .0032$, $\omega = .1356 \times 10^{-4}$, $\alpha = .0881$, and $\beta = .87$. Here we assumed $\phi = 0$. These parameter values were chosen on the basis of the same data as for Figure 15.1; see p. 233 in Rémillard (2013) for details.

Comparing Figures 15.1 and 15.2, one can see that the range of values is larger for the GARCH model than for the Black–Scholes model. While the Black–Scholes AR(1) model assumes that the returns are independent and identically distributed, the GARCH extension makes the more realistic assumption that the return at time k depends on the performance of the stock at earlier dates. As a result of this serial dependence, the prediction intervals for the future return values are not all identical; they vary as a function of time. However, as can be seen from Figure 15.2, they stabilize after a while. In other

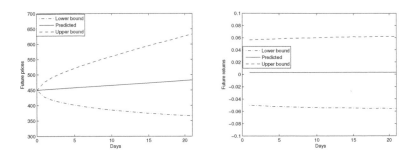

FIGURE 15.2: Prediction intervals for future prices (left) and returns (right) for the next 21 trading days predicted by the GARCH model when the current price is $450.

words, the distribution of the long-term return values gradually stabilizes or converges to a so-called stationary distribution. In particular, the long-term prediction becomes a constant, which is usually the mean of the stationary distribution.

In practice, it sometimes happens that series of financial returns move quite a distance from the mean of their stationary distribution. However, they will eventually come back to it. This so-called "return-to-the-mean" property is a consequence of the famous ergodic theorem, established by the American mathematician George David Birkhoff. It is therefore important for modelers and financial analysts to consider models that have a (unique) stationary distribution. It is not always possible to produce a formula for this distribution but for GARCH models, say, it is relatively easy to simulate it using computer-intensive methods. One could, for example, use the Monte Carlo Markov Chain approach (MCMC) discussed by Jeffrey Rosenthal in Chapter 6.

15.2.3 Choice of Distribution for the Innovations

Another important aspect of the model building process is the choice of distribution for the innovations $\varepsilon_1, \varepsilon_2, \ldots$ The predictive properties of the model very much depend on this choice. The Normal curve is only one option. Another popular choice is the Student t distribution, which has fatter tails, meaning that its range of likely values is larger than for the Normal. This choice is often dictated in practice by the presence in the series of a larger proportion of extreme observations than we would expect under the Normal paradigm.

Figure 15.3 shows the impact of replacing the Normal distribution by a Student t_5 in the GARCH model described above. As one can see, the prediction intervals are now wider. Suppose, for example, that the current price of one Apple share is $450 and that we wish to predict its price in 21 days, based on the GARCH model. If the Normal distribution is used, there is a

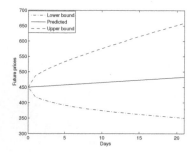 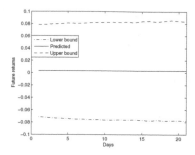

FIGURE 15.3: Prediction intervals for future prices (left) and returns (right) for the next 21 trading days predicted by the GARCH model with Student t_5 innovations when the current price is $450.

99.7% chance that the price will fall between $366 and $633. If a Student t_5 distribution is used, however, there is about a 99% chance that it will fall in this interval. The 99.7% prediction interval for the price is now much wider; it extends from $352 to $658!

What principles can guide us in choosing an appropriate distribution for the innovations? A major stumbling block is that for models defined by equation (15.1), the variables $\varepsilon_1, \varepsilon_2, \ldots$ cannot be observed or measured except in the rare (not to say unrealistic) cases where all model parameters are known. We are thus faced with a situation where we cannot even plot a simple histogram of these innovations to guide us in the choice of their distribution.

The solution to this problem is to replace the unobservable innovations by proxies, that is, quantities that can be computed from the data and which closely resemble or imitate the innovations. The best proxy candidates are the so-called model residuals, denoted $e_1, e_2, \ldots$ The kth residual can be computed as

$$e_k = \frac{R_k - \hat{\mu}_k}{\hat{\sigma}_k}.$$

In this formula, $\hat{\mu}_k$ and $\hat{\sigma}_k$ are estimated values of μ_k and σ_k, respectively. Histograms and other graphical representations of the residuals can then be plotted to help us choose an appropriate distribution.

At this stage, it is tempting to go one step further by trying to apply standard statistical tests to the residuals in order to check various assumptions about the innovations ε_k. As it happens, however, conventional tests were developed for variables that are truly observable and independent of one another. This is not the case for residuals and, as many statisticians have begun to realize, this sometimes makes the tests non-operational, because when residuals are used, the distributions of the statistics are different from those when the mean and standard deviation are known, and not estimated. For details, see Bai (2003) or Ghoudi and Rémillard (1998, 2004), among others. Therefore, new methodologies adapted to residuals must be developed.

15.2.4 Challenge of Model Validation

Today, the vast majority of articles in financial engineering, including doctoral dissertations, involve financial data modeling. The major journals not only require that the models be thoroughly investigated from a theoretical point of view, but also that they be tried out in applications and that their characteristics be showed to fit the data at hand.

Similarly, financial institutions need to validate the assumptions of the models they rely on to develop new financial products. For obvious reasons, it is increasingly difficult to convince investors of the potential of a new financial product if its relevance and the adequacy of the models on which it is based cannot be justified properly. It is imperative, therefore, to understand the effect of the use of residuals on statistical tests and procedures, and in particular to develop new ways of validating financial models. This is currently the subject of much fundamental research in statistics, both in Canada and abroad.

In order to validate the adequacy of a proposed model for a given dataset, a data analyst must be able to perform the following tasks:

1. *Detect change points in the distribution of innovations over time.* Change points are important characteristics of financial data. In particular, a change in the distribution is likely to occur when the market crashes. As a result, it might be unrealistic to assume that over a long period of time, the distribution of innovations is the same. This is why statistical techniques have to be used to detect change points. However, remember that the data show serial dependence, so change point tests would be applied to residuals. Recent results suggest that this can be done (Rémillard, 2012).

2. *Find a plausible model for the distribution of returns and estimate the parameters as precisely as possible.* Suppose for instance that a GARCH model has been fitted under the assumption that the innovations are Normally distributed. A specification test must then be used to check whether this assumption is supported by the data; see, e.g., Ghoudi and Rémillard (2013) and references therein.

3. *Check whether the innovations are serially independent.* Testing this hypothesis using the residuals is not an easy task. In fact it can be very difficult, especially for GARCH models; see Berkes et al. (2003), among others. Canadians have contributed to the development of new statistical techniques to handle this problem (Genest et al., 2007). Their approach works for simple models but more research is needed to develop appropriate techniques for more complex models like GARCH.

4. *If necessary, model dependence between time series.* When several financial series are involved, as is often the case in practice (options on several securities, etc.), one must also model the dependence between innovations in each series and test the adequacy of the dependence model for these multivariate data. Copulas can be used to this end (Embrechts et al., 2002), but much additional work will be required to develop their use to its full potential. For more information on this topic, see Chapter 4 by Christian Genest and Johanna Nešlehová, and Chapter 8 of Rémillard (2013) for applications in a financial engineering context.

15.3 Applications to Financial Engineering

Once financial data have been modeled correctly, one can go a step further and implement the model. In this section, we will review briefly three important financial applications for these models: portfolio management, option pricing, and risk management.

15.3.1 Portfolio Management

Following Bachelier (1900), the importance of the Normal distribution in finance was reaffirmed by American economist Harry Markowitz. In his doctoral dissertation, Markowitz (1952) used a weighted sum of Normally distributed random variables to model the return of a portfolio. More precisely, if $\omega_1, \ldots, \omega_d$ represent the fractions of the wealth invested in $d \geq 2$ possibly risky assets with returns $R_1, \ldots, R_d$, the return of the portfolio with weights $\omega_1, \ldots, \omega_d$ is then

$$P_\omega = \sum_{j=1}^{d} \omega_j R_j.$$

Note here that we are assuming $w_1 + \cdots + w_d = 1$; in other words, all the money must be invested in the assets (one of them could be a bank account). The weights are usually positive, but they could be negative too. A negative weight corresponds to a strategy called short-selling; it consists of cashing the present value of the asset without selling it, but the asset must be bought at the end of a given period. It is basically like borrowing the value of a stock.

In financial markets, short-selling is often used to reduce the risk in a portfolio. To illustrate, suppose that at one point in time, an investor has in hand one share of Apple listed at \$450 and that he chooses to be "short" 10 shares of Microsoft listed at \$30 per share. This means that he will cash immediately $10 \times \$30 = \$300$, thereby reducing his current investment to \$150.

The weights associated with this investment strategy are $w_1 = 450/150 = 3$ for Apple and $w_2 = -300/150 = -2$ for Microsoft. This way, $w_1 + w_2 = 3 - 2 = 1$. In return, however, the investor will need to buy 10 shares of Microsoft at some future point in time, whatever its price may be at the time. One advantage of this strategy is that if the values of the two stocks go down, the investor can still make money if Apple performs better than Microsoft. For example, if both stocks decrease by \$1, the value of the portfolio will be \$459. From the borrowed \$300 obtained by short-selling Microsoft, you just have to reimburse \$290.

In the above example, $w_1 = 3$ and $w_2 = -2$ are only one possible choice of weights. Can we do better? Using the standard deviation as a measure of the risk of a portfolio P_w, Markowitz looked at the problem of minimizing this risk for a given average return; he also considered the complementary problem of maximizing the expectation $R_w = \mathrm{E}(P_w)$ for a given level of risk. Assuming only that the weights sum up to 1, Markowitz was able to solve this optimization problem using elementary calculus. It can be shown that the variance (i.e., the square of the standard deviation) of the portfolio P_w is given by

$$\sigma_w^2 = \sum_{i=1}^{d}\sum_{j=1}^{d} w_i w_j \sigma_i \sigma_j \rho_{ij}. \tag{15.2}$$

Therefore, the problem is to minimize σ_w^2, subject to the constraints

$$\sum_{j=1}^{d} w_j = 1 \quad \text{and} \quad \sum_{j=1}^{d} w_j \mu_j = \mu.$$

Here μ_j and σ_j are respectively the mean and standard deviation of return R_j, μ is the target mean of the portfolio, and ρ_{ij} is the so-called Pearson correlation coefficient between the returns R_i and R_j, which measures the degree of dependence between them. This leads to the concept of efficient frontier, which is the graph of the maximal expected return as a function of the standard deviation. For this and subsequent work on portfolio management, Markowitz was awarded the Nobel Prize in Economics in 1990.

For example, using the portfolio with weights $w_1 = 3$ and $w_2 = -2$, we end up with an average return of .2627 and a standard deviation of .0514, as marked by a square on the graph of the efficient frontier in Figure 15.4. This portfolio is optimal for a standard deviation of .0514, but not for a standard deviation of .03, as seen from the graph.

The field of portfolio management, which basically consists in finding optimal weights of a portfolio changing from period to period, is a very active field of research from the statistical point of view. The issue of non-Normal returns is particularly relevant. Furthermore, portfolio managers may be interested either in maximizing a utility function, which represents the value of wealth of the investor, or in minimizing a risk measure, such as the standard deviation.

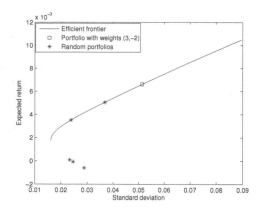

FIGURE 15.4: Graph of the efficient frontier, position of the portfolio with weights $\omega_1 = 3$ and $\omega_2 = -2$, and positions of some portfolios with random weights.

Among the problems of interest for statisticians, one can look at constraints on the weights, discrete time and continuous time models, etc. The computation of the optimal solution for a portfolio composed of a large number of assets is also quite challenging. Several statisticians working in Canadian universities have contributed to the field; see Watier (2003), Vaillancourt and Watier (2005), Labbé and Heunis (2007), and Elliott and Siu (2009), to name a few.

15.3.2 Option Pricing

Going back once again to the case of Apple, assume that we have in hand a theoretical model that has been validated with data. The next challenge then lies is the determination of a "fair" price for this option. A price is considered fair if neither the buyer nor the seller of the option have a chance of making money while being sure of not losing any money in the process.

The most common way of determining a fair price for an option consists of choosing an equivalent probability distribution, called the equivalent martingale measure, under which the actualized value of the asset in the future is a martingale. A martingale has the property that given the current value, the expected value at any future time is the same as the current value. To understand the concept of martingale, suppose for an instant that the value of an asset would correspond to a gambler's fortune. To say that the fortune is a martingale means that there is no optimal way for the gambler to decide when to stop playing; the average return is the same, whatever the player's strategy.

The existence of an equivalent martingale measure guarantees that there exists a fair price. In practice, the choice of the equivalent martingale measure is dictated by the previous real market prices of the option. Then the future prices of any option can be theoretically determined.

For example, the value of the call option described at the beginning of Section 15.2 is $18.04 if we assume that the Black–Scholes model is correct and the annual interest rate is 2%. As it happens, there is only one equivalent martingale measure in this case. Assuming that the model is correct, we can compute the chances that a buyer will not exercise the call option. This probability is approximately 22%. In this case, the buyer will actually lose $18.10, which is the value of $18.04 in two months, assuming an annual interest rate of 2%.

When the price of the stock is above the strike price of $460, the buyer will also lose money if the future price is lower than $460 + $18.04 = $478.04. According to the model, this happens 33% of the time. Overall, the negative gain (or loss) turns out to be −$15 while the average positive gain is $63.

A histogram of the distribution of the net gain is given in Figure 15.5. Note that the average net gain is $37.03. Therefore, here it seems that the buyer has a net advantage over the seller of the option. However, if the seller could trade continuously on Apple, he could generate exactly the payoff of the option, i.e., what is due to the seller at maturity, by building a portfolio composed only of cash and a fraction of the asset. This was shown in the article of Black and Scholes (1973). It means that the position of the seller is not risky at all, while the position of the buyer is quite risky. In addition, if the price paid for the option is larger than the fair price $18.04, there is a way for the seller to make money for sure. If the price paid for the option is less than $18.04, then there is a way for the buyer to make money for sure.

Given an equivalent martingale measure, one cannot find in general an explicit formula for an option price. However, the option price being an average value, or expectation, one can use simulation algorithms to price it. This was first proposed by Canadian-based finance Professor Phelim Boyle (1977). For other applications of Monte Carlo simulation algorithms in finance, see, among others, the books by American mathematician Paul Glasserman (2004) and Canadian statistician Don McLeish (2011) from the University of Waterloo. Monte Carlo methods are particularly effective to demonstrate to potential investors that a proposed product or strategy is interesting for them, as was done in the previous example.

15.3.3   Risk Management

Risk management is another key field of applications of statistical methods in financial engineering. The importance of the field is such that some even believe that risk assessment should be integrated into the business student curriculum (*The Economist*, 2012). An excellent reference on risk management and statistical methods is the book by McNeil et al. (2005).

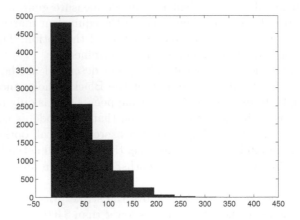

FIGURE 15.5: Histogram of the net gain for buying the option obtained by simulating 10,000 values of the stock price at maturity.

Since 1988, the Basel Committee on Banking Supervision has issued three supervisory accords — Basel I, II and III — that comprise recommendations on banking regulations. These recommendations are being implemented by many financial institutions worldwide. In particular, the Basel II Accord, initially published in June 2004, called for for the computation of risk measures and capital requirements to guard against three important sources of risks: risks of default of debt payments of borrowers (credit risk), risks of losses on the markets (market risk), and risk of loss resulting from inadequate or failed internal processes, people and systems, or from external events (operational risk).

At the heart of risk management is the issue of how to measure risk. Beginning with Markowitz (1952), the standard deviation was the measure of risk in use for a long period. Since the Basel Accords, the so-called Value-at-Risk (VaR for short) has replaced the standard deviation as a measure of market risk. The VaR corresponds to the quantile (usually of order 99.9%) of a loss, meaning that 99.9% of the losses should be smaller than the VaR. Stimulated by the Basel I Accord, Artzner et al. (1999) proposed axioms that should be satisfied by what they called "coherent risk measures." Today, the issue of risk measurement is by no means settled and many researchers, including statisticians, continue to work on this problem.

One of the main challenges in risk management is how to model individual risk factors and account for their interdependence. Two areas of modern statistics are relevant to this problem: copula modeling and extreme-value theory. The latter is particularly important for market and operational risks; see, e.g., Roncalli (2004). As its name indicates, extreme-value theory deals with the modeling of extremal events, such as sample maxima or losses exceeding high thresholds. Statistical tools developed in that context allow researchers to extrapolate beyond the range of observable data. Such extrapolation is required for the computation of risk measures and financial reserves, and extreme-value techniques are particularly useful because loss distributions typically have considerably fatter tails than the omnipresent Normal distribution.

Extreme-value theory has been an integral part of probability and statistics for over 50 years. Several Canadian statisticians have contributed to it, including UBC Professor Harry Joe and my colleague at HEC Montréal, Debbie Dupuis, but also my frequent coauthors, Christian Genest and Johanna Nešlehová, currently both at McGill. Yet some people are still oblivious to these developments. In his book which popularized the expression "Black Swans," Lebanese American essayist Nassim Nicholas Taleb (2007) even goes as far as blaming statisticians for not studying extremal events!

Another important risk management challenge is modeling times until default in the context of credit risk. Surprisingly perhaps, models and techniques developed for the analysis of survival data in medical studies come in handy; see, e.g., the book by University of Waterloo professor Jerry Lawless (2003). One specific twist of credit risk modeling is that defaults do not occur often, and in fact you would prefer that they never occur! Simplifying assumptions typically need to be made in order to model default times of firms for which no defaults were ever observed. The articles by Merton (1974) and Jarrow et al. (1997) count among the seminal contributions to this issue. Modeling default times on mortgages and credit cards is also quite a challenge. Yet another problem is how to account for dependence between default times. In fact, such dependencies played a major role in the so-called subprime crisis that led to a financial crisis and a subsequent recession in 2008. In that context, the risk of "contagion," producing interdependence of defaults, was underestimated by financial institutions and rating agencies.

Risk management is a fledgling field of research. For example, as this chapter is being written, liquidity risk is a hot and emerging topic. This risk has not yet been precisely articulated but in broad terms, it refers to the fact that at some point, there are too few buyers or sellers. How best to measure liquidity risk is a subject of debate and finding appropriate ways of estimating it is one of the future challenges; see, e.g., Jarrow et al. (2012).

15.4   Final Comment

The recent financial crisis has put the use of statistics in finance in a spotlight. The use and abuse of mathematical models and statistical methodology has been the object of much polemic in the popular press. For example, former

French Prime Minister Michel Rocard wrote, in the 2–3 November 2008 edition *Le Monde*,

"On reste trop révérencieux à l'égard de l'industrie de la finance et de l'industrie intellectuelle de la science financière. Des professeurs de maths enseignent à leurs étudiants comment faire des coups boursiers. Ce qu'ils font relève, sans qu'ils le sachent, du crime contre l'humanité."

The response of senior French probabilists and academicians Jean-Pierre Kahane and Marc Yor is well worth reading (Kahane et al., 2009). In the English language press, British journalist Felix Salmon (2009, 2012) wrote

"A formula in statistics, misunderstood and misused, has devastated the global economy." [The formula in question is the so-called bivariate Gaussian copula; again, see Chapter 4 by Genest and Nešlehová.]

Are statistical techniques really so dangerous? Of course not! It is the duty of statisticians to educate financial practitioners and make them aware of the limits of their models. The same is true in every other area of application of statistical sciences.

Acknowledgments

I am grateful to Christian Genest for his time and effort in helping me finalize this chapter. I would also like to thank the editor and the associate editor for their helpful comments. This research was funded in part by the Natural Sciences and Engineering Research Council of Canada and by the Fonds de recherche du Québec – Nature et technologies.

About the Author

Bruno Rémillard is a professor of financial engineering at HEC Montréal since 2001. He is a graduate of Université Laval (BSc, MSc) and Carleton University (PhD). His research interests cut across probability, statistics and financial engineering, and he has consulted extensively on finance and investment, helping to develop and implant new quantitative methods for both alternative and traditional portfolios. He is the author of three books, including *Statistical Methods for Financial Engineering* (Chapman & Hall/CRC, 2013).

Bibliography

Artzner, P., Delbaen, F., Eber, J.-M., and Heath, D. (1999). Coherent measures of risk. *Mathematical Finance*, 9:203–228.

Bachelier, L. (1900). Théorie de la spéculation. *Annales scientifiques de l'École normale supérieure, 3ᵉ série*, 17:21–86.

Bai, J. (2003). Testing parametric conditional distributions of dynamic models. *The Review of Economics and Statistics*, 85:531–549.

Berkes, I., Horváth, L., and Kokoszka, P. (2003). Asymptotics for GARCH squared residual correlations. *Econometric Theory*, 19:515–540.

Black, F. and Scholes, M. (1973). The pricing of options and corporate liabilities. *Journal of Political Economy*, 81:637–654.

Boyle, P. (1977). Options: A Monte Carlo approach. *Journal of Financial Economics*, 4:323–338.

Elliott, R. J. and Siu, T. (2009). On Markov-modulated exponential-affine bond price formulae. *Applied Mathematical Finance*, 16:1–15.

Embrechts, P., McNeil, A. J., and Straumann, D. (2002). Correlation and dependence in risk management: Properties and pitfalls. In *Risk Management: Value at Risk and Beyond (Cambridge, 1998)*, pp. 176–223. Cambridge University Press, Cambridge.

Genest, C., Ghoudi, K., and Rémillard, B. (2007). Rank-based extensions of the Brock–Dechert–Scheinkman test for serial dependence. *Journal of the American Statistical Association*, 102:1363–1376.

Genest, C. and Rémillard, B. (2004). Tests of independence or randomness based on the empirical copula process. *Test*, 13:335–369.

Ghoudi, K. and Rémillard, B. (1998). Empirical processes based on pseudo-observations. In *Asymptotic Methods in Probability and Statistics (Ottawa, ON, 1997)*, pp. 171–197. North-Holland, Amsterdam.

Ghoudi, K. and Rémillard, B. (2004). Empirical processes based on pseudo-observations. II. The multivariate case. In *Asymptotic Methods in Stochastics*, volume 44 of *Fields Institute Communications*, pp. 381–406. American Mathematical Society, Providence, RI.

Ghoudi, K. and Rémillard, B. (2013). Comparison of specification tests for GARCH models. *Computational Statistics & Data Analysis*, in press.

Glasserman, P. (2004). *Monte Carlo Methods in Financial Engineering*. Springer, New York.

Jarrow, R. A., Lando, D., and Turnbull, S. M. (1997). A Markov model for the term structure of credit risk spreads. *The Review of Financial Studies*, 10:481–523.

Jarrow, R. A., Protter, P., and Roch, A. F. (2012). A liquidity-based model for asset price bubbles. *Quantitative Finance*, 12:1339–1349.

Kahane, J.-P., Talay, D., and Yor, M. (2009). Finance, politique et mathématiques, quels liens? *Image des Maths, CNRS*, May 2009.

Labbé, C. and Heunis, A. J. (2007). Convex duality in constrained mean-variance portfolio optimization. *Advances in Applied Probability*, 39:77–104.

Lawless, J. F. (2003). *Statistical Models and Methods for Lifetime Data*, Second Edition. Wiley, Hoboken, NJ.

Markowitz, H. (1952). Portfolio selection. *The Journal of Finance*, 7:77–91.

McLeish, D. L. (2011). *Monte Carlo Simulation and Finance*. Wiley, New York.

McNeil, A. J., Frey, R., and Embrechts, P. (2005). *Quantitative Risk Management: Concepts, Techniques and Tools*. Princeton University Press, Princeton, NJ.

Merton, R. C. (1974). On the pricing of corporate debt: The risk structure of interest rates. *The Journal of Finance*, 29:449–470.

Rémillard, B. (2012). *Nonparametric Change Point Problems Using Multipliers*. Technical Report, SSRN Working Paper Series No. 2043632.

Rémillard, B. (2013). *Statistical Methods for Financial Engineering*. Chapman & Hall, London.

Roncalli, T. (2004). *La gestion des risques financiers*. Economica, Paris.

Salmon, F. (2009). Recipe for disaster: The formula that killed Wall Street. *Wired Magazine*, 17.03.

Salmon, F. (2012). The formula that killed Wall Street. *Significance*, 9(1):16–20.

Taleb, N. N. (2007). *The Black Swan: The Impact of the Highly Improbable*. Penguin, London.

The Economist (2012). Academic view: Underestimating risk. October 31st Edition.

Vaillancourt, J. and Watier, F. (2005). On an optimal multivariate multiperiod mean-variance portfolio. *Mathematical Reports of the Academy of Science, The Royal Society of Canada*, 27:92–96.

Watier, F. (2003). *Optimisation stochastique dans le cadre de la gestion de portefeuilles*. Doctoral dissertation, Université de Sherbrooke, Sherbrooke, QC.

16

Making Personalized Recommendations in E-Commerce

Mu Zhu

University of Waterloo, Waterloo, ON

Many practical problems in our Internet Age can benefit from ideas in statistics. In this chapter, I briefly tell the story of how two statistical ideas can be applied quite naturally to one such problem.

16.1   Introduction

The problem that I will focus on is that of making personalized recommendations in e-commerce. We encounter personalized recommendations everywhere. If you buy a book or watch a movie online, e.g., from `www.amazon.com` or `www.netflix.com`, their recommender systems will suggest a few other books or movies that they "think" you might also like. Table 16.1 contains a hypothetical example, created to illustrate the main ideas. It shows how four different customers would have rated four different books on the scale of 0–100 (see Remark 16.1). In reality, at any given time, the customers will only have revealed their preferences on a limited number of books. For example, they may have purchased a few and explicitly rated a few others. Therefore, we will pretend that only some of the ratings in Table 16.1 are available, while others — in particular, those marked by "?" — are missing, and the recommender system must predict them based on the observed entries. If the predicted rating is high, a recommendation can then be made. In what follows, we will refer more generally to "users" and "items" rather than just "customers" and "books."

The two statistical ideas that I have alluded to are: regression and shrinkage. Linear regression is one of the most widely used statistical techniques. The idea has been around since at least the early 1800s, and some of its early champions included such mathematical giants as Carl Friedrich Gauß. It postulates that a target variable of interest, y, is a linear function of a number of

TABLE 16.1: Illustrative example: Four users rate four books. *Moby Dick* refers to the novel *Moby Dick; or, The Whale* by Melville (1851). *Dreams* refers to the book *The Interpretation of Dreams* by Freud (1913). *Species* refers to the book *The Origins of Species* by Darwin (1859). *Relativity* refers to the book *Relativity* by Einstein (1916). A question mark (?) indicates that the corresponding rating is treated by various methods as if it were missing/unobserved and to be predicted. Predictions closer to these entries here are deemed more accurate.

	Moby Dick	*Dreams*	*Species*	*Relativity*
Alice	90	~~70~~ ?	30	10
Bob	90	70	~~30~~ ?	10
Cathy	10	~~30~~ ?	70	90
David	10	30	~~70~~ ?	90

other variables $x_1, \ldots, x_d$ plus some random noise ϵ,

$$y = \beta_0 + \beta_1 x_1 + \cdots + \beta_d x_d + \epsilon,$$

and uses data to estimate the linear function — in particular, the coefficients in front of each x_j, $j \in \{1, \ldots, d\}$. Shrinkage estimation (James and Stein, 1961) is a much newer idea, but it has been central to modern statistical practice. For example, while a "natural" way to predict the batting averages of baseball players in a new season is to base the predictions on each player's historical average (y_i), a much better way (see, e.g., Efron and Morris, 1977) is to shrink each y_i toward the overall mean of all players, i.e.,

$$\bar{y} = \frac{1}{n} (y_1 + \cdots + y_n).$$

Remark 16.1. In practice, we will not usually have ratings of such fine resolution. In many cases, we can only obtain rough indications of whether a customer likes or dislikes a certain book, e.g., a binary indicator of whether the customer has purchased it or read a few of its reviews. Explicit ratings are possible, but they rarely go beyond a five-point scale such as "terrible," "bad," "fair," "good," and "excellent." However, I have chosen the finer-than-usual scale deliberately, so that I can demonstrate more easily the differences of various methods. On such a small example (4 customers × 4 books), the predictions made by various methods would differ by too little if I only used a rough scale.

16.2   Nearest Neighbors

There are many different ways to predict the missing entries in Table 16.1; see, for example, a recent review by Feuerverger et al. (2012) and references therein. The nearest neighbors approach (see, e.g., Koren, 2008) is perhaps the most intuitive. For example, to predict Alice's rating (r_A) of Freud's *The Interpretation of Dreams*, we can consider observed ratings of this book — in this case, those given by Bob ($r_B = 70$) and by David ($r_D = 30$) — and ask: whose preferences, Bob's or David's, do we expect to be closer to those of Alice's? Suppose $s(A, B)$ and $s(A, D)$ measure the similarities between Alice and Bob, and between Alice and David, respectively. Then, we can predict r_A to be

$$\left\{ \frac{s(A, B)}{s(A, B) + s(A, D)} \right\} \times r_B + \left\{ \frac{s(A, D)}{s(A, B) + s(A, D)} \right\} \times r_D,$$

a weighted average of Bob's and David's ratings, each weighted by their respective similarities to Alice. The similarity between two users can be inferred from items that they have already rated in common. In this case, Alice and Bob have both rated Melville's *Moby Dick* and Einstein's *Relativity*, and their ratings of these items are highly similar. On the other hand, Alice and David have both rated the same two items as well, but their ratings of these items are much less similar. It appears, therefore, that $s(A, B) > s(A, D)$. To give a concrete numeric example, suppose the similarity measure $s(\cdot, \cdot)$ were specified in such a way that $s(A, B) = .75 > .25 = s(A, D)$. Then, our prediction of r_A would be

$$\left(\frac{.75}{.75 + .25} \right) \times 70 + \left(\frac{.25}{.75 + .25} \right) \times 30 = 60.$$

The choice of the similarity measure, $s(\cdot, \cdot)$, clearly plays a major role in this approach, but I will not go into the mathematical details here.

16.3   Matrix Factorization

A slightly more abstract approach — the matrix factorization approach (see, e.g., Koren et al., 2009) — became popular as a result of the highly publicized Netflix Prize (www.netflixprize.com). The data in Table 16.1 form a user-item rating matrix, R, typically with many missing entries. In general, suppose there are n users and m items ($n = m = 4$ in Table 16.1). After accounting for both user-effects and item-effects (more on this below), the matrix factorization approach aims to factor the matrix R into the product of two low-rank matrices,

$$R \approx PQ^\top = \underbrace{\begin{pmatrix} p_1^\top \\ \vdots \\ p_n^\top \end{pmatrix}}_{n \times K} \underbrace{\begin{pmatrix} q_1 & \cdots & q_m \end{pmatrix}}_{K \times m}, \tag{16.1}$$

where p_u, q_i are vectors in $\mathbb{R}^K$ and $K \ll \min(n, m)$. The user- and item-effects refer to the fact that some users are more difficult to please, while some items are better liked in general than others. After removing the overall mean of R, these effects can be estimated by the row-means and column-means of R, and are typically removed as well prior to performing the factorization (16.1). In other words, the matrix factorization approach aims to estimate latent coordinates, $p_u, q_i \in \mathbb{R}^K$, respectively for each user (u) and for each item (i), such that the user-item rating (r_{ui}) can be approximated — modulo user- and item-effects — by $p_u^\top q_i$, a measure of how closely aligned the user- and item-coordinates are.

If the coordinates are two-dimensional (i.e., $K = 2$), then this process will literally give us a map of all users and items (Figure 16.1, I). In order to make recommendations to a user, all we have to do is to first locate the user on the map, and then recommend items that are close by (see Remark 16.2). For $K > 2$, the idea is exactly the same, except that the map is high-dimensional. In practice, the parameter K is determined empirically, but generally should be chosen so that the total number of parameters being estimated ($nK + mK$) is considerably smaller than the total number of observed ratings. The key, of course, lies in our ability to create such a user-item map. This can be accomplished by solving a regularized optimization problem, but I will, again, omit the mathematical details.

Remark 16.2. One may notice that, in Figure 16.1 (I), Bob is much farther away from *Moby Dick* than Alice is, even though they both have given it the same rating of 90 (Table 16.1). This is because user- and item-effects have been removed prior to matrix factorization. Based on available ratings, Bob appears to be less critical than Alice is — in particular, Bob's average rating is $(90 + 70 + 10)/3 \approx 56.67$, whereas Alice's average rating is $(90 + 30 + 10)/3 \approx 43.33$. That's why a book has to be much closer to Alice for her to give it a high rating, but it doesn't have to be as close to Bob for him to give it an equally high rating. The same explanation applies to Cathy and David.

16.4   Matrix Completion

Lately, an even more abstract approach — the matrix completion approach — has attracted some attention as well. The idea is to fill in the missing

(I)

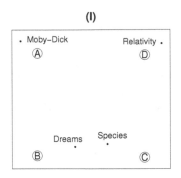

(II)

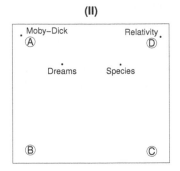

(III)

FIGURE 16.1: Illustrative example: Maps produced by different matrix factorization methods, after having removed user- and item-effects. A = Alice; B = Bob; C = Cathy; D = David. (I) A map of users and items, without incorporating any content information. (II) A map of users and items, incorporating content information by the shrinkage approach. (III) A map of users and items' content features, incorporating content information by the regression approach.

entries of R in such a way that the completed matrix — call it $\widehat{R}$ — is as low-rank as possible. Mathematically, this amounts to solving a constrained rank minimization problem (Candès and Recht, 2009). The rationale behind rank minimization is similar to that behind the matrix factorization approach: we believe that user-preferences are driven by only a few key factors; therefore, the rank of the rating matrix cannot be very high. Thus, the two approaches — matrix factorization and matrix completion — share a common philosophical underpinning, but they differ in that the matrix factorization approach is more explicit about the nature of the low-rankness. Rank minimization by itself is an NP-hard problem, essentially meaning that there is currently no way to guarantee an exact solution except when the size of the problem is very small. However, recent theoretical advances (e.g., Candès and Tao, 2005; Candès and Recht, 2009; Recht et al., 2010) have established that, under certain conditions, we can obtain the same solution by solving a much easier, convex optimization problem instead, replacing $\text{rank}(\widehat{R})$ with $\|\widehat{R}\|_*$, the so-called "nuclear norm" of $\widehat{R}$ (see Remark 16.3). The mathematical details here are very technical, and we will definitely stay away from them in this chapter.

Remark 16.3. For those with enough background to appreciate why the so-called ℓ_1-norm has been so important for high-dimensional problems in statistics (see, e.g., Tibshirani, 1996; Donoho, 2006), the nuclear norm of $\widehat{R}$ is defined as

$$\|\widehat{R}\|_* = \sum_k |\sigma_k|^1,$$

where $\sigma_1, \sigma_2, \ldots$ are the singular values of $\widehat{R}$. Recall that $\text{rank}(\widehat{R}) = \sum_k |\sigma_k|^0$. Hence, if we write $\mathbf{v} = (\sigma_1, \sigma_2, \ldots)^\top$ as the vector stacking all the singular values together, then $\text{rank}(\widehat{R})$ and $\|\widehat{R}\|_*$ are just the ℓ_0- and ℓ_1-norms of $\mathbf{v}$, respectively. Therefore, nuclear-norm minimization is to rank minimization what the lasso (Tibshirani, 1996) is to subset selection. For additional information about the lasso, see Chapter 5 by Rob Tibshirani.

16.5   Content-Boosted Matrix Factorization

Sometimes, we may have additional content information about the items. For example, Table 16.2 contains some features that can be used to describe the four books listed in Table 16.1. According to this table, Melville's *Moby Dick* shares three features in common with Freud's *The Interpretation of Dreams*, but only one with Darwin's *The Origin of Species*. Clearly, such information may explain why some users prefer certain items to others. In our illustrative example, for instance, the ratings are highly predictable from the items' content features — any user's ratings of any two items are always $20 \times (4 - Z)$ points apart if the two items share Z content features in common ($Z = 0, 1, 2$, or 3). In reality, the content features will not have such a strong bearing on the users' ratings, but they are still more likely than not to be at least partially informative. If so, they can (and should) be exploited to enhance the predictions of the recommender system. We have proposed two different ways (Forbes and Zhu, 2011; Nguyen and Zhu, 2013) to incorporate such content information into the matrix factorization approach, which we discussed two paragraphs ago. Our proposals are natural applications of the two statistical ideas that I mentioned at the beginning of this chapter.

For example, we can bias two items' coordinates to be close to each other if they share at least a certain number of common features (Figure 16.1, II).

TABLE 16.2: Illustrative example: Content information about the four books in Table 16.1. Italicized words are used as abbreviated descriptions of each feature in Figure 16.1, III.

	Moby Dick	*Dreams*	*Species*	*Relativity*
Themes of *conflict*	√	√	√	×
Elements of *moral* philosophy	√	√	×	×
Darkness of human nature	√	√	×	×
Other *empirical* evidence	×	×	√	√
Logical rigor	×	×	√	√
A grand new *theory*	×	√	√	√

Suppose that each item i is associated with a binary content vector a_i (e.g., a column in Table 16.2, taking "√" as 1 and "×" as 0). Then,

$$\mathcal{S}_c(i) = \{i' : \ a_i^\top a_{i'} \geq c\}$$

is the set of all items that share at least c common features with i. In the iterative procedure to estimate (p_u, q_i), this added bias amounts to *shrinking* the coordinates of each item (q_i) at every step toward the mean coordinates of those that share enough common features with it — that is, shrinking q_i toward

$$\sum_{i' \in \mathcal{S}_c(i)} \frac{q_{i'}}{|\mathcal{S}_c(i)|},$$

where $|\mathcal{S}_c(i)|$ denotes the size of the set $\mathcal{S}_c(i)$. We call this the shrinkage approach.

Alternatively, we can force an item's coordinates to depend on the item's content features by means of a *regression* relationship, i.e.,

$$q_i = Ba_i. \tag{16.2}$$

In a K-dimensional user-item map, each item i has K coordinates. For each item i, Equation (16.2) actually encodes K simultaneous regression relations,

$$q_i(k) = \sum_j B(k,j)a_i(j), \tag{16.3}$$

one for each coordinate k. Under the constraint (16.2), the problem becomes one of estimating (p_u, B) rather than (p_u, q_i). This can be accomplished by making a relatively small change to the iterative procedure for estimating (p_u, q_i). We call this the regression approach.

For our illustrative example (Tables 16.1–16.2), predictions made by these different matrix factorization approaches (all using $K = 2$) are given in Table 16.3. By directly comparing with Table 16.1, we can see that incorporating content information contained in Table 16.2, whether by shrinkage or by regression, has indeed led to more accurate predictions of the four "missing" entries (emboldened in Table 16.3).

TABLE 16.3: Illustrative example: Predictions made by different matrix factorization methods using $K = 2$. (I) Without incorporating any content information. (II) Incorporating content information by the shrinkage approach. (III) Incorporating content information by the regression approach.

		Moby Dick	*Dreams*	*Species*	*Relativity*
(I)	Alice	89	**48**	30	10
	Bob	90	70	**51**	11
	Cathy	11	**52**	70	90
	David	10	30	**49**	89
(II)	Alice	89	**63**	30	10
	Bob	90	70	**37**	11
	Cathy	11	**37**	70	90
	David	10	30	**63**	89
(III)	Alice	90	**75**	30	10
	Bob	90	70	**25**	10
	Cathy	10	**25**	70	90
	David	10	30	**75**	90

An interesting by-product of the regression approach is that each column of the matrix B — a vector in $\mathbb{R}^K$ — can be interpreted as the latent coordinates for each corresponding content feature. To see this, notice that Equation (16.3) can be interpreted as

(kth coordinate of item i) =

$$\sum_j (k\text{th coordinate of feature } j) \times \mathbf{1}(\text{item } i \text{ has feature } j),$$

where $\mathbf{1}(E)$ is an indicator function taking on the values of 1 or 0 depending on whether E is true or false. Therefore, not only can we create a user-item map to facilitate personalized recommendations, we can also put content features onto the same map (Figure 16.1, III), and gain fresh insight about the content features themselves. For example, using data from `http://allrecipes.com/` and treating ingredients as content features of recipes, we have found "mozzarella" and "firm tofu" to be similar ingredients, but "cottage cheese" and "Swiss cheese" to be dissimilar ones; using the "MovieLens 100K" data from `http://www.grouplens.org/` and treating genres as content features of movies, we have found that "action" movies are more similar to "science fiction" movies than to "war" movies, whereas "war" movies are closer to "animation" movies than to "action" movies (Nguyen and Zhu, 2013).

16.6   Discussion

We are currently contemplating how to incorporate content information into the matrix completion approach. As mentioned earlier, the matrix completion approach is based on the premise that the completed matrix should be low-rank, without being explicit about the nature of the low-rankness. The lack of an explicit parameterization makes the kind of extensions we have proposed to the matrix factorization approach elusive, and it appears to us that a different paradigm is needed altogether. There are certainly many opportunities for statisticians to make contributions.

Acknowledgments

The author's research is supported by the Natural Sciences and Engineering Research Council of Canada. Although the illustrative example (Tables 16.1–16.2) has been created by the author specifically for this chapter, the research being highlighted is based on joint work with his former undergraduate students: Peter Forbes, now a doctoral student at the University of Oxford, and Jennifer Nguyen, now a master's student at University College London. The author thanks the editor for inviting him to contribute this chapter to the special volume. He also thanks the editor, an associate editor, and two anonymous reviewers for their comments and suggestions.

About the Author

Mu Zhu is an associate professor of statistics at the University of Waterloo. A *Phi Beta Kappa* graduate of Harvard University, he obtained his PhD from Stanford University. His primary research interests are machine learning, multivariate analysis, and health care informatics. In 2012–13, he served as president of the Business and Industrial Statistics Section of the Statistical Society of Canada.

Bibliography

Candès, E. J. and Recht, B. (2009). Exact matrix completion via convex optimization. *Foundations of Computational Mathematics*, 9:717–772.

Candès, E. J. and Tao, T. (2005). Decoding by linear programming. *IEEE Transactions on Information Theory*, 51:4203–4215.

Donoho, D. L. (2006). For most large underdetermined systems of linear equations the minimal ℓ_1-norm solution is also the sparsest solution. *Communications on Pure and Applied Mathematics*, 59:797–829.

Efron, B. and Morris, C. (1977). Stein's paradox in statistics. *Scientific American*, 236:119–127.

Feuerverger, A., He, Y., and Khatri, S. (2012). Statistical significance of the Netflix challenge. *Statistical Science*, 27:202–231.

Forbes, P. and Zhu, M. (2011). Content-boosted matrix factorization for recommender systems: Experiments with recipe recommendation. In *Proceedings of the Fifth ACM Conference on Recommender Systems*, RecSys'11, pp. 261–264, New York.

James, W. and Stein, C. M. (1961). Estimation with quadratic loss. In *Proceedings of the Fourth Berkeley Symposium on Mathematical Statistics and Probability, Vol. I*, pp. 361–379. University California Press, Berkeley, CA.

Koren, Y. (2008). Factorization meets the neighborhood: A multifaceted collaborative filtering model. In *Proceedings of the 14th ACM SIGKDD International Conference on Knowledge Discovery and Data Mining*, KDD'08, pp. 426–434, ACM, New York.

Koren, Y., Bell, R., and Volinsky, C. (2009). Matrix factorization techniques for recommender systems. *Computer*, 42:30–37.

Nguyen, J. and Zhu, M. (2013). Content-boosted matrix factorization techniques for recommender systems. *Statistical Analysis and Data Mining*, 6:286–301.

Recht, B., Fazel, M., and Parrilo, P. A. (2010). Guaranteed minimum-rank solutions of linear matrix equations via nuclear norm minimization. *SIAM Review*, 52:471–501.

Tibshirani, R. J. (1996). Regression shrinkage and selection via the lasso. *Journal of the Royal Statistical Society, Series B*, 58:267–288.

17

What Do Salmon and Injection Drug Users Have in Common?

Laura L. E. Cowen

University of Victoria, Victoria, BC

Wendell O. Challenger, and Carl J. Schwarz

Simon Fraser University, Burnaby, BC

Many populations consist of elusive members that can make these populations difficult to study using standard sampling methods. These populations (e.g., salmon, large and small mammals, birds, injection drug users, sex workers) are impossible to census and most "regular" sampling techniques produce biased estimates (estimates that are, on average, too large or too small) of demographic processes (e.g., birth, death, immigration, and emigration) and underestimate the size of the population when detectability (elusiveness) is ignored. Capture-recapture has evolved to deal with this situation and highlights the pivotal role statistics can play in aiding general scientific understanding, of which many Canadian researchers are at the forefront. Capture-recapture studies are used to estimate the population size, determine if the population is growing, determine the rate of population change, estimate the rate of survival, determine if individuals move from one location to another, and/or determine the rate of movement between locations.

17.1 Methodology Background

The fundamental problem associated with elusive populations is that individuals are less than 100% detectable, which reduces our ability to fully enumerate members of the population or to fully resolve individual fates. For example, if we are interested in the number of injection drug users, we may attempt to enumerate the population at some point in time by a census of known locations with a large number of users. However, estimates based on direct counts will consistently underestimate the true population size because injection drug users face social stigma and lack trust in health researchers, often making them elusive to surveys.

To gain an accurate estimate of the total number injection drug users we need to inflate the number we observe by a factor proportional to the rate that individuals were missed. If we had an estimate of the detection rate (p), an estimator of the population size N at a single point in time would be

$$\hat{N} = n/\hat{p}, \qquad (17.1)$$

where n is the number of injection drug users we did observe and $\hat{p}$ is the estimate of the detection rate (the circumflex above p and N indicates these are estimates of the true detection rate p and the true population size N respectively). An estimated detection rate of $\hat{p} = .1$ would indicate we only observed 10% of the population during enumeration, therefore we can expect the true number of users would be roughly 10 times our observed count.

Estimates of demographic processes such as births, deaths, initiation (an individual begins injection drug use), cessation (an individual ends injection drug use), immigrations (a user that moved to a new city), and emigrations (a user moving from a city) will also have to be similarly adjusted. This can be of critical importance because accurate estimates of demographic processes are required to define policy for at-risk human populations or to manage wildlife populations.

At the heart of capture-recapture methodology is the explicit handling of imperfect detection. This is done through the application of "marks" (unique identifiers, typically referred to as "tags") and sampling protocols that attempt to recapture a subset of marked individuals. The use of uniquely marked individuals allows previously captured individuals to be identified in subsequent sampling occasions, which in turn allows for the proportion of missed individuals to be estimated. The exact method of how detection rates are estimated depends on the sampling protocol and modeling assumptions, but all capture-recapture models share this simple premise. Today, a wealth of tag types (Figure 17.1) and sampling protocols exist, supplemented by an equally rich set of modeling frameworks and associated software available to practitioners. In this exposition, we will only be scratching the surface of the type of research done, but we hope to demonstrate the real-world applicability of capture-recapture type studies.

17.2 Closed Population Models and Two-Sample Experiments

The simplest study design that allows for the estimation of the detection rate and of the population size is a two-sample experiment. Two samples are obtained from the population. Individuals captured on the first occasion are

FIGURE 17.1: A variety of tag types exist, from external tags that are read (a-c) to internal tags that transmit a unique code (d). All tags perform the same basic task of uniquely identifying individuals. Image Credits: Smithsonian Environmental Research Center (a), Mickey Samuni-Blank (b), The Pomacea Project, Inc. (c), Kintama Research Services Ltd (d).

marked and returned to the population. The second sample of the population again captures individuals, but now the sample contains a mixture of previously marked individuals and individuals captured for the first time. There are three potential fates of individuals after the first sample time (Figure 17.2): they are alive and are captured, they are alive and not captured, or they are dead.

Of all the possible fates, individuals that are alive and in the study area provide information about detection efficiency. In some circumstances, it is possible to choose the sampling occasions so that the assumption of population closure holds (closure implies that negligible births, deaths, immigration, or emigration events have occurred). Under closure, individuals identified in the first sample, but not the second, can be assumed to have been missed (Figure 17.2). Thus, under closure, the proportion of "recaptures" (observing the same individual again) provides a direct estimate of the detection rate p, thereby allowing for estimation of N via (17.1).

If n_1 represents the total individuals in the first sample and n_2 the total number caught in the second sample (with m_2 representing the number caught in both samples, or the number of marked-individuals captured at occasion 2) we can estimate the true population size N. Assuming that all individuals are equally catchable at each sampling occasion, the proportion recaptured on

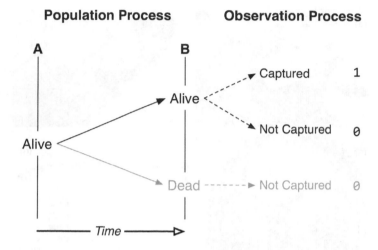

FIGURE 17.2: Possible fates in a two-sample study. Individuals captured and marked at the first time point may be alive and captured, alive and not captured or dead at the second sampling point. When sampling is done close in time, it can be assumed that no deaths occurred, removing death as a possible outcome (greyscale). Observed data for an individual is a "1" if captured and a "0" if not captured.

the second occasion should on average be the same as the proportion initially marked. In other words,

$$\frac{n_1}{N} = \frac{m_2}{n_2},$$

which can be rearranged to give

$$\hat{N} = \frac{n_1 n_2}{m_2}. \tag{17.2}$$

For example, Xu et al. (2013) had $n_1 = 254$ injection drug users in the first sample, $n_2 = 250$ in the second, and $m_2 = 19$ in both. Thus $\hat{N} = 254 \times 250/19 = 3342$ individuals. Using our earlier form (17.1), we would substitute m_2/n_1 for $\hat{p}$. While an estimate of N is our primary interest, estimating p is also important, despite often being referred to as a "nuisance" parameter. By accounting for detection, we are able to derive an estimate of N, which we would not be able to do with a simple count. While we have taken the first steps of accounting for detection, we must also be careful that p is an appropriate description of the detection process and that we did not make any mistakes reading tags, nor that individuals lose their tags between sampling occasions.

The described estimator is termed a Lincoln–Petersen type estimator and was originally developed to estimate the population of France in 1783 (Laplace, 1786). Today it is one of the most widely used two-sample capture-recapture estimators by wildlife ecologists and epidemiologists. Despite the simple study design, much of the current work involves relaxing or dealing with violations to various modeling assumptions. For example, the assumption that all individuals are equally catchable is often violated, which is referred to as heterogeneity in catchability. Normally, there is not sufficient information to detect or to adjust for heterogeneity in the simple two-sample experiment; however, for both injection drug users and migrating salmon, we can make use of additional information to understand and to adjust for heterogeneity. Chapter 18 by Louis-Paul Rivest and Sophie Baillargeon takes a detailed look at heterogeneity.

17.2.1 Victoria's Injection Drug Users

Xu et al. (2013) first "captured" individuals in Victoria, British Columbia, in 2003 through a national cross-sectional survey designed to track changes in the prevalence of HIV and hepatitis C. Participants were recruited in Victoria through AIDS Vancouver Island's needle exchange program and at shelter services run by the Victoria Cool Aid society. Posters, flyers, word of mouth, and contact with Vancouver Island Health Authority staff were other methods used for recruitment. Individuals were associated with a unique identifier — a combination of birth date and initials. A second survey was completed in 2005 and individuals whose identifiers matched those from the first sample were considered to be "recaptured." A standard Lincoln–Petersen approach resulted in estimates that were much higher than the accepted population size estimate of Stajduhar et al. (2004) obtained from the client load of AIDS Vancouver Island's needle exchange program.

Xu et al. (2013) also looked at two other methods to account for heterogeneity in capture probability by incorporating covariates (information from other variables) using methods of Huggins (1989); and by using (hidden) mixture models (Pledger, 2000) that assume the existence of groups of individuals with different catchabilities (but does not rely on additional covariate information). In the Huggins model, capture probabilities varied by the sex of an individual and whether or not an individual had been captured before, through the use of the function

$$\log\left(\frac{p_{ij}}{1 - p_{ij}}\right) = \beta_0 + \beta_1 \text{sex}_i + \beta_2 z_{ij},$$

where p_{ij} is the capture probability for individual i on occasion j, and z_{ij} is equal to one if individual i was captured before occasion j and zero otherwise. As the sex of unobserved individuals is unknown, one can consider only individuals captured at least once.

In Pledger's model, we assume that individuals belong to groups with different capture probabilities, but group membership is latent (or unobservable). Capture probabilities are allowed to vary by time, behavior and/or group. Here capture probability p_{jab} is the probability of capture for an individual i at occasion j who has behavior type b in group a. It is modeled as

$$\log\left(\frac{p_{jba}}{1 - p_{jab}}\right) = \mu + \tau_j + \beta_b + \eta_a + (\tau\beta)_{jb} + (\tau\eta)_{ja} + (\beta\eta)_{ba} + (\tau\beta\eta)_{jba},$$

where b equals one if individual i was not caught before occasion j and two otherwise, τ_j is the effect of time, β_b is the effect of behavior, and η_a is the effect of group a. Parameters are typically estimated using maximum likelihood methods.

Ultimately the three estimates from the different models were very similar (with confidence intervals of between 2246 and 5078 injection drug users) and all three differed drastically from the accepted population size estimate of 1500–2000 individuals (Stajduhar et al., 2004). The Vancouver Island Health Authority will use these new estimates to secure resources and plan harm reduction programs such as fixed-site needle exchanges to meet the health care needs of the injection drug user population. In turn, implementation of these services will help control the transmission of HIV and hepatitis C, which will benefit the population at large.

17.2.2   Rotary Screw Traps and the Time Stratified Petersen Estimator

Rotary screw traps have long been used as a method to sample salmonoid smolts during outward migration (Figure 17.3). These devices capture an unknown proportion of the migrating smolts as they move downstream. In one study design, fish are captured at an up-river trap, marked, and released. Some of the marked fish are recaptured at a second trap downstream giving an estimate of the capture efficiency (Schwarz and Dempson, 1994). For example, if 100 smolts are marked at the upstream trap and only 10 are recaptured at the downstream trap, then the downstream trap is estimated to have a 10% capture efficiency. Creative approaches using a single trap can also be constructed by capturing, then moving captured smolts upstream and releasing them, which allows for recaptures to occur at the single trap as the smolts resume their downstream migration; see, e.g., Macdonald and Smith (1980).

This outward migration occurs over long periods of time (i.e., weeks to months) necessitating multiple periods of time when smolts are captured, tagged, and released. The efficiency of the trap can vary dramatically across the study as water levels and other environmental variables (e.g., water temperature) change. This results in a violation of the assumption of homogenous catchability. Consequently, the simple approach of pooling all releases and recoveries over the entire duration of the study and using a simple Lincoln–

FIGURE 17.3: Rotary screw traps have long been used by biologists to capture juvenile salmonoids migrating downstream in medium- and large-sized streams. *Image Credit*: FISHBIO.

Petersen estimator (17.2) to estimate population size could result in a biased estimate (Arnason et al., 1996).

Temporal stratification (breaking up the study into sub-studies or groups over time) has been the standard approach used to deal with changing catchability. Strata (an individual sub-study or group) are constructed such that roughly uniform catchability can be expected (e.g., weekly strata), separate population size estimates are then derived for each stratum, and then estimates from each stratum are combined to derive an overall population size estimate. The stratified-Petersen estimator has a long history in terrestrial surveys (Schaefer, 1951; Chapman and Junge, 1956; Darroch, 1961), but suffered from the constraint that the number of release and recovery strata needed to be equal. Arnason et al. (1996) developed user-friendly software to implement this estimator that is still in use today and Plante et al. (1998) relaxed this constraint of equal strata.

This method has been used, for example, to study the problem of "five million missing salmon" that occurred in the Fraser River watershed in 1991. A discrepancy occurred between the estimate of the number of adult fish allowed to escape the fishery and the numbers adult spawners observed on the spawning ground. This discrepancy was thought to be due to a statistical artefact of naively applying the simple pooled-Petersen estimator when catchability differed for each stratum. Schwarz and Taylor (1998) applied a stratified-Petersen estimator to account for the heterogeneity in catchability, and indeed showed that the missing fish were just a statistical artefact due to not incorporating variable catchability into the estimator.

While the stratified-Petersen proved to be an important advancement for Canadian fishery science, it was not without limitations. Heterogeneity in both catchability and degree of mixing can be extreme, requiring large numbers of strata to adequately deal with the changes in heterogeneity. A large number of strata often results in poor precision for the estimate of sample size due to small stratum sample sizes combined with a large number of model parameters that must be estimated. In doing so an unfortunate trade-off between bias and precision is created.

One way of avoiding this trade-off is to add additional structure to the models that can reduce the effective number of parameters used to describe the problem (e.g., adjacent strata should have similar parameter values). One approach used information about smolt travel times to provide information about catchability (Schwarz and Dempson, 1994); another approach used environmental factors to model recapture probabilities (Plante et al., 1998). A more modern approach (Bonner and Schwarz, 2011) assumed the number of smolts passing the traps over time can be expected to occur in a smooth and predictable manner (e.g., increasing over time to a peak, followed by a decrease) and used splines to model this process. They also used Bayesian methods to share information from neighboring strata about the capture rates in a particular stratum.

Furthermore the use of smooth functions (splines) has additional advantages such as the ability to provide abundance estimates for strata that may be missing due to crew illness, high water flows, or other causes. The approach of Bonner and Schwarz (2011) also removes the need to make arbitrary decisions when defining strata, as the model is self-selecting so that with poor data, more extensive sharing of information among strata takes place, while with rich data, very little sharing of information among strata takes place. A user-friendly R package (BTSPAS) was also developed (Bonner and Schwarz, 2012) and is in use extensively on the west coast of North America.

17.2.3   Extending the Closed Population Models

The two-sample experiment can be generalized with additional sampling occasions and by allowing for known subgroups (e.g., sex) in the population. The catchability of individuals can now vary over time, because of behavior, or because of intrinsic heterogeneity. For example, small rodents may become "trap-happy" (they like the bait) or "trap-shy" (they dislike being handled) after being captured, thus subsequent capture probabilities may be affected. Trap stations may be dispersed over a geographical grid and because of home range limitations, animals captured in the northwest section of a trapping grid cannot be captured in the southeast portion of a trapping grid. Larger animals may be more visible than smaller animals and more difficult to recapture. While the closure assumption is still paramount, the additional information in larger experiments allows for quite complex modeling to take place to deal with a variety of these other complications.

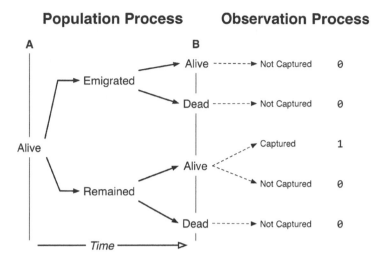

FIGURE 17.4: Open population models need to consider more fates than closed population models. Additional fates include: death or emigration and immigration or births (not shown). Observed data is a "1" if captured and a "0" if not captured.

17.3   Open Populations

In most populations of interest, some amount of birth, death, immigration or emigration is likely occurring between sampling occasions and these populations are considered "open." In the case of Cassin's Auklets on Triangle Island, British Columbia, Bertram et al. (2000) studied annual survival rates between 1994 and 1997. With a multi-year open population study there are a number of possible fates that must be considered (Figure 17.4). Because survival rate (rather than population size) was of primary interest, Bertram et al. (2000) used the Cormack–Jolly–Seber (CJS) model (Cormack, 1964; Jolly, 1965; Seber, 1965) and considered both age of the bird and capture (netting) location in their model, thus controlling for some amount of capture heterogeneity.

The data obtained in these studies are typically a sequence of digits (termed a capture history) that provide a concise summary of whether or not an individual was encountered during the course of the study. A "1" indicates an individual was observed, and a "0" indicates the individual was unobserved. For example, the capture history {1010} indicates a four sample experiment, where the individual was observed (alive) on the first and third sample times. The zero observed at time 2 indicates that the individual was not detected, but is known to be alive due to the subsequent detection. The missed detection may have resulted from the individual not being present in the study area during sampling (due to temporary emigration) or the result of individual detectability being less than 100% (i.e., the individual was not captured). The missed detection provides information on the detection rate at sample time 2, assuming emigration did not happen. The final zero at sample time 4 is less informative — now a third possibility (death) must be considered, in addition to imperfect detection or emigration as the other possibilities (Figure 17.5).

The key to the analysis of such a capture-recapture dataset is to construct a probability for each possible capture history. For example, a probability expression for the history {1010} takes the form

$$\phi_1(1 - p_2)\phi_2 p_3\{1 - \phi_3 + \phi_3(1 - p_4)\},$$

where ϕ_t represents the probability of surviving between times t and $t + 1$, and p_t is the probability of capture at time t. Notice that after the third sample time, the last term in the expression represents the probability of death, $1 - \phi_3$, or survival but no detection on the last occasion, $\{\phi_3(1 - p_4)\}$. These probability statements are produced for each individual under study and are combined to create a model whereby survival can be estimated (using maximum likelihood methods for example). Biological hypotheses can be investigated by looking at several plausible models. For example, Bertram et al. (2000) studied models where capture rates varied over time, or netting location, or both.

For more complicated studies, the "digits" in the capture-history can be generalized to record other information, such as the breeding state and location of the individual (e.g., "A^b" could indicate the individual was observed breeding (b) on site A).

If population size is of interest, then the Jolly–Seber model (Jolly, 1965; Seber, 1965) is typically employed where entrance to the population must also be modeled. Schwarz and Seber (1999) provide a thorough review of this model.

17.3.1   Novel Applications — Pacific Ocean Shelf Tracking

Open-population models are arguably the most popular approach and can be applied to a diverse array of situations. While much of our discussion of capture-recapture has focused on physical captures, many studies relax this constraint by using a variety of methods to "capture" individuals. For example, once a tag is applied (requiring capture), subsequent recaptures can consist of non-invasive approaches such as resightings (e.g., reading bird bands from a distance) or recoveries (e.g., through hunting or fishery harvest). Furthermore, physical capture can be avoided in some circumstances by relying on naturally occurring markings (e.g., whale fluke patterns), camera traps, or DNA sampling (e.g., using hair or feces) to sample individuals. Other unique implementations can be creative constructions of what constitutes a "capture

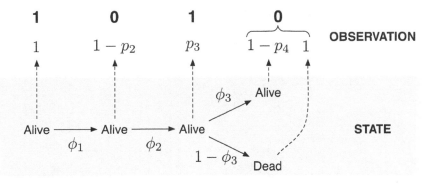

$$P(1010) = \phi_1 \left(1 - p_2\right) \phi_2 p_3 \left\{1 - \phi_3 + \phi_3 \left(1 - p_4\right)\right\}$$

FIGURE 17.5: Fate diagram and corresponding capture history probability expression for the capture history $\{1010\}$ under an open population scenario. The expression is conditioned on first capture (the probability of observing a "1" is 1), with survival (ϕ) and capture (p) probability parameters used to describe the possible states. Living individuals may be captured with probability p, or missed with probability $1 - p$, while deceased are missed ("0") with probability 1.

occasion." The use of electronic tagging in the study of migrating marine animals, such as salmon, is a good example of this type of approach.

Traditional non-electronic tags have been used to tag fish or other marine organisms; however, these provide only crude estimates of animal movements and often require intensive sampling to recover a small proportion of tags, resulting in low precision estimates of model parameters. Electronic tags now exist that are small enough to implant into juvenile salmon (Figure 17.1d), providing rich datasets from a much smaller sample size. These tags carry a power source allowing them to actively transmit information for extended periods of time. Transmissions are in the form of a coded signal, which contain a unique identifier, allowing individuals to be identified as they come within proximity of specialized passive receivers. Acoustic tags transmit sound waves, which can be detected across salt and freshwater; while radio tags may only be detected in freshwater due to saltwater signal attenuation.

Similar to the rotary screw trap studies, capture histories can be constructed by setting up multiple listening lines of receivers along known migration routes (Figure 17.6). Migration pathways of many anadromous fish species, such as Pacific salmon, are well known and occur in a directed manner. As such, tagged fish can be expected to transverse the listening lines in a predictable manner, creating well defined capture histories, with intervals

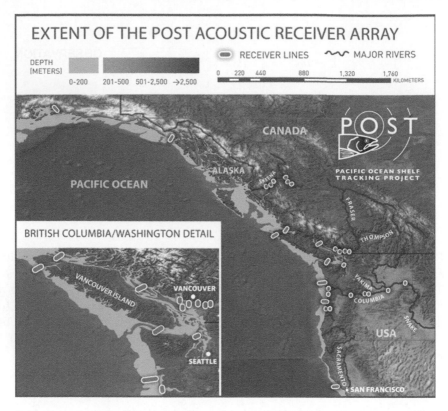

FIGURE 17.6: The POST project deployed acoustic listening lines along the continental shelf and along several major rivers, spanning from California to Alaska. This allowed acoustically tagged salmon, such as Fraser River sockeye, to be tracked during parts of the oceanic and river phases of migration. Reproduced from Jackson (2011).

between "capture" occasions representing physical segments, rather than time periods.

A large-scale research program that employed this type of approach was the Pacific Ocean Shelf Tracking (POST) program. A field program of the decade-long Census of Marine Life, from 2003 through to 2012, the POST program deployed a series of acoustic listening lines along the continental shelf, and in select rivers of the West Coast of North America. The POST array spans nearly 3,000 km from Alaska, through British Columbia, to California and can track migrating salmon through the river and along the ocean shelf. While the receiver lines can detect a variety of species — 18 species as of 2011 (Jackson, 2011) — much of the published work has focused on Steelhead trout and Pacific salmon such as chinook and sockeye; see, e.g., Crossin et al. (2007),

Welch et al. (2003), and Welch et al. (2009). Studies combining the POST array, as well as radio tagging studies restricted to the Fraser River, have been instrumental in furthering our understanding of some factors impacting Pacific salmon. An example of this has been the insights gained into the mortality problem that has recently emerged for late-run Fraser River sockeye.

Fraser River sockeye are divided into roughly four management groups based on the timing of their spawning migrations. Late-run sockeye typically enter the Fraser River last, at the end of summer, after spending a period of time (historically six weeks) holding, off the mouth of the Fraser River. Sometime around 1995 this holding period was drastically reduced and stocks within this timing group began experiencing severe spawning mortality problems somewhere within the Fraser River (Cooke et al., 2004). For some late-run stocks the mortality was as high as 90% (Hinch et al., 2012). The changes in river migration timing were also unprecedented, with holding periods reduced by three weeks and even eliminated in some years. Due to the sampling techniques used (abundance estimates at river entry and at the spawning grounds), the exact timing and location of the in-river mortality was not known.

Driven by concern over these events, and the inability of more standard sampling approaches to provide the necessary information, a series of large-scale capture-recapture telemetry experiments using both acoustic (POST) and radio technologies were carried out. POST listening lines were used to allow researchers to track the movement and timing of individually tagged sockeye from as far away as the Gulf of Alaska to just off the spawning ground; see, e.g., Crossin et al. (2009). Due to the sequential nature of the migration route, studies could use standard capture-recapture approaches to analyze the data.

This finer scaled approach allowed for much clearer understanding of the major mortality mechanisms. Using sockeye radio-tagged in the ocean it was found that the date of river entry strongly predicted subsequent in-river survival (English et al., 2005). After removing fishery-related mortality it was also found that this mortality followed a smooth function related to date of entry, with earlier entrants exhibiting the highest levels of in-river mortality. Sockeye that displayed little to no ocean holding prior to entering the river had virtually no prospects of arriving on the spawning grounds. Temperature exposure (and subsequent freshwater disease expression) was thought to be driving this mortality. This suspicion was confirmed by the discovery that fish held in warm water incurred much higher mortality levels during the subsequent river migration than fish held in cool water (Crossin et al., 2008). The mechanism was also further established in follow-up observational studies; see, e.g., Mathes et al. (2010). Interestingly, some of the early entrants that do survive to the spawning ground appear to hold in the coolest portions of their natal lakes after migrating up the river (Mathes et al., 2010; Hinch et al., 2012).

Furthermore, by combining telemetry with physiological sampling at the time of capture, studies have shown that sockeye may already be compromised prior to river entry (Cooke et al., 2006; Miller et al., 2011) and that early-entry behavior may be triggered by the physiological and energetic status of the fish (Cooke et al., 2006). All of these scientific insights relied on being able to track individual movements and subsequent survival. None of this would be possible without capture-recapture methodology.

17.4   Additional Complications

As we have seen, complications to standard mark-recapture studies typically arise when assumptions of the models are violated. We discussed some possible sources of heterogeneity affecting capture probabilities, but there are many other sources of heterogeneity that can be incorporated. Capture rates might vary with individual characteristics that differ each sample time such as body mass or wing length. Bonner and Schwarz (2006) showed how this information can be incorporated for meadow voles whose body mass was measured on each sampling occasion. The difficulty here is that when a vole was not captured, the value of body mass for that sample time was missing.

Similarly, individuals in different locations or breeding states might have different catchabilities. In this context, capture-recapture can also be used to determine patterns of migration through the use of a multistate model (Arnason, 1972, 1973; Schwarz et al., 1993). Cowen et al. (2009) incorporated the location of an individual into a multistate model to estimate the fishery exploitation rates of yellowtail flounder on the Grand Bank of Newfoundland.

Another important capture-recapture assumption is that individuals behave independently of one another. This is often violated in waterfowl that form pair-bonds and share fates. Challenger (2010) considered the problem in a Harlequin duck population in Hinton, Alberta, and was able to estimate the degree to which one individual was associated with its pair in terms of survival and catchability.

Finally, tag loss is assumed to be negligible in standard capture-recapture studies. When individuals are double-tagged, tag loss can be estimated and accounted for in the models; see Cowen and Schwarz (2006), Cowen et al. (2009) and Xu et al. (2014). Cowen and Schwarz (2006) studied walleye in Mille Lacs, Minnesota, that had been double-tagged and found tag loss rates changed over time, which in turn affected population size estimates from standard methods. Tag loss in radio-telemetry studies comes in the form of battery failure. Cowen and Schwarz (2005) studied chinook smolts traveling down the Columbia River, WA and were able to account for battery failure by incorporating information about battery lifetimes into the model, thus producing more accurate survival estimates for the smolts.

17.5   Concluding Remarks

Capture-recapture experiments continue to provide new problems for Canadian researchers. This is particularly true with the arrival of new technologies for monitoring animals that promise to provide information about animal locations and interactions in real time. With these new technologies, methodological research will be shifting from problems with a sparsity of data to an overabundance of data, often collected on very fine time (daily) and large spatial (over continents) scales, or collected from different sources (e.g., radio collars; DNA) that will have to be integrated together. These data will require development of models with many thousands of parameters, all of which are highly interrelated.

Capture-recapture methodological development has quickly been integrated into mainstream ecological fields. A short illustrative list of the application of capture-recapture methods to Canadian wildlife populations is: Cassin's and Rhinoceros Auklet survival (Bertram et al., 2000), humpback whales abundance (Smith et al., 1999), Black Brant site fidelity Reed et al. (1998), Norway rat eradication (Drever, 1997), population change in Ancient Murrelets (Gaston and Descamps, 2011), and grizzly bear abundance (Mowat and Strobeck, 2000).

About the Authors

Laura L. E. Cowen is an associate professor of statistics at the University of Victoria. She received an MMath from the University of Waterloo and a PhD in statistics from Simon Fraser University. She is a statistical ecologist with a primary research focus in capture-recapture methodology. Her earlier background in biological science has led to collaborations with seabird ecologists, fisheries scientists, microbiologists, anthropologists, and ophthalmologists.

Wendell O. Challenger was formerly a biologist before focusing on statistics for his PhD. Now a statistical ecologist, his primary research interests are on capture-recapture methodology with an emphasis on accommodating implementation issues associated with field studies. He has recently completed a postdoctoral fellowship working on the acoustic tracking of salmon.

Carl J. Schwarz is a professor of statistics at Simon Fraser University. He graduated from the University of Manitoba in 1985 and joined SFU in 1994. He specializes in statistical methods for ecology, notably capture-recapture and multi-list techniques for estimating abundance, survival, movement and other demographic parameters of wildlife populations. He is a former editor of the *Journal of Agricultural, Biological, and Environmental Statistics.*

Bibliography

Arnason, A. N. (1972). Parameter estimates from mark-recapture experiments on two populations subject to migration and death. *Researches on Population Ecology,* 13:97–113.

Arnason, A. N. (1973). The estimation of population size, migration rates and survival in a stratified population. *Researches on Population Ecology,* 15:1–8.

Arnason, A. N., Kirby, C. W., Schwarz, C. J., and Irvine, J. R. (1996). *Computer Analysis of Marking Data from Stratified Populations for Estimation of Salmonid Escapements and the Size of Other Populations.* Canadian Technical Report of Fisheries and Aquatic Sciences no 2106.

Bertram, D. F., Jones, I. L., Cooch, E. G., Knechtel, H. A., and Cooke, F. (2000). Survival rates of Cassin's and Rhinoceros Auklets at Triangle Island, British Columbia. *The Condor,* 102:155–162.

Bonner, S. J. and Schwarz, C. J. (2006). An extension of the Cormack–Jolly–Seber model for continuous covariates with application to *Microtus pennsylvanicus. Biometrics,* 66:1256–1265.

Bonner, S. J. and Schwarz, C. J. (2011). Smoothing population size estimates for time-stratified mark-recapture experiments using Bayesian p-splines. *Biometrics,* 67:1498–1507.

Bonner, S. J. and Schwarz, C. J. (2012). *BTSPAS: Bayesian Time Stratified Petersen Analysis System.* R Package Version 2012.0219.

Challenger, W. O. (2010). *Modeling Uncertainty and Heterogeneity in Mark-Recapture and Occupancy Experiments.* Doctoral dissertation, Simon Fraser University, Burnaby, BC.

Chapman, D. G. and Junge, C. O. (1956). The estimation of the size of a stratified animal population. *The Annals of Mathematical Statistics,* 27:375–389.

Cooke, S. J., Hinch, S. G., Crossin, G. T., Patterson, D. A., English, K. K., Shrimpton, J. M., Kraak, G. V. D., and Farrell, A. P. (2006). Physiology of individual late-run Fraser River sockeye salmon (*Oncorhynchus nerka*) sampled in the ocean correlates with fate during spawning migration. *Canadian Journal of Fisheries and Aquatic Sciences,* 63:1469–1480.

Cooke, S. J., Hinch, S. G., Farrell, A. P., Lapointe, M. F., Jones, S. R. M., Macdonald, J. S., Patterson, D., Healey, M. C., and Van Der Kraak, G. (2004). Abnormal migration timing and high en route mortality of sockeye salmon in the Fraser River, British Columbia. *Fisheries,* 29:22–33.

Cormack, R. M. (1964). Estimates of survival from the sighting of marked animals. *Biometrika*, 51:429–438.

Cowen, L. L. E. and Schwarz, C. J. (2005). Capture-recapture studies using radio telemetry with premature radio-tag failure. *Biometrics*, 61:657–664.

Cowen, L. L. E. and Schwarz, C. J. (2006). The Jolly–Seber model with tag loss. *Biometrics*, 62:699–705.

Cowen, L. L. E., Walsh, S. J., Schwarz, C. J., Cadigan, N., and Morgan, J. (2009). Estimating exploitation rates of migrating yellowtail flounder (*Limanda ferruginea*) using multistate mark-recapture methods incorporating tag loss and variable reporting rates. *Canadian Journal of Fisheries and Aquatic Sciences*, 66:1245–1256.

Crossin, G. T., Hinch, S. G., Cooke, S. J., Welch, D. W., Batten, S. D., Patterson, D. A., Kraak, G., Shrimpton, J. M., and Farrell, A. P. (2007). Behaviour and physiology of sockeye salmon homing through coastal waters to a natal river. *Marine Biology*, 152:905–918.

Crossin, G. T., Hinch, S. G., Cooke, S. J., Welch, D. W., Patterson, D. A., Jones, S. R. M., Lotto, A. G., Leggatt, R. A., Mathes, M. T., Shrimpton, J. M., Van Der Kraak, G., and Farrell, A. P. (2008). Exposure to high temperature influences the behaviour, physiology, and survival of sockeye salmon during spawning migration. *Canadian Journal of Fisheries and Aquatic Sciences*, 86:127–140.

Crossin, G. T., Hinch, S. G., Welch, D. W., Cooke, S. J., Patterson, D. A., Hills, J. A., Zohar, Y., Klenke, U., Jacobs, M. C., Pon, L. B., Winchell, P. M., and Farrell, A. P. (2009). Physiological profiles of sockeye salmon in the Northeastern Pacific Ocean and the effects of exogenous GnRH and testosterone on rates of homeward migration. *Marine and Freshwater Behaviour and Physiology*, 42:89–108.

Darroch, J. N. (1961). The two-sample capture-recapture census when tagging and sampling are stratified. *Biometrika*, 48:241–260.

Drever, M. C. (1997). *Ecology and Eradication of Norway Rats on Langara Island, Queen Charlotte Islands*. Doctoral dissertation, Simon Fraser University, Burnaby, BC.

English, K. K., Koski, W. R., Sliwinski, C., Blakley, A., Cass, A., and Woodey, J. C. (2005). Migration timing and river survival of late-run Fraser River sockeye salmon estimated using radiotelemetry techniques. *Transactions of the American Fisheries Society*, 134:1342–1365.

Gaston, A. J. and Descamps, S. (2011). Population change in a marine bird colony is driven by changes in recruitment. *Avian Conservation and Ecology*, 6.

Hinch, S. G., Cooke, S. J., Farrell, A. P., Miller, K. M., Lapointe, M., and Patterson, D. A. (2012). Dead fish swimming: A review of early migration and high premature mortality in adult Fraser River sockeye salmon *Oncorhynchus nerka*. *Journal of Fish Biology*, 81:576–599.

Huggins, R. M. (1989).   On the statistical analysis of capture experiments. *Biometrika*, 76:133–140.

Jackson, G. D. (2011).   The development of the Pacific Ocean Shelf Tracking Project within the decade long census of marine life. *PLoS ONE*, 6:e18999. doi: 10.1371/journal.pone.0018999.g006.

Jolly, G. M. (1965). Explicit estimates from capture-recapture data with both death and immigration-stochastic model. *Biometrika*, 52:225–247.

Laplace, P. S. (1786). Sur les naissances, les mariages et les morts. *Histoire de l'Académie Royale des Sciences, Année 1783*, 693.

Macdonald, P. D. M. and Smith, H. D. (1980). Mark-recapture estimation of salmon smolt runs. *Biometrics*, 36:401–417.

Mathes, M. T., Hinch, S. G., Cooke, S. J., Crossin, G. T., Patterson, D. A., Lotto, A. G., and Farrell, A. P. (2010). Effect of water temperature, timing, physiological condition, and lake thermal refugia on migrating adult Weaver Creek sockeye salmon (*Oncorhynchus nerka*). *Canadian Journal of Fisheries and Aquatic Sciences*, 67:70–84.

Miller, K. M., Li, S., Kaukinen, K. H., Ginther, N., Hammill, E., Curtis, J. M. R., Patterson, D. A., Sierocinski, T., Donnison, L., Pavlidis, P., Hinch, S. G., Hruska, K. A., Cooke, S. J., English, K. K., and Farrell, A. P. (2011). Genomic signatures predict migration and spawning failure in wild Canadian salmon. *Science*, 331:214–217.

Mowat, G. and Strobeck, C. (2000). Estimating population size of grizzly bears using hair capture, DNA profiling, and mark-recapture analysis. *Journal of Wildlife Management*, 64:183–193.

Plante, N., Rivest, L.-P., and Tremblay, G. (1998). Stratified capture-recapture estimation of the size of a closed population. *Biometrics*, 54:47–60.

Pledger, S. (2000).   Unified maximum likelihood estimates for closed capture-recapture models using mixtures. *Biometrics*, 56:434–442.

Reed, E. T., Cooch, E. G., Goudie, R. I., and Cooke, F. (1998). Site fidelity of Black Brant wintering and spring staging in the Strait of Georgia, British Columbia. *The Condor*, 100:426–437.

Schaefer, M. B. (1951). Estimation of size of animal populations by marking experiments. *United States Fish and Wildlife Service, Fishery Bulletin*, 69:190–203.

Schwarz, C. J. and Dempson, J. B. (1994). Mark-recapture estimation of a salmon smolt population. *Biometrics*, 50:98–108.

Schwarz, C. J., Schweigert, J. F., and Arnason, A. N. (1993). Migration rates using tag-recovery data. *Biometrics*, 49:177–193.

Schwarz, C. J. and Seber, G. A. F. (1999). Estimating animal abundance: Review III. *Statistical Science*, 14:427–456.

Schwarz, C. J. and Taylor, C. G. (1998). Use of the stratified-Petersen estimator in fisheries management: Estimating the number of pink salmon (*Oncorhynchus gorbuscha*) spawners in the Fraser River. *Canadian Journal of Fisheries and Aquatic Sciences*, 55:281–296.

Seber, G. A. F. (1965). A note on the multiple recapture census. *Biometrika*, 52:249–259.

Smith, T. D., Allen, J., Clapham, P. J., Hammond, P. S., Katona, S., Larsen, F., Lien, J., Mattila, D., Palsboll, P. J., Sigurjonsson, J., Stevick, P. T., and Oien, N. (1999). An ocean-basin-wide mark-recapture study of the North Atlantic humpback whale (*Megaptera novaeangliae*). *Marine Mammal Science*, 15:1–32.

Stajduhar, K., Poffenroth, L., Wong, E., Archibald, C., Sutherland, D., and Rekart, M. (2004). Missed opportunities: Injection drug use and HIV/AIDS in Victoria, Canada. *International Journal of Drug Policy*, 15:171–181.

Welch, D. W., Boehlert, G. W., and Ward, B. R. (2003). POST: The Pacific Ocean Salmon Tracking Project. *Oceanologica Acta*, 25:243–253.

Welch, D. W., Melnychuk, M. C., Rechisky, E. R., Porter, A. D., Jacobs, M. C., Ladouceur, A., McKinley, R. S., and Jackson, G. D. (2009). Freshwater and marine migration and survival of endangered Cultus Lake sockeye salmon (*Oncorhynchus nerka*) smolts using POST, a large-scale acoustic telemetry array. *Canadian Journal of Fisheries and Aquatic Sciences*, 66:736–750.

Xu, Y., Cowen, L. L. E., and Gardner, C. (2014). Group heterogeneity in the Jolly–Seber-Tag-Loss model. *Statistical Methodology*, 17:3–16.

Xu, Y., Fyfe, M., Walker, L., and Cowen, L. L. E. (2013). Estimating the number of injection drug users in Greater Victoria, BC, using capture-recapture methods. *Harm Reduction Journal*, in revision.

18

Capture–Recapture Methods for Estimating the Size of a Population: Dealing with Variable Capture Probabilities

Louis-Paul Rivest and Sophie Baillargeon
Université Laval, Québec, QC

18.1 Estimating Abundance with Marked Units

Demographers predict that the Canadian population will reach the 35 million mark in 2013; this represents a 4% increase over a 5-year period. This is a projection of the results of the 2011 census that shows that the Canadian population is growing at a steady pace. In a similar way, the size of an animal population is a basic demographic characteristic that reflects its general well-being. Consider, for instance, the George River caribou herd in Northern Québec and Labrador. Thirty years ago, it was among the most expansive groupings of large mammals on the planet, with more than 600,000 animals. Now the herd has dwindled and is monitored closely. As this is a migratory animal, estimating the herd's size is a challenge. The current method uses animals marked with GPS collars to locate and photograph the large groups that form during the short nordic summer. A complete enumeration is not possible and final estimates are calculated by "entering the number of caribou found by biologists into an algorithm that determines the herd's size" as explained in a recent article in a Montréal newspaper. The statistical model behind this algorithm is detailed in Rivest et al. (1998) and several ways to account for missed animals are available. The estimate for the summer 2012 inventory is 27,600 caribou and protective measures, such as reducing hunting permits, have been taken.

The caribou inventory provides an example of an ingenious method developed by statisticians and biologists for estimating the size of a wildlife population using marked animals. According to Lecren (1965), marked animals have been used in ecology since the start of the 20th century. This consists of tagging animals with a unique identifier that allows for identification on subsequent captures or, more generally, encounters. Nowadays, advances in molecular biology broaden the scope of applications of capture-recapture techniques as they permit capturing and identifying animals through DNA collected in feces or hair snags. One distinguishes investigations of open populations, whose main objective is the estimation of survival rates, from those of closed populations that focus on population sizes; see Williams et al. (2002). This work is concerned with the latter; it assumes that the number of deaths and of births is negligible during the survey period and that the size of the population is, for all practical purposes, constant. Capture-recapture is the main method for estimating the abundance of animals not easily visible on the ground, such as the harvest mouse population discussed in this work. The previous chapter, by Cowen, Challenger, and Schwarz, describes some other applications of capture-recapture techniques.

Let N be the unknown size of the population of interest. Its estimation by capture-recapture typically involves the capture and the recapture of animals over a short period. The study discussed in the next sections has $t = 14$ weekly capture occasions. At the start, a grid of live traps is laid out in the population habitat. All the traps are visited on each capture occasion. When an animal is captured for the first time, it is marked with a unique identifier and released. Thus, when it is recaptured, an animal bearing a mark can be identified and linked with its previous captures. At the end of the study, each animal has its own capture history; it consists of 14 entries, equal to either 0 or 1, where a 1 in position i means that the animal has been captured at occasion i. Covariates, such as sex and body weight, are sometimes recorded as they might be related to the number of times that an animal is captured. The probability of being captured is often the same for all capture occasions. In such a case, the complete capture history is not useful, only the total number of captures is reported. The dataset is then given by $\{(c_j, x_j) : j = 1, \ldots, n\}$ where n is the number of animals caught at least once, c_j is the number of captures for animal j, and x_j represents covariates measured on that animal. Animals with the capture history $(0, \ldots, 0)$ are never caught and their number is unknown; the goal of the analysis is to estimate that number.

In ecology, species richness can be estimated using standard capture-recapture techniques for closed populations. In this context, a unit is a species and a capture occasion is an observation station. A unit is observed at a station if it is seen or heard at that station; the goal is to estimate the total number of species in an area, including those that went undetected. In this example, the species detected at an observation station are a partial or incomplete list of all the species in the area. Viewing a capture occasion as an incomplete list of units makes capture-recapture techniques also applicable to elusive human populations. For instance, the total number of illegal drug users in a city can be estimated through partial lists of users provided by police stations, drug treatment services and community centers. In a similar way, partial lists of patients coming from hospital records, drug prescriptions and doctors' offices provide the basis for estimating the number of people suffering from a particular disease in an administrative region. When applied to administrative records, capture-recapture methods are called multiple system estimation; linking the records from various lists together in order to create capture histories raises interesting statistical issues. Nowadays the wide availability of administrative data broadens the application areas of capture-recapture techniques. For example, multiple system estimation is used extensively by the Human Rights Data Analysis Group to evaluate human rights violations, such as killings, in troubled areas; see Harrison (2012) for a discussion of the methodological challenges involved.

Capture-recapture also applies in investigations without clearly defined capture occasions. A famous example, due to McKendrick (1926), attempts to estimate the number of households affected by cholera in an Indian village from a list containing the name and the address of persons suffering from cholera. The addresses identify households and the list allows to determine the number of members affected, or "captured," by cholera in each one. In some disease free households the cholera might be latent and appear later; the goal is to estimate the number of such households. In this example a household is captured when at least one of its member is diagnosed with cholera. There are no well defined capture occasions and "continuous captures" describes such a dataset.

Capture-recapture models attempt to extrapolate, to the whole population, observations made on captured units. This works as long as the units captured are representative of the whole population. However, consider a simple situation where the population consists of two groups, say 1 and 2. If the units in Group 1 are much easier to capture than those in Group 2, then capture-recapture estimates are derived mostly from Group 1. They do not apply to the whole population unless the units caught in Group 2 are given more weight to account for their underrepresentation. Indeed if the "unit level covariate" group could be recorded, then the solution would be to estimate the size of the two groups separately. When some units are easier to catch than others, the capture probabilities are said to be variable or heterogeneous. This is quite common. When studying animals, the size of the animal's home range and its weight might be positively associated with its capture probability while the detection of a species depends on the importance of its habitat in the study area. In epidemiological applications, the severity of an illness is positively correlated with the probability of appearing in a partial list of patients. Such heterogeneity in capture probability makes the estimation of N difficult as the data might not be representative of the whole population. The solution is to measure "unit level covariates," such as the group identifier alluded to earlier, and to use them to model the capture probabilities before estimating N. Even when this is done, the covariates might fail to explain all of the heterogeneity and the estimate of N might be biased. The goal of this chapter is to review the statistical methods available to detect heterogeneity in capture probabilities and to account for it when estimating N. These methods are presented through the analysis of two datasets that are introduced in the next section.

18.2   Datasets

The first dataset considered here has been obtained from a study of the harvest mouse (*Micromys minutus*) conducted at the Wulin Recreation Area in Shei-Pa National Park, Taiwan, in the summer of 2008. Such small mammal populations are investigated in ecology since they give handy indicators of the impact of human interventions, such as wood cutting and reforestation, on wildlife habitats. The data are presented in Stoklosa and Huggins (2012) and Stoklosa et al. (2011); there are $t = 14$ capture occasions. A total of $n = 142$ different mice have been captured over the 14 occasions. The number of captures per mouse ranged from 1 to 10, and the body weight of a mouse was found to be related to the capture probabilities. Other covariates such as sex and hindfoot measurement were available; however, they were unrelated to the capture probabilities and are not considered here. This dataset is available in the R package PL.popN (Stoklosa, 2012) in the form $\{(c_j, x_j) : j = 1, \ldots, n\}$, where $c_j \geq 1$ is the number of times that mouse j is captured and x_j is the weight of mouse j. Leaving the weight variable aside, the data can be presented in aggregated form as $f_1 = 68$, $f_2 = 29$, $f_3 = 16$, $f_4 = 8$, $f_5 = 10$, $f_6 = f_7 = 4$, $f_8 = 2$, $f_{10} = 1$, and $f_9 = f_{11} = f_{12} = f_{13} = f_{14} = 0$, where f_i is number of animals captured i times, that is the number of occurrences of $c_j = i$ among the 142 mice that have been captured.

The second dataset is an instance of continuous captures. It concerns illegal immigrants in four large cities in the Netherlands over a one-year period. These immigrants cannot be effectively expelled and are therefore liable to be captured several times by the police; see van der Heijden et al. (2003) for a detailed presentation. We use the portion of the data available in Supplement C to Böhning and van der Heijden (2009). It features $n = 799$ illegal immigrants apprehended once or more. The covariates possibly related to their detection are their sex, a dichotomized age (above or below 40 years old), their country of origin with five possible values, and their reason for being arrested. The maximum number of encounters was 6 and the aggregated data gives $f_1 = 686$, $f_2 = 90$, $f_3 = 18$, $f_4 = 3$, $f_5 = f_6 = 1$.

In these two examples, the goal is to estimate the total population size N that includes the units that were missed. In the mouse example, N may provide an evaluation of the quality of the site's habitat while in the illegal immigrant example it gives an indication of the extent of this social problem in the Netherlands. The plots and the calculations presented in Section 18.3 have been carried using Rcapture, an R package for capture-recapture models presented in Baillargeon and Rivest (2007).

18.3 Estimation of Population Size Using Aggregated Data

This section reviews methods to estimate the population size N without using the unit level covariates. The first comprehensive presentation of models and methods for doing so is the monograph *Statistical Inference from Capture Data for Closed Animal Populations* by Otis et al. (1978) along with CAPTURE, its companion software. It distinguishes three types of effects that need to be addressed when estimating a population size, namely a time effect, a behavioral effect, and an heterogeneity effect. This work focuses on the latter, as it is also a concern in multiple system estimation of human populations and in studies featuring continuous captures.

Otis et al. (1978) used the acronym M_h to denote the heterogeneity model while M_0 represents the homogeneity model. Only estimates of N derived from statistical models for the data are considered. The jackknife estimator, used by Otis et al. (1978), does not fall in this category as it does not rely on a particular model. The modern statistical methodology on the estimation of a population size emphasizes fitting a model as an important intermediate step in the analysis. Models providing a good fit to the data are expected to give reliable estimates of population sizes. We start with the homogeneity model M_0 in the next section.

18.3.1 Homogeneity Model

For a capture-recapture dataset featuring t capture occasions, the homogeneity model postulates that there is a positive probability p for a unit to be captured at a given occasion and that it is the same for all units and all occasions. The number of times C that a unit is captured is a random variable with what is known as a binomial distribution with parameters p and t for all units in the population. The probability that C is equal to i is then

$$\Pr(C = i) = \binom{t}{i} p^i (1 - p)^{t-i}, \quad i \in \{0, \ldots, t\}.$$

Under this model the probability that a unit is not captured is $(1 - p)^t$, thus on average $N(1 - p)^t$ will be missed in the study, and an estimate of N is given by n plus an estimate of $N(1 - p)^t$.

To obtain a numerical value for $N(1 - p)^t$, we fit the homogeneity model to the data $\{f_i : i = 1, \ldots, t\}$. Given that C has a binomial distribution, the expectation (or predicted value) for f_i is given by

$$\mathrm{E}(f_i) = N \binom{t}{i} p^i (1 - p)^{t-i}.$$

or

$$E(f_i) = E_i = \binom{t}{i} e^{\gamma + i\beta}, \quad i \in \{1, \dots, t\}, \tag{18.1}$$

where $\gamma = \log\{N(1 - p)^t\}$ is the logarithm of the predicted frequency for the missed units and $\beta = \log\{p/(1 - p)\}$ is called the logit of the capture probability p. For fixed values of (γ, β) the discrepancy between the observed values f_i and the predicted values E_i is measured by what statisticians term the deviance, given by

$$G^2 = 2 \sum_{i=1}^{t} \{f_i \log(f_i/E_i) - (f_i - E_i)\}. \tag{18.2}$$

Note that the value of G^2 is small when the E_i closely approximate the observed frequencies f_i. The values of (γ, β) for which G^2 is minimum are called the maximum likelihood estimates of the parameters and are denoted $(\hat{\gamma}, \hat{\beta})$. These are the values that will be used to estimate N,

$$\hat{N} = n + e^{\hat{\gamma}}. \tag{18.3}$$

In statistics, model (18.1) is called a generalized linear model. Besides providing the point estimates $(\hat{\gamma}, \hat{\beta})$, it also gives variance estimates $(v(\hat{\gamma}), v(\hat{\beta}))$ as measures of uncertainty. The sampling variance of $\hat{N}$ can then be estimated by

$$v(\hat{N}) = e^{\hat{\gamma}} + v(\hat{\gamma}) e^{2\hat{\gamma}}; \tag{18.4}$$

see Rivest and Lévesque (2001) for more discussion. In Tables 18.1–18.3, the standard error (SE) is the square root of the variance estimate $v(\hat{N})$ in (18.4) and the coefficient of variation (CV), defined as the ratio $SE/\hat{N}$, gives a standardized measure of precision. Roughly, estimates with a CV smaller than 5% are considered as being accurate while CVs larger than 20% are considered imprecise.

For continuous captures, the distribution for the number of captures under the homogeneity model is the Poisson distribution,

$$\Pr(C = i) = \frac{1}{i!} \lambda^i e^{-\lambda}, \quad i \in \{0, 1, \dots\},$$

where λ, the "encounter rate," is the average number of detections per unit time. For the illegal immigrant data, λ represents the average number of captures per year. In the homogeneity model, λ is assumed to be the same for all units in the population. One has

$$E(f_i) = E_i = \frac{1}{i!} e^{\gamma + i\beta}, \quad i \in \{1, \dots, t_{\max}\}, \tag{18.5}$$

where $t_{\max}$ is the largest i for which $f_i > 0$, $\gamma = \log(Ne^{-\lambda})$ and $\beta = \log \lambda$. Once the generalized linear model for (18.5) is fitted, the estimates of N and of its variance are calculated using Equations (18.3) and (18.4).

TABLE 18.1: Homogeneity and lower bound (LB) models fitted to the harvest mouse and to the illegal immigrant data.

	Harvest Mouse Data			Illegal Immigrant Data		
	M_0	LB	M_0 ($t_0 = 3$)	M_0	LB	M_0 ($t_0 = 3$)
$\hat{N}$	158	216	195	2,765	3,413	3,002
SE	5.0	24.2	14.4	205.6	344.1	243.8
CV	3.1%	11.2%	7.4%	7.4%	10.1%	8.1%
G^2	81.9	—	2.6	22.1	—	6.2
df	12	—	1	4	—	1

One method used to assess whether a model fits the dataset $\{f_i\}$ is to compare the deviance statistic G^2 for the fitted model with its degrees of freedom. The number of degrees of freedom is t (or $t_{\max}$), the number of f_i-values, minus the number of parameters in the model. A model is said to fit well if its deviance is small, in particular smaller than its degrees of freedom. For model M_0, the deviance has $t - 2$ degrees of freedom. Table 18.1 gives the deviances of the homogeneity model obtained for the two datasets of Section 18.2. The two deviances are much larger than their degrees of freedom indicating a bad fit. Thus the M_0 estimates of N for the two datasets, 158 and 2,765, should be discarded.

18.3.2 A Probability Plot and a Lower Bound Estimate

Under model (18.1), $\log\{\mathrm{E}(f_i)/\binom{t}{i}\}$ is a linear function of i. Thus a simple binomial probability plot to investigate this assumption contains the points $(i, \log\{f_i/\binom{t}{i}\})$ for all i for which $f_i > 0$. If the plot is approximately linear, then the homogeneity model fits well. For continuous captures, a Poisson probability plot considers the points $(i, \log(i!f_i))$ for the positive f_i's. It should be linear, up to sampling errors, if the homogeneity model fits well; see Hoaglin (1980) for more discussion. The probability plots for the two datasets are shown in Figure 18.1 together with lines, $\hat{\gamma} + i\hat{\beta}$, representing the fitted homogeneity models. The two plots are not linear and the lines for the homogeneity model do not represent the points with large i-values well. This agrees with the conclusion of Section 18.3.1 that the homogeneity model is inadequate.

The homogeneity model fails when the capture probability p changes from one unit to the next. This feature is included in the statistical model by considering p as a random variable rather than a fixed parameter. In this case $\Pr(C = i)$ is proportional to

$$\binom{t}{i}\mathrm{E}\{p^i(1 - p)^{t-i}\},$$

where $\mathrm{E}(\cdot)$ gives an average value over the units in the population. One can

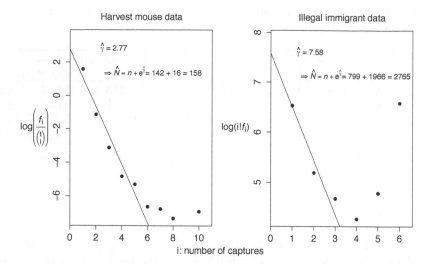

FIGURE 18.1: Probability plots for the two datasets with graphical representations of the fit of the homogeneity model.

show that when p is random $\log\{E(f_i)/\binom{t}{i}\}$ is a convex up function of i. Thus if the shape of the the points $(i, \log\{f_i/\binom{t}{i}\})$ for $i = 1, \ldots, t$ is convex up, then the capture probability may vary from one unit to the next. The same is true of the Poisson probability plot; a convex up plot indicates heterogeneity in encounter rates. The two probability plots in Figure 18.1 are convex up, suggesting the presence of heterogeneity in these two datasets. We now consider the construction of alternative estimates that address the issue of heterogeneity.

A lower bound estimate can be derived by considering the probability plots in Figure 18.1. It is constructed by calculating the smallest possible value of f_0 which preserves the convexity of these plots, when the point $(0, \log(f_0))$ is added. The value of f_0 can be determined geometrically, as illustrated in Figure 18.2. This yields $\hat{f}_0 = (t-1)f_1^2/(2tf_2)$ as an estimate for the number of missed units for a dataset with t capture occasions; the lower bound estimate of N is

$$\hat{N} = n + \frac{(t-1)f_1^2}{2tf_2}, \qquad (18.6)$$

as first derived in Chao (1987); see also Rivest and Baillargeon (2007) for the derivation of a variance estimate. Letting t tend to infinity gives the lower bound estimate for continuous captures, $\hat{f}_0 = f_1^2/(2f_2)$. In Table 18.1, the columns LB give lower bound estimates $\hat{N} = 216$ and $\hat{N} = 3{,}413$ for the two datasets. They are 30% larger than the M_0 estimates. This highlights the severe bias of the latter when there is heterogeneity in the data. See Hwang and Huggins (2005) and Rivest (2008) for further discussions of this point.

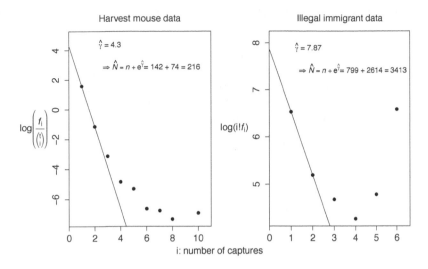

FIGURE 18.2: Geometric construction of the lower bound estimate.

The lower bound estimate can be calculated using formula (18.3) when the parameters (γ,β) are estimated by solving equations that set f_1 and f_2 equal to their predicted values, E_1 and E_2, under M_0 as given in equations (18.1) and (18.5). It makes sense to take out the units caught several times when estimating f_0, especially when the capture probabilities vary between units, because these units might not be representative of the ones that were missed. This increases the variance, however. In Table 18.1, the coefficients of variation (CV) of the two LB estimates are larger than 10%. Observe that in Figure 18.1 the probability plots are nearly linear for $i = 1, 2, 3$; this suggests adding f_3 to f_1 and f_2 to estimate f_0. Bringing f_3 in the estimation amounts to estimating the parameters in (18.1) or (18.5) by fitting the homogeneity model using only the units caught $t_0 = 3$ times or less. This is model M_0 $(t_0 = 3)$ in Table 18.1, where t_0 stands for the maximum number of captures for a unit to be kept when fitting (18.1). This model should give an estimate for N close to the true population size if the units caught three times or less are representative of the units that were missed. The heterogeneity would then be caused by units caught four times or more. They might correspond to a small trap-happy group whose behavior is not representative of the units that were missed.

In Table 18.1, bringing f_3 in the estimation of N reduces both $\hat{N}$ and its standard errors. For the illegal immigrant data the deviance of 6.2 with $df = 1$ indicates that this model does not fit, while for the mouse data the fit is acceptable. The next section discusses estimates of N obtained with models that deals with the heterogeneity by incorporating the random nature of p and λ in the model construction.

18.3.3   Parametric Models with Random Capture Probabilities

In this section, a distribution F_θ describes the variations of the probability of capture p in the population, where θ is a vector of unknown parameters. The choices for F_θ considered here are either the Normal or distributions leading to simple models for the data $\{f_i\}$. As seen in Section 18.3.2, the expectation of the number of units caught i times is

$$\mathrm{E}(f_i) = \binom{t}{i} N E_p\{p^i(1-p)^{(t-i)}\},$$

where the expectation $E_p(\cdot)$ is taken with respect to the distribution F_θ of p. In this context, (18.1) becomes

$$\mathrm{E}(f_i) = E_i = \binom{t}{i} e^{\gamma + \phi_\theta(i)}, \quad i \in \{1, \ldots, t\}, \tag{18.7}$$

where γ is, as before, the logarithm for the predicted number of missed animals and $\phi_\theta(i)$ is a convex function of i that depends on F_θ. The parameters (γ, θ) are estimated by minimizing the deviance (18.2), as in Section 18.3.1. Equations (18.3) and (18.4) are used to estimate N and its sampling variance. For continuous captures, a heterogeneity in encounter rate is modeled by assuming that λ is a random variable with a distribution F_θ and the same construction applies.

The Normal distribution, $\mathcal{N}(\beta, \sigma^2)$, where $\theta = (\beta, \sigma^2)$ are unknown parameters, is often used to model the random capture probabilities and the random encounter rates. This is considered by Coull and Agresti (1999) for the discrete case and by Bunge and Barger (2008) for continuous captures. In Table 18.2, the estimates $\hat{N}$ obtained with Normal random effects are 284 and 7,224 for the harvest mouse and the illegal immigrant examples. The latter estimate is quite large, as it is equal to 2 times the LB estimate of Table 18.1; its high CV of 23% is also noteworthy. Both models fit well as their deviances G^2 are smaller than the degrees of freedom (df). Thus the heterogeneity in capture probability could be described with a Normal distribution.

A simple model can be derived by assuming that, in (18.7), $\phi_\theta(i)$ has a simple linear form such as $i\beta + \psi(i)\tau$ where $\theta = (\beta, \tau)$ are unknown parameters and ψ is a known convex function of i. Lindsay (1986) and Rivest and Baillargeon (2007) show that there exists a distribution F_θ such that $\phi_\theta(i)$ calculated with F_θ gives $i\beta + \psi(i)\tau$ for several convex functions ψ such as $\psi(i) = 2^i - 1$, $\psi(i) = i^2/2$, and $\psi(i) = -\log(3.5 + i) + \log(3.5)$. The corresponding models are called the Poisson2, the Darroch and the Gamma3.5, respectively. Typically one has $\hat{N}(\text{Gamma3.5}) > \hat{N}(\text{Darroch}) > \hat{N}(\text{Poisson2})$, and these estimates allow us to investigate the impact of various specifications of F_θ on the magnitude of $\hat{N}$. For continuous captures, Bunge and Barger (2008) found that Poisson mixtures constructed with a finite mixture of the

TABLE 18.2: Heterogeneity models fitted to the harvest mouse and to the illegal immigrant data.

| | Harvest Mouse Data | | | |
	Normal	Poisson2	Darroch	Gamma3.5
$\hat{N}$	284	158	204	350
SE	44	5.1	15	63.9
CV	15.5%	3.2%	7.4%	18.2%
G^2	7.5	77.4	13.9	5.1
df	11	11	11	11
	Illegal Immigrant Data			
	Normal	Poisson2	Darroch	Gamma3.5
$\hat{N}$	7,224	3,374	5,629	9,425
SE	1,651	305	956	2,628
CV	22.9%	9%	17%	27.9%
G^2	.9	3.8	.7	1.0
df	3	3	3	3

negative exponential distribution often provide a very good fit to the aggregated data $\{f_i\}$.

Table 18.2 presents the estimates $\hat{N}$ obtained with these simple models and Figure 18.3 gives the probability plots introduced in Section 18.3.2 and the fitted convex functions, $\hat{\gamma} + \phi_{\hat{\theta}}(i)$, for the four models of Table 18.2. All models appear to fit the data well, except possibly the Poisson2 model for the harvest mouse data. For the harvest mouse data, the fit of the Normal and the Gamma3.5 model are similar, both in Figure 18.3 and considering the deviances of Table 18.2. There is a 20% difference in the two values of $\hat{N}$, however. For the immigrant data, the four models presented in Table 18.2 fit well since their deviances are small. Threefold variations in the value of $\hat{N}$ are observed however and it seems difficult to obtain a definitive estimate for the population size. Figure 18.3 and Table 18.2 illustrate findings of Huggins (2001) and Link (2003): the data at hand does not permit us to identify the distribution F_θ for the random capture probability p (or the random detection rate λ) and the value of $\hat{N}$ depends critically on F_θ. By using auxiliary variables to account for the variations in p and λ, one expects to narrow down the range of possible values for N. This is discussed briefly in the next section.

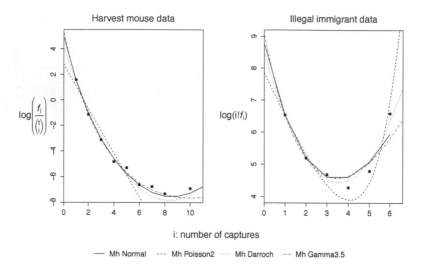

FIGURE 18.3: Graphical representation of the fit of various models for M_h.

18.4   Modeling Capture Probabilities with Unit Level Covariates

This section uses explanatory variables x, measured on each unit caught in the study, to model the capture probability p or the encounter rate λ. Using subscript j for the unit, the dataset is $\{(c_j, x_j) : j = 1, \ldots, n\}$ where x_j is a $d \times 1$ vector. The logit and the log link functions are used to express the capture probability p_j and the encounter rate λ_j in terms of the covariate x_j, viz.

$$p_j = \frac{e^{x_j^\top \beta}}{1 + e^{x_j^\top \beta}} \quad \text{and} \quad \lambda_j = e^{x_j^\top \beta},$$

where β is a $d \times 1$ vector of unknown regression parameters. If p_j and λ_j were known, then N would be estimated using a so-called Horvitz–Thompson estimator, viz.

$$\hat{N} = \sum_{j=1}^{n} \frac{1}{\pi_j},$$

where π_j is the probability of being caught at least once, $\pi_j = 1 - (1 - p_j)^t$ for discrete captures and $\pi_j = 1 - e^{-\lambda_j}$ for continuous captures. By weighting each unit captured by its inverse probability of being caught, one constructs a good estimate of N.

In practice, β is unknown and numerical values for this parameter have to be determined. As in Section 18.3.1, we want to calculate maximum likelihood estimates. This is done by maximizing what is known as the likelihood function. It is constructed by considering the distribution of the number of captures c_j. For discrete captures, the number of captures has a binomial distribution as discussed in Section 18.3.1. Since only units captured at least once are recorded in the dataset, we say that c_j has a binomial distribution truncated at 0. In a similar way, for continuous captures, c_j has a Poisson distribution truncated at 0.

The maximization of the likelihood for discrete captures is carried out with statistical software for truncated binomial regression while one would use a truncated Poisson regression for continuous captures. See Huggins (1989) and van der Heijden et al. (2003) for a detailed presentation of the calculations underlying the estimation of N and of its variance in these two cases. Note, however, that for these complex models the maximization of the likelihood does not produce a statistic, such as the deviance in Section 18.3.1, that could be used to investigate the fit of these models.

If, once the covariates are accounted for, there is still some residual heterogeneity in the data, then the estimate of N might be negatively biased, as in Section 18.3.1. A sensitivity analysis can be carried out by using only the units captured less than t_0 times to estimate the parameter β. As argued in Section 18.3.2, such an estimate should be more representative of the units that were never captured. If the estimate of N is the same as that obtained with the whole dataset, then residual heterogeneity should not be a concern. Such an analysis has been proposed by Böhning and van der Heijden (2009) for continuous captures. To our knowledge, this has not been considered for discrete captures; only models using all the captured units are used to estimate β, regardless of the number of captures c_j.

Table 18.3 gives estimates of N calculated using unit level covariates. Preliminary statistical analyses, not reported here, found that the covariates had a significant effect on p and λ. The capture probability of a mouse increases with its body weight and the encounter rate varies according to the immigrant's country of origin. For both datasets, the pattern of increasing $\hat{N}$ with decreasing t_0 suggests that the covariates do not explain all the heterogeneity in capture probabilities. Since they use all the data to estimate β, the $t_0 = 14$ and $t_0 = 6$ estimates might underestimate the true population sizes in the two examples. Unit level covariates narrow down the range of possible values for N as compared with Table 18.2; note, however, that bringing in the covariates does not seem to reduce the standard error of $\hat{N}$ for estimates having the same magnitude.

The best size estimates of N, among all those considered here, are the ones of Table 18.3 with small values of t_0. However, their CVs are relatively high, especially for the illegal immigrant data, and the lower bound estimate of Table 18.1 might be a better estimate for this dataset. It would be interesting to analyze the data with models featuring both unit level covariates and a

TABLE 18.3: Models with unit-level covariates fitted to the mouse and to the illegal immigrant data.

	Harvest Mouse Data			Illegal Immigrant Data		
	$t_0 = 14$	$t_0 = 4$	$t_0 = 2$	$t_0 = 6$	$t_0 = 3$	$t_0 = 2$
$\hat{N}$	176	197	249	4,883	5,166	5,545
SE	9.4	14.5	40.4	1,114	1,140	1,267
CV	5.4	7.3	16.2	22.8	22.1	22.9

random unit effect. For continuous captures, assuming a gamma distribution for the residual unit effect leads to the zero truncated negative binomial as a distribution for the number of captures C; see Cruyff and van der Heijden (2008). When t is finite, the specification of models for C featuring both fixed covariates and a residual unit effect is more complex and has not, to our knowledge, been investigated.

18.5   Discussion

Capture-recapture methods for closed populations answer a very basic question: How many? This work has treated two examples involving an animal population and an elusive human population, as developments of the statistical methods for this type of data have been motivated by applications in these two areas. Nowadays capture-recapture techniques are applied in a variety of fields. They are, for instance, used in genetic research to estimate the size of a gene pool and in computer engineering to estimate the number of bugs in a new software. Each new application must address the question of heterogeneity: are some units more likely to be captured than others? A failure to deal with this problem may result in a severe underestimation of the population size. This work has presented some of the statistical tools that have been developed to cope with this difficulty and has applied them on two examples.

About the Authors

Louis-Paul Rivest is a professor and the holder of a Canada Research Chair in Statistical Sampling and Data Analysis at Université Laval. He studied mathematics and statistics at the Université de Montréal and at McGill University. His research interests include capture-recapture models, survey sampling, and geometrical and multivariate statistics. He founded Laval's *Service de consultation statistique* and has been involved throughout his career in collaborative research with biologists, foresters, engineers, and social scientists. He was president of the Statistical Society of Canada in 2000–01 and received the SSC Gold Medal for research in 2010.

Sophie Baillargeon is a research associate and lecturer in statistics at Université Laval, where she completed an MSc degree in 2005. She has produced several R software packages that implement methodology developed by the Laval Statistics Group.

Bibliography

Baillargeon, S. and Rivest, L.-P. (2007). The Rcapture package: Loglinear models for capture-recapture in R. *Journal of Statistical Software*, 19. http://www.jstatsoft.org/v19/i05, Rcapture CRAN URL: http://CRAN.R-project.org/package=Rcapture.

Böhning, D. and van der Heijden, P. G. M. (2009). A covariate adjustment for zero-truncated approaches to estimating the size of hidden and elusive populations. *The Annals of Applied Statistics*, 3:595–610.

Bunge, J. and Barger, K. (2008). Parametric models for the number of classes. *Biometrical Journal*, 50:971–982.

Chao, A. (1987). Estimating the population size for capture-recapture data with unequal catchability. *Biometrics*, 43:783–791.

Coull, B. A. and Agresti, A. (1999). The use of mixed logit models to reflect heterogeneity in capture-recapture studies. *Biometrics*, 55:294–301.

Cruyff, M. J. L. F. and van der Heijden, P. G. M. (2008). Point and interval estimation of the population size using a zero-truncated negative binomial regression model. *Biometrical Journal*, 50:1035–1050.

Harrison, A. (2012). Counting the unknown victims of political violence: The work of the human rights data analysis group. *Human Rights and Information Communications Technologies: Trends and Consequences of Use.*

Hoaglin, D. C. (1980). A Poissonness plot. *The American Statistician*, 34:146–149.

Huggins, R. M. (1989). On the statistical analysis of capture experiments. *Biometrika*, 76:133–140.

Huggins, R. M. (2001). A note on the difficulties associated with the analysis of capture-recapture experiments with heterogeneoous capture probabilities. *Statistics & Probability Letters*, 54:147–152.

Hwang, W.-H. and Huggins, R. M. (2005). An examination of the effect of heterogeneity on the estimation of population size using capture-recapture data. *Biometrika*, 92:229–233.

Lecren, E. D. (1965). A note on the history of mark-recapture population estimates. *Journal of Animal Ecology*, 34:453–454.

Lindsay, B. G. (1986). Exponential family mixture models (with least-squares estimators). *The Annals of Statistics*, 14:124–137.

Link, W. A. (2003). Nonidentifiability of population size from capture-recapture data with heterogeneous detection probabilities. *Biometrics*, 59:1123–1130.

McKendrick, A. G. (1926). Application of mathematics to medical problems. *Proceedings of the Edinburgh Mathematical Society*, 44:98–130.

Otis, D. L., Burnham, K. P., White, G. C., and Anderson, D. R. (1978). *Statistical Inference from Capture Data on Closed Animal Populations*, volume 62 of *Wildlife Monographs*. Wildlife Society.

Rivest, L.-P. (2008). Why a time effect often has a limited impact on capture-recapture estimates in closed populations. *The Canadian Journal of Statistics*, 36:75–84.

Rivest, L.-P. and Baillargeon, S. (2007). Applications and extensions of Chao's moment estimator for the size of a closed population. *Biometrics*, 63:999–1006.

Rivest, L.-P., Couturier, S., and Crépeau, H. (1998). Statistical methods for estimating caribou abundance using postcalving aggregations detected by radio telemetry. *Biometrics*, 54:865–876.

Rivest, L.-P. and Lévesque, T. (2001). Improved log-linear model estimators of abundance in capture-recapture experiments. *The Canadian Journal of Statistics*, 29:555–572.

Stoklosa, J. (2012). *PL.popN: Population Size Estimation*. R Package Version 1.2. http://CRAN.R-project.org/package=PL.popN.

Stoklosa, J. and Huggins, R. M. (2012). A robust P-spline approach to closed population capture-recapture models with time dependence and heterogeneity. *Computational Statistics & Data Analysis*, 56:408–417.

Stoklosa, J., Hwang, W.-H., Wu, S.-H., and Huggins, R. M. (2011). Heterogeneous capture-recapture models with covariates: A partial likelihood approach for closed populations. *Biometrics*, 67:1659–1665.

van der Heijden, P. G. M., Bustami, R., Cruyff, M., Engbersen, G., and van Houwelingen, H. C. (2003). Point and interval estimation of the population size using the truncated Poisson regression model. *Statistical Modeling*, 3:305–322.

Williams, B. K., Nichols, J., and Conroy, M. J. (2002). *Analysis and Management of Animal Populations*. Academic Press, San Diego, CA.

19

Challenges in Statistical Marine Ecology

Joanna Mills Flemming and Christopher A. Field
Dalhousie University, Halifax, NS

As a coastal nation, it is important for Canada that we understand the oceans around us and how they and their inhabitants are responding to changes induced either directly or indirectly by human activity. Within biology, marine ecology is the scientific study of marine-life habitat, populations, and interactions among organisms and the surrounding environment.

19.1 Introduction

Marine ecologists ask questions at various scales ranging from worldwide populations to the behavior of individual species or animals. Statistics are essential for answering these questions: from describing the study design and data collection, to providing models and methodology appropriate for answering the question(s) of interest, and finally in making available the necessary expertise to fully understand and interpret results.

In marine ecology, the scientific questions often concern the abundance of a particular species, how that abundance may depend on habitat or other species and how the abundance changes over space and/or time in response to environmental or human-induced changes. What are the features of marine ecological data that make them challenging statistically? First, data often include substantial random variation and other inaccuracies at least in part due to the difficulty of making accurate measurements of abundance in the marine environment. For instance, it is challenging to estimate the number of cod in a particular fishing zone and consequently we must often use surrogate and rather imprecise measures. Often data also vary quite dramatically over both space and time and series are commonly short and sparse. Furthermore, data on marine species or systems of interest usually need be linked to environmental measurements that are typically collected on different scales.

Given a clear marine ecological question of interest, the next step is to translate this into statistical terminology such that the study design, data collection and ensuing model framework are effective. Statistical models are used to represent important features of the ecological processes of interest. Such models are mathematical representations; for example, they might describe the changing numbers (abundance) of fish in a particular region of interest. There is a symbiotic relationship between the model and the data collected. That is, the model must be complex enough to capture the essential features of the process but simple enough that it can be fitted to the data. Typically a model will have some unknown quantities, referred to as states or parameters which we wish to estimate from the data. For instance if we want to fit a straight line to a simple two-dimensional dataset, then the slope and intercept are often the parameters of interest. We aim for states and parameters that have simple ecological interpretations, can be reasonably well estimated and for which we are able to assess the accuracy of estimates.

In this chapter, we describe two research projects that utilize novel statistical methodologies to answer important scientific questions. The goal of the first project is to describe how a collection of fish populations is changing over time in response to fishing and other sources of mortality. This is critical to sustaining some of our most valuable renewable resources.

The second project presents a statistical model for obtaining accurate estimates of abundance of the critically endangered hammerhead shark. Although both projects are concerned with estimating abundance of marine species, they require the development of very different statistical methods and rely on vastly different types of data. Further details are given below.

Young fish are very vulnerable to competition, predation, hydrography, temperature, and a host of other environmental factors that result in extraordinarily high levels of mortality. Their varying ability to survive and grow ultimately determines how many fish become part of the adult population (the part of the total population than can be legally fished). Understanding the dynamics of their growth period is therefore critical to understanding fish population dynamics and the management of fisheries resources. Yet, despite a century of fruitful investigation, much uncertainty remains in the understanding of stock replenishment. Here the focus is on the scientific question of whether the number of Atlantic cod joining the adult population has changed over time and, where observed, are such changes similar to those of proximate populations. The data used were extracted from a quality controlled global fish stock assessment database (Ricard et al., 2012) that was recently developed at Dalhousie University.

In the second project we note that for many endangered marine species, the only available data related to their abundance are counts of when they are caught unintentionally in a fishery (i.e., as bycatch). These data typically involve a large number of zero counts (indicating that none were caught as bycatch in a particular haul) and very few positive counts (obtained if one or more are caught as bycatch in a haul) and are based on hauls that occur within trips which are made by each of a number of vessels. Using these data, we need to answer the important question of whether we can accurately estimate the probability of bycatch as this is an important first step in attempting to protect endangered marine species (Lewison et al., 2004). The second project uses bycatch data on the hammerhead shark, obtained by Julia Baum—see Baum (2007), for further details—from the US National Marine Fisheries Service Pelagic Observer Program (http://www.sefsc.noaa.gov/pop.jsp). In the spring of 2013 these sharks, which are commercially valuable and whose numbers are thought to have been declining dramatically in recent years, were given added protection by CITES (the Convention on International Trade in Endangered Species of Wild Fauna and Flora) whose aim is to ensure that international trade in specimens of wild animals (and plants) does not threaten their survival.

19.2   Sustaining North Atlantic Cod Stocks

A fish stock is a group of fish of the same species that live in the same geographic area and mix enough to breed with each other when mature. Stock assessments provide fisheries managers with the information that is used in the regulation of a fish stock. Data used in stock assessments can be classified as fishery-dependent or fishery-independent. Fishery-dependent data are collected from both commercial and recreational fishing activities. There is a variety of methods for obtaining fishery-dependent data. The most common approach is to use records of the amount of fish sold with the numbers typically reported in total weight. Another common mode for acquiring fishery-dependent data is through portside sampling of the catch of both recreational and commercial fishermen. Other less common methods for obtaining data are through the use of onboard observers, self-reporting, telephone surveys, and vessel-monitoring surveys. Fishery-independent data are obtained in the absence of any fishing activity with the majority of these data collected by government agencies. A wide variety of methods are used to acquire fishery-independent data and sampling equipment can include trawls, seines, acoustic and/or video surveys. Stock assessments are often conducted using both fishery-dependent and fishery-independent data.

Traditional approaches to predicting the number of fish that will be present in a stock either in the current year or in future years have relied on a statistical model by Ricker (1954). Typically we refer to fish as belonging to a particular cohort or year class and want to estimate the number of new fish entering the cohort (recruits) based on the number of parent fish (spawners). We label cohort or year by t and want to predict the number of recruits in year t based on the number of spawners at time $t - \tau$, where τ represents the years required for a fish to mature. Estimates of the number of spawners and recruits for 10 different cod stocks corresponding to fishing regions off the coast of Nova Scotia and Newfoundland from a new quality controlled global fish stock assessment database (Ricard et al., 2012) developed at Dalhousie University were used. To write down the Ricker model for a collection of J stocks use $j \in \{1, \dots, J\}$ as the stock label and t as the year label, where $t \in \{t_{j,0}, t_{j,0} + 1, \dots, T_j\}$ (the start $t_{j,0}$ and end years T_j are stock-specific since the start and end points of the assessments differ by stock with some stretching back close to a century while others begin within the last 30 years). The Ricker model can then be written as

$$R_{j,t} = \alpha_j S_{j,t-\tau_j} \exp(-\beta_j S_{j,t-\tau_j} + \epsilon_{j,t}),$$

where $R_{j,t}$ is the number of recruits in stock j, in year t at age τ_j (age of maturity can depend on the particular stock) and $S_{t-\tau_j}$ is the number of spawners. When suitably scaled, α_j is referred to as the maximum reproductive rate, which is a measure of how productive the stock is at recruiting fish to the adult population and we see that as α_j increases the number of recruits increases. β_j is a parameter describing density-dependence and can be thought of as the inverse of the carrying capacity of the environment of the stock. As the carrying capacity increases, β_j decreases and the number of recruits increases due to the minus sign in front of it. The $e^{\epsilon_{j,t}}$ are Log-Normally distributed errors and represent all other factors which might affect recruitment. The model can be linearized by dividing through by the number of spawners and taking the natural logarithm, viz.

$$\log\left(\frac{R_{j,t}}{S_{j,t-\tau_j}}\right) = a_j - \beta_j S_{j,t-\tau_j} + \epsilon_{j,t}. \tag{19.1}$$

The natural logarithm of the ratio of recruits to spawners is termed survival and $a_j = \log(\alpha_j)$ is referred to as the productivity. In order to understand how productivity changes over time and how productivity in one stock is related to another, one can allow a_j to to vary over time leading to a so-called state space model (SSM). A SSM involves two equations: the first is the process equation which describes the underlying (yet unobserved) state that one would like to model, namely the productivity $a_{j,t}$; the second is the measurement equation relating the unobserved states to the observed measurements. Here Equation (19.1) is the measurement equation (with a_j replaced by $a_{j,t}$). For the process equation, a random walk on $a_{j,t}$ is used. That is, the productivity at time t is determined by the productivity at time $t-1$ plus a Normally distributed random error $\eta_{j,t}$ termed a process error (which can be either positive or negative) and is expressed as

$$a_{j,t} = a_{j,t-1} + \eta_{j,t}.$$

SSMs similar to the above have been fitted in the past but only to one stock at a time; see, e.g., Peterman et al. (2003) and Dorner et al. (2008). The method (Minto et al., 2013) allows one to simultaneously model multiple stocks (e.g., stocks in different geographic regions). That is, the multivariate SSM allows us to simultaneously estimate trends and correlations in productivity for multiple stocks. The formulation includes a correlation matrix (Q) describing the correlation of productivity across stocks. Where stocks are behaving similarly we would expect positive values in Q and conversely where they behave differently negative values would be expected. The method allows one to (i) test if productivity has changed over time; (ii) determine how productivity is correlated across stocks; and (iii) interpret the scale of productivity changes where they exist.

In order to estimate the states and parameters in the multivariate SSM, the Kalman Filter is used; see Harvey (1991) and Petris et al. (2009) for a description. SSMs are capable of dealing with missing values by interpolating using the process equation. This provides an opportunity to estimate historical trends in productivity for a given stock during a time period in which no data exist.

19.2.1   Results

By applying the multivariate SSM to 10 cod (*Gadus morhua*) stocks from different regions in the northwest Atlantic, we see that productivity has varied markedly through time. Furthermore, it was found that many stocks display lower productivity today, compared with historical averages. This indicates some fundamental changes in stock biology, or alternatively, in their environment. In the northwest Atlantic (Figure 19.1), the southern stocks of Georges Bank (GB), the Gulf of Maine (GOM), and the Southwestern Scotian Shelf (SWSS) are currently close to a historic minimum (based on hindcasts for Georges Bank and the Gulf of Maine). The Eastern Scotian Shelf (ESS), Southern (SGOSL) and Northern Gulf of St. Lawrence (NGOSL) stocks have potentially higher recent productivity than that observed or predicted at the earliest time points. All three stocks displayed markedly elevated productivity in the late 1970s, early 1980s. The Southern Newfoundland (SNEW) stock displayed increasing productivity to the 1980s and has been declining since. The Labrador Northern Newfoundland (LABNNEW) stock displayed constant productivity until the late 1980s, when it dipped precipitously prior to the cod fishing moratorium in 1992. A similar but more consistently decreasing trend was observed for the Southern Grand Banks (SGRAND) stock, again with a precipitous decline in the early 1990s. Both stocks are currently at depressed productivity levels, although forecasts suggest slow though uncertain productivity increases in both. The Flemish Cap (FLEM) stock displays differing productivity trends to those elsewhere in the northwest Atlantic; while other stocks reached a peak in productivity in the early 1980s, the Flemish Cap displayed a trough, similarly in the mid- to late-1990s.

Note that in Figure 19.1 results obtained by fitting a SSM to each stock individually are shown in light gray. A comparison of these estimated states with those of the multivariate SSM (shown in dark gray) shows similar trends for most stocks with overlapping confidence intervals. The confidence intervals

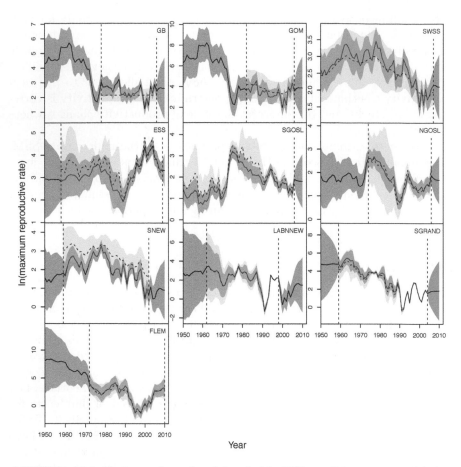

FIGURE 19.1: Estimated productivity (with 95% confidence intervals) from single-stock (light gray) and our multivariate (dark gray) fits to 10 Northwest Atlantic cod stocks. Dashed vertical lines represent the earliest and latest data points for each stock. Pre-dashed line fits are hindcasts. Post-dashed lines are forecasts.

are tighter for the multivariate SSM, thereby illustrating that one is better able to infer the productivity dynamics by borrowing information from other stocks. The method also generates hindcasts and forecasts that are not available if we consider each stock separately. In fact, single-stock models have long been viewed as too simplistic (Larkin, 1977) and there has been an increasing emphasis on ecosystem-based management (Link et al., 2011).

The correlation of the time-varying productivity across stocks showed marked patterns in the northwest Atlantic (Figure 19.2). Some strong correlations exist, e.g., the Georges Bank, Gulf of Maine and Southwestern Scotian

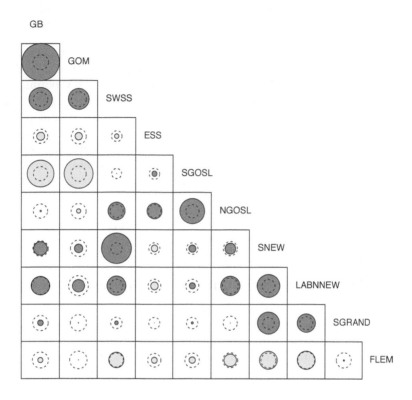

FIGURE 19.2: Estimated correlations in productivity for 10 Northwest Atlantic cod stocks. The strength of the correlation is represented by the radius of the circle, the direction by the color (light gray representing negative correlations and dark gray positive). Radii greater than that of the dashed circles are significant according to a statistical test.

Shelf stocks and further north, the Labrador Northern Newfoundland, Southern Newfoundland and Southern Grand Banks stocks displaying positive correlations. The Southwestern Scotian Shelf also displays positive correlations with distant stocks such as the Northern Gulf of St. Lawrence, Southern Newfoundland and Labrador Northern Newfoundland. Equally noticeable, however, is the isolation of several individual stocks from those adjacent. For example, the Eastern Scotian Shelf stock displays no or weakly negative correlations with the nearby stocks of Gulf of Maine and the Georges Bank. Similarly, the Southern Gulf of St. Lawrence stock displays no or negative correlation with all but the adjacent Northern Gulf stock (Figure 19.2). Outside the positive correlations already highlighted, the Southern Grand Banks displays weak correlations with all other stocks. Interestingly, the Flemish Cap displays weakly negative correlations with all stocks (Figure 19.2).

These statistical methods have made it possible to describe what is happening to productivity both within and between northwest Atlantic cod stocks. We found that productivity of northwest Atlantic cod, a key population parameter, varies markedly through time as well as across stocks and that these relationships change with geographic distance, identifying similar productivity regimes among neighboring stocks. The observed trends displayed some synchronicity across stocks, but also smaller scale variability that is not readily explained by current ecological hypotheses. The next step is to work with marine ecologists to attempt to understand why these ecological patterns are emerging.

19.3   Conserving the Endangered Hammerhead Shark

Marine species conservation is an important goal of ecologists. For many endangered species the only available data related to their abundance are counts of when they are caught unintentionally in a fishery; see Figure 19.3. We refer to this as bycatch data and here examine that of Hammerhead sharks obtained from the US National Marine Fisheries Service Pelagic Observer Program. Hammerhead sharks have garnered much attention recently due to CITES, which took decisive action in March of 2013 to halt their further decline. These bycatch data are complicated in that we have counts of bycatch of Hammerhead sharks recorded for each of a collection of hauls occurring during each of a number of trips by particular vessels. Such data are difficult to model with standard statistical methods for two main reasons: first, they involve a preponderance of hauls for which no Hammerheads are recorded (zero counts, see Figure 19.4); second, hauls are nested, i.e., they occur within trips which are nested within vessels, and this structure results in dependencies in the data that must be incorporated into the model structure for proper inference. Data with such features are commonly referred to by statisticians as clustered count data with many zero observations and are typical of endangered species data, particularly in marine environments.

In Cantoni et al. (2013) a statistical model was formulated for clustered count data with many zero observations. The formulation allows one to describe the bycatch of hammerhead sharks in terms of two random processes: one process determines the presence or absence of sharks in a haul, and in those hauls for which sharks are present, a second process determines the number of sharks. It deals with the clustering (also known as dependence) resulting from hauls made as part of the same trip and possibly trips made

FIGURE 19.3: A hammerhead shark as bycatch.

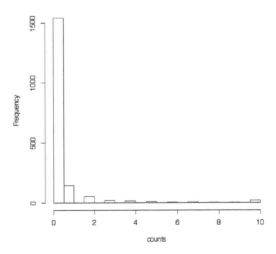

FIGURE 19.4: Histogram of the counts of hammerhead sharks caught as bycatch.

by the same vessel by introducing random effects. For instance, a trip random effect is an attempt to measure the effects on the bycatch which are particular to that trip and measure the extent to which the bycatch on one haul on the trip is correlated with the bycatch from another haul on the same trip. Parameters are introduced which control whether these random effects appear in one or both parts of the model. The two parts of the model are independent if these parameters equal zero and dependent otherwise.

Here we describe the model by supposing that there are c trips, with each trip i containing n_i, $i \in \{1, \ldots, c\}$, hauls (for simplicity we are ignoring the effect of vessel here). On the jth haul of the ith trip, one observes a count of Hammerhead shark bycatch, y_{ij}, along with other pieces of information, referred to as covariates. These include the year and season, the average hook depth, the area where the fishing is occurring as well as some measure of fishing effort. Some of these covariates will affect the presence or absence of a hammerhead while others will affect the number of hammerheads given their presence. Hence two (possibly overlapping) sets of covariates are allowed. Denote the values of the covariates in the jth haul of the ith trip as $\mathbf{x}_{ij}$ and $\mathbf{z}_{ij}$, $j \in \{1, \ldots, n_i\}$, $i \in \{1, \ldots, c\}$. The $\mathbf{x}_{ij}$ are the covariates which are anticipated to affect the presence or absence of a hammerhead while the $\mathbf{z}_{ij}$ are those affecting the number of hammerheads given their presence. This so-called hurdle model specifies that the counts y_{ij} are independent (given the random effects) with probabilities of the form:

$$
\Pr(Y_{ij} = y_{ij} \mid \mathbf{u}_i, \mathbf{v}_i) =
\begin{cases}
1 - p(\mathbf{x}_{ij}, \mathbf{u}_i) & y_{ij} = 0, \\[2mm]
p(\mathbf{x}_{ij}, \mathbf{u}_i) f\{y_{ij}, \lambda(\mathbf{z}_{ij}, \mathbf{u}_i, \mathbf{v}_i)\} & y_{ij} = 1, 2, 3, \ldots
\end{cases}
$$

where p is the probability of observing a positive count (i.e., "crossing the hurdle"), $f\{y_{ij}, \lambda(\mathbf{z}_{ij}, \mathbf{u}_i, \mathbf{v}_i)\}$ is the density of a distribution defined on the positive integers with parameter λ which is a function of the covariates and the random effects $\mathbf{u}_i$ and $\mathbf{v}_i$. Both $\mathbf{u}_i$ and $\mathbf{v}_i$ represent trip random effects with $\mathbf{u}_i$ being the effect of a trip on the presence/absence of sharks and $\mathbf{v}_i$ being the effect on the number given presence.

In many cases, it is of great interest to predict the effects of trips and/or vessels and other quantities that are functions of them. For example, we may be interested in determining which vessels tend to have more shark bycatch or which trips tend to result in more bycatch. We may also be interested in predicting the probability of bycatch for a particular haul, trip and vessel combination $\Pr(Y_{ij} > 0 | \mathbf{u}_i)$, or the expected number of sharks comprising bycatch (given non-zero bycatch) $E(Y_{ij} | Y_{ij} > 0, \mathbf{u}_i, \mathbf{v}_i)$ or, the expected abundance $E(Y_{ij} | \mathbf{u}_i, \mathbf{v}_i)$. We may also be interested in the analogous quantities aggregated over all trips made by a particular vessel or also across all vessels; for instance, the probability of bycatch, the expected bycatch given non-zero bycatch, or the expected bycatch.

This approach was used to model 1825 bycatch counts made on 292 different trips, with from 1 to 21 hauls per trip. Each of the $c = 292$ trips was treated as a cluster; vessels potentially add another layer of clustering though not incorporated here. For these data, we had 85% zeros and the positive counts ranged from 1 to 46. The covariates thought to potentially influence both the probability of bycatch and its magnitude that were considered are: the year (YEAR), average or median hook depth (AVGHKDEP, from 6.40 to 182.88 meters), area (4 = South Atlantic Bight and 5 = Mid Atlantic Bight), and season (SEASON, with 464 observations in autumn, 543 in spring, 525 in summer and 293 in winter). The catch effort was measured using the logarithm of the number of hooks (TOTHOOK, ranging from 25 to 1548). Baum et al. (2003) provide a more complete description of these data.

19.3.1 Results

The fitted model showed that there are more variables which are significant in the abundance part than in the presence-absence part. This is quite common in ecological problems; more factors affect the abundance given presence than whether we observe a positive count or not. The variable YEAR is significant in both parts and has a negative effect. This means that sharks are observed less often through time, and even when they are present they are less abundant. With each additional year the odds of observing a positive count reduce by 5.7%. SEASON too plays a role in both parts, with spring and winter significantly different from autumn. The catch effort, log(TOTHOOK), does not affect the presence-absence, but is significant in the abundance part. The hook depth also affects the abundance given presence significantly. Furthermore, a new parameter that was included in the model formulation demonstrated that the two parts of the model were dependent on each other. This sort of information is critical for informing management decisions, since, by incorrectly fitting the two parts of the model separately, e.g., one could potentially underestimate the extent to which their abundance is decreasing through time.

Figures 19.5 and 19.6 show the predicted probability of presence $\Pr(Y_{ij} > 0|\mathbf{u}_i)$ and the conditional expectations $\mathrm{E}(Y_{ij}|Y_{ij} > 0, \mathbf{u}_i, \mathbf{v}_i)$ with their corresponding confidence intervals for the first five trips (identifers as shown). We can interpret these results quite precisely. For example, in Figure 19.5, we see that for trip 01A019 there are essentially two different groups of predictions. It turns out that this is due to two different values of AVGHKDEP for this trip.

In Figure 19.6, we observe that the length of the confidence intervals can be quite variable. For trip 01A019, the two much larger intervals correspond to a different combination of AVGHKDEP and log(TOTHOOK), which is significant in the abundance part of the model and therefore has an impact on the estimation of $\mathrm{E}(Y_{ij}|Y_{ij} > 0)$. Similarly, the longer interval for trip 01A029 corresponds to the sole observation of this cluster with a different value of AVGHKDEP. The smaller variations in length are due to the differences in log(TOTHOOK).

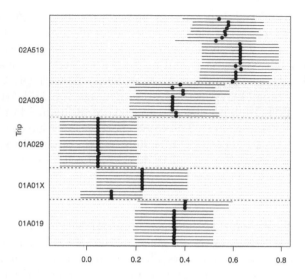

FIGURE 19.5: Confidence intervals for the probability of presence of hammerhead shark for the first five trips.

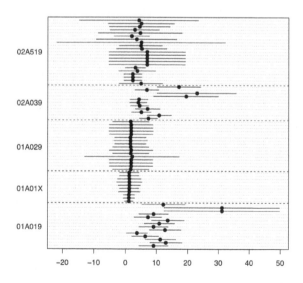

FIGURE 19.6: Confidence intervals for the expected number of hammerhead sharks present (given that bycatch has occurred) for the first five trips.

Tools to estimate abundance and predict important quantities like the probability of presence of critical endangered species have been much anticipated. We have demonstrated here how much can be learned in this regard from bycatch data. Future research will involve combining this information with measures of fishing intensity so as to attempt to fully describe abundance in regions of interest.

19.4 Conclusions

As we have illustrated with these two projects, we are able to gain insight into important questions in marine ecology by the careful construction of statistical models. As with many datasets in marine ecology, the structure is complex and there is also dependence among the observations which we need to accommodate in our models.

The development of the statistical model is determined by the scientific question of interest and the available data. The parameters to be estimated need to have clear ecological interpretations for the scientific question at hand. The confidence bounds on the parameters provide a measure of the accuracy of the estimates and information on the strength of the underlying ecological process in the data.

Our overall aim is to provide statistical models and tools necessary to understand the marine environment. This is especially important as data collection technologies continue to rapidly evolve leading to more sophisticated and complex data. For Canada, it is essential to understand how the oceans are changing, particularly the Arctic, as we continue to experience climate change.

Acknowledgments

The authors are very grateful to Coilin Minto, Boris Worm, Eva Cantoni, and Alan Welsh, with whom the first author collaborated on the projects discussed herein. This research has been supported by the Natural Sciences and Engineering Research Council of Canada.

About the Authors

Joanna Mills Flemming is an associate professor of statistics at Dalhousie University. She studied computer science and statistics at the University of Guelph and Dalhousie, and worked in industry for 18 months before returning to academia to spend two years as a postdoctoral fellow at the Université de Genève. Her research interests center on the development of statistical methodology for data exhibiting spatial and/or temporal dependencies, with a particular interest in marine ecology and more broadly, environmental science. She is an associate editor for *The Canadian Journal of Statistics*.

Christopher A. Field is a professor emeritus of statistics at Dalhousie University, where he spent his career. He holds degrees in mathematics and statistics from Dalhousie and Northwestern University. His research interests include robust statistics and small sample asymptotics plus collaborative work with scientists in many disciplines. He served as president of the Statistical Society of Canada in 1992–93 and was awarded the 2006 SSC Gold Medal for research. He is a fellow of the American Statistical Association and an honorary member of the SSC.

Bibliography

Baum, J. (2007). *Population- and community-level consequences of the exploitation of large predatory marine fishes.* Doctoral dissertation, Dalhousie University, Halifax, NS.

Baum, J. K., Myers, R. A., Kehler, D. G., Worm, B., Harley, S. J., and Doherty, P. A. (2003). Collapse and conservation of shark populations in the Northwest Atlantic. *Science*, 299:389–392.

Cantoni, E., Mills Flemming, J., and Welsh, A. (2013). A novel approach to the introduction and prediction of random effects in hurdle models. *Biometrika*, submitted.

Dorner, B., Peterman, R. M., and Haeseker, S. L. (2008). Historical trends in productivity of 120 Pacific pink, chum, and sockeye salmon stocks reconstructed by using a Kalman filter. *Canadian Journal of Fisheries and Aquatic Sciences*, 65:1842–1866.

Harvey, A. (1991). *Forecasting, Structural Time Series Models and the Kalman Filter.* Cambridge University Press, Cambridge.

Larkin, P. A. (1977). An epitaph for the concept of maximum sustained yield. *Transactions of the American Fisheries Society*, 106:1–11.

Lewison, R. L., Crowder, L. B., Read, A. J., and Freeman, S. A. (2004). Under-standing impacts of fisheries bycatch on marine megafauna. *Trends in Ecology and Evolution*, 19:598–604.

Link, J. S., Bundy, A., Overholtz, W. J., Shackell, N., Manderson, J., Duplisea, D., Hare, J., Koen-Alonso, M., and Friedland, K. D. (2011). Ecosystem-based fisheries management in the Northwest Atlantic. *Fish and Fisheries*, 12:152–170.

Minto, C., Mills Flemming, J., Britten, G. L., and Worm, B. (2013). Productivity dynamics of Atlantic cod. *Canadian Journal of Fisheries and Aquatic Sciences*, in press.

Peterman, R. M., Pyper, B. J., and MacGregor, B. W. (2003). Use of the Kalman filter to reconstruct historical trends in productivity of Bristol Bay sockeye salmon (*Oncorhynchus nerka*). *Canadian Journal of Fisheries and Aquatic Sciences*, 60:809–824.

Petris, G., Petrone, S., and Campagnoli, P. (2009). *Dynamic Linear Models with* R. Springer, Berlin.

Ricard, D., Minto, C., Baum, J. K., and Jensen, O. P. (2012). Examining the knowledge base and status of commercially exploited marine species with the ram legacy stock assessment database. *Fish and Fisheries*, 13:380–398.

Ricker, W. (1954). Stock and recruitment. *Journal of the Fisheries Research Board of Canada*, 11:559–623.

20

Quantifying the Human and Natural Contributions to Observed Climate Change

Francis W. Zwiers
University of Victoria, Victoria, BC

Gabriele C. Hegerl
University of Edinburgh, Edinburgh, Scotland

Xuebin Zhang, Qiuzi Wen
Environment Canada, Downsview, ON

20.1   Introduction

The natural climate is characterized by enormous amounts of chaotic variability that we experience as weather on short time scales and as year-to-year (and longer) differences in climate characteristics. Not only do we experience variations in temperature, precipitation, wind speed and so on within the course of a few days, but we also experience a summer, for example, that may be warm or cool, dry or wet, or be otherwise distinguished from other summers. This natural chaotic variability is driven by the energy from the Sun, which is absorbed by the Earth and re-emitted to space in the form of heat. Much of the weather and climate variability we experience results from the transfer of energy from places where it is absorbed to places where it is ultimately radiated back to space. This variability occurs spontaneously, and would occur even in the absence of any external influences on the climate, such as that from volcanic eruptions. Thus climate scientists often refer to this variability as natural internal climate variability.

In the absence of external influences on the climate, the amount of incoming solar energy is balanced by an equal amount of outgoing heat. External influences which affect that balance are called *forcing* agents. One group of forcing agents that has been recognized since the time of Joseph Fourier (Fourier, 1824) is greenhouse gases such as carbon dioxide. Increasing the greenhouse gas concentration in the atmosphere makes it less transparent to the passage of the infrared radiation (heat) that is trying to escape to space, which implies that the climate should become warmer. The atmospheric concentration of carbon dioxide has increased markedly since the beginning of the industrial revolution, increasing from about 280 ppm in 1750 to about 400 ppm today. This rise is principally from fossil fuel use and changes in land use, such as deforestation. Concentrations of other important greenhouse gases, such as nitrous oxide and methane, have also increased. Other external influences that affect the Earth's energy balance include changes in the Earth's orbit around the Sun, changes in the Sun's energy output, volcanic eruptions that affect the amount of reflective aerosol present in the stratosphere (which reduces the amount of solar energy that reaches the Earth's surface), human induced changes in atmospheric aerosol concentrations and changes in land surface properties related to development activities (e.g., due to urbanization and conversion to agricultural uses).

The climate that we experience can be thought of as a combination of the effects of a suite of external forcing agents, including human induced greenhouse gas increases, and natural internal climate variability. Invariably, a conceptual or physical model of some kind is used to estimate the expected effects of those forcing agents. Subsequently, a range of statistical techniques is used to attempt to detect these estimated signals in climate observations, and ultimately, to attribute causes to observed changes.

20.2   Scope of the Problem

Before beginning, it is worth considering some of the details of the climate observations in which we might look for evidence of the effects of increasing greenhouse gases or other external forcing factors. For example, the global mean surface temperature has warmed by approximately .7°C over the period 1901–2010, and we might want to determine whether external influences on the climate played a role in that warming. To do so convincingly, it will be necessary to consider not just the trend in global mean temperature, but also the evolution of the spatial pattern of warming over time, to be sure that the observed change was not of natural origin.

The warming pattern that is expected from human influence on the climate is distinct from natural patterns of internally caused surface temperature variability. Based on physical considerations we expect that the forcing from human influences on the climate should lead to more warming over land than over oceans and more warming at high northern latitudes than elsewhere due to feedback processes operating in the Arctic that help to amplify the climate response to human influences. We also expect that warming will evolve somewhat differently in different places because some forcing agents, such as aerosols, have regional as well as global scale effects. Thus the pattern of expected change is complex, with spatial features that evolve slowly over time.

Many studies therefore consider only decade-to-decade changes in surface temperature. For example, Jones et al. (2013) considered the evolution of decadal mean surface temperature patterns of the 11-decade period 1901–2010. The decadal surface temperature patterns are often based on decadal mean temperatures in regions of approximately 250,000 km^2 (areas about the size of United Kingdom that span 5° of latitude and longitude). If it were possible to calculate a decadal surface temperature average for every such region on the face of the Earth for every decade in the 11-decade period, there would be a total of 28,512 decadal means (11 decades × 2592 regions). Even considering that in many regions decadal averages cannot be calculated because insufficient observations are available to estimate a regional mean for that decade, there are nevertheless thousands of decadal values. Thus the observed space-time pattern of surface temperature changes has very high dimension when organized as a space-time data vector. Since the expected pattern of change is spatially smooth, most studies reduce the spatial dimension of the problem by retaining the equivalent of about 25 subcontinental regional means to represent the large-scale surface temperature pattern for each decade, which still implies an observed space-time data vector of length 275.

Thus the sheer size of the problem poses a challenge. While the observed pattern of change is simplified substantially by considering only decadal means over very large areas, we are still faced with the fact that the presence or absence of a human signal in this very large scale pattern of observed change needs to be assessed relative to a background of considerable natural internal climate variability. However, it is not possible to estimate the characteristics of the natural internal climate variability directly from observations since the historical observations are presumably affected by external forcing. Moreover, the pattern of change that we are interested in involves most of the available historical instrumental surface temperature observations; temperature readings from thermometers become increasingly sparse as you go back into the 19th century, with readings available in the 18th century in only a few isolated locations. The approaches that have been developed in statistical science for modeling the behavior of high dimensional space-time processes are generally not used (Cressie and Wikle, 2011) because most involve assumptions about the properties of the natural internal variability that are not well justified (e.g., assumptions such as isotropy in space, which can hardly be expected given the land-sea distribution of the Earth). Approaches that take advantage of some of the expected properties of the covariance matrix, which describes how variations at different times and places are related (Rajaratnam et al., 2008; Guillot et al., 2013) are also only rarely used at the moment, although there are exceptions, such as Ribes and Terray (2013).

Instead, estimates of the covariance matrix of the background internal climate variability are usually based on climate simulations obtained from climate models that have been run under control conditions (i.e., with constant atmospheric composition, no episodic volcanic influences, and no variation in solar output). In some cases, it has been possible to use climate models to simulate several thousand years of the evolution of the climate under such conditions, thus providing up to 100, or possibly more, simulated realizations of how the climate might have evolved in the absence of external influences since 1901. These 100 or so pseudo-realizations of the unperturbed climate are used to calculate a sample covariance matrix that estimates the true covariance matrix of the internal variability.

20.3   A Weighted Regression Approach

A general approach in detection and attribution studies is to regard an observed climate change Y as a linear combination of externally forced signals X and residual internal climate variability ϵ (Hegerl and Zwiers, 2011), viz.

$$Y = X\beta + \epsilon, \tag{20.1}$$

where vector Y is a filtered version of the observed record, the columns of matrix X contain estimates of the effects of the external forcings that are under investigation, and β is a vector of regression coefficients that adjust the amplitudes of those expected patterns of change so that the linear combination matches the observations as closely as possible.

Typically, X contains no more than three columns (e.g., representing the responses to greenhouse gas changes, the combined effects of other anthropogenic influences such as aerosol emissions and land surface changes, and the combined effects of natural external influences from explosive volcanic eruptions and changes in solar output). Each column in X is estimated from a set of climate model simulations in which the climate model has been run with the forcing that corresponds to that column. For example, the first column in X might contain estimates of expected surface temperature changes that are derived from climate model simulations run with the observed history of greenhouse gas concentration changes, the second column might be based on additional runs using the observed history of aerosols and land surface changes, and perhaps a third might be based on runs with the observed history of volcanic eruptions and changes in solar output.

In Figure 20.1, for example, which is based on the assessment of Hegerl et al. (2007), we suppose that the observed 1901–2005 change in surface temperature (Figure 20.1(a), representing the vector Y) can be represented as a combination of the effects of two kinds of forcing, which are represented by Figures 20.1(b) and 20.1(c). Figure 20.1(b) shows an estimate of the effects of human induced forcing (greenhouse gases, aerosols, land use effects, etc.) and natural external forcing (solar output changes and volcanic forcing) combined, and could be considered as a candidate for the first column of matrix X, say X_1. Figure 20.1(c) shows an estimate of the effects of natural external forcing only, and could be considered as a candidate for the second column of

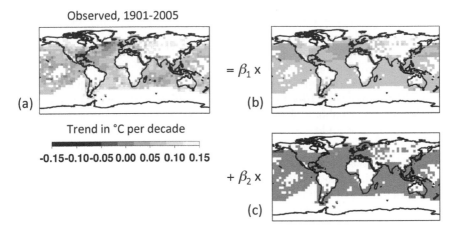

FIGURE 20.1: After Hegerl and Zwiers (2011), based on Hegerl et al. (2007): Schematic for detection and attribution. The observed change (panel (a), shown here as the pattern of temperature change over the twentieth century) is assumed to be composed of a linear combination of the effects of two kinds of external forcing plus the effect of internal climate variability. Panel (b) shows an estimate from climate models of the effects of all forcings combined, which includes human induced forcings and natural external forcings. Panel (c) shows an estimate from climate models of the effects of natural forcings only. See text for further details. Units are °C.

the matrix, say X_2. The amplitudes of the two patterns of forced change are adjusted with the coefficients β_1 and β_2 that make up the vector of regression coefficients β so that the sum best matches the observed changes. Note that in climate science, these regression coefficients are also often referred to as scaling factors.

It is assumed that the effects of human induced forcings and natural forcings add linearly. Thus the figure supposes, as does Equation (20.1), that

$$Y = \beta_1 X_1 + \beta_2 X_2 + \epsilon,$$

where
$$X_1 = X_{\text{ANT}} + X_{\text{NAT}}, \quad X_2 = X_{\text{NAT}},$$

with X_{ANT} and X_{NAT} representing the effects of human induced (ANThropogenic) and NATural external forcings respectively. Equivalently, this also means that $Y = \beta_1 X_{\text{ANT}} + \beta_2' X_{\text{NAT}} + \epsilon$, where $\beta_2' = \beta_1 + \beta_2$. This small algebraic manipulation is required because climate modeling centers have most frequently produced separate ensembles of simulations with the anthropogenic and natural forcings combined and with natural forcing only, in preference to separate ensembles of simulations with anthropogenic forcings only and natural forcings only.

Depending upon the study, vector Y and the columns of X may be of dimension of 275 or so, representing the decade-by-decade evolution of the large-scale spatial pattern of changes in a climate variable such as surface temperature over the 110-year period from 1901 to 2010, where each decadal spatial pattern has about 25 spatial components in each decade; see, e.g., Jones et al. (2013). It should be noted that the climate model output is spatially and temporally complete (there are no missing values), whereas the observations are often missing, both in space and in time. Figure 20.1(a) shows linear trends in surface temperature over the period 1901–2005. Trends are shown for $5° \times 5°$ latitude by longitude regions that are judged to have sufficient observations to allow the reliable estimation of trends. Grey areas in Figure 20.1(a) indicate locations where trends cannot be estimated due to insufficient data being available. This includes some large parts of the global oceans where historically, there was little ship traffic (up until recently, most ocean surface temperature readings were collected by commercial ships). Also, there are data voids in both polar regions and over several large land areas (particularly in South America, Africa, and parts of Asia) either due to an insufficient observing network, because archived paper records have not yet been digitized, or, in some cases, because national meteorological services do not have the mandate or resources to disseminate the data that they gather. As shown in Figures 20.1(b)–(c), the climate model output is therefore made to be missing at the same times and places as the observed data so that a like with like comparison can be made between the observations and the models.

The scaling factors (regression coefficients) β are often estimated by the so-called weighted least squares method as

$$\hat{\beta} = (X^\top C^{-1} X)^{-1} X^\top C^{-1} Y = (\tilde{X}^\top \tilde{X})^{-1} \tilde{X}^\top \tilde{Y}.$$

Here the matrix $\tilde{X}$ and vector $\tilde{Y}$ are the weighted versions of the signal patterns and observations, such that after weighting, the estimator of the regression coefficients looks like that used in the ordinary least squares method. This weighting reduces the uncertainty of the scaling factors $\hat{\beta}$ (Hasselmann, 1997; Allen and Tett, 1999), although in some detection and attribution problems, that benefit is difficult to realize because of difficulty in determining a complete set of weights; see, e.g., Hegerl et al. (1996) and Polson et al. (2013). The weighting is based on an estimate of the covariance matrix of the climate's natural internal variability, which as explained above, is derived from a limited set of climate model control runs. Thus the weights themselves are uncertain. Climate scientists account for that uncertainty by using a second set of climate model control simulations to estimate the natural internal variability of $\tilde{Y}$ (Hegerl et al., 1997; Allen and Stott, 2003) and consequently, the uncertainty of the regression coefficients $\hat{\beta}$.

Having fitted regression model Equation (20.1) to the observations, climate scientists ask essentially three questions.

1. Is the residual variability $\hat{\epsilon} = Y - X\hat{\beta}$ consistent with the estimates of natural internal climate variability obtained from climate model control simulations? If not, then the rest of the analysis is drawn into question since the criteria that will be used to evaluate questions about the regression coefficients depend on estimates of internal variability from climate models. One approach that has been used in such cases is to proceed to questions (ii) and (iii) after arbitrarily inflating the climate model simulated variability by a factor of 2 or more.

2. Are the estimated regression coefficients $\hat{\beta}$ significantly greater than zero? If not, then there is insufficient evidence to support the idea that observations reflect the estimated effects of external forcing that are described in matrix X.

3. If there is evidence that the expected effects of external forcing are reflected in the observations, then does their magnitude correspond to that which is expected based on our theoretical understanding of the physical processes that link forcing to its effects and given known uncertainties, for example, in the magnitude of forcing and its influence on climate?

We will come back to these inference questions in Section 20.6, but before doing so, it is important to describe the role of the climate models in detection and attribution studies in a bit more detail.

20.4   Role of the Global Climate Models

Most recent detection and attribution work uses signal patterns, often called fingerprints, that vary in time, space, or both. In the case of space and time, you can think of the fingerprints as a time series of patterns of changes that are expected at different times from a given forcing (or set of forcings combined). Figures 20.1(b)–(c) illustrate two spatial fingerprints (expected patterns of change over a 105-year period), but you could also easily imagine a time series of patterns connected together, describing the expected change decade-by-decade.

The fingerprints are almost always obtained from comprehensive climate models, consisting of coupled atmospheric and oceanic models that simulate winds and currents and are based on the principles of thermo- and fluid-dynamics, together with sea ice, land surface, and ice-sheet models, and often also, interactive vegetation, carbon cycle and chemistry components. While the physical and chemical processes represented by these models can be described mathematically in a relative concise manner, the equations that comprise the models are far too complex to be solved mathematically. Hence, the equations are "discretized" and solved via large-scale computer calculations (runs) using some of the largest computers on the planet. By discretized we mean that the equations are approximated in such a way that the atmosphere and ocean are represented by sets of cubes covering the Earth and extending upward through the atmosphere and downward through the depths of the ocean, the land surface is represented by a set of tiles that covers the Earth (or several layers of tiles to represent how soil properties vary with depth), and time passes by in fixed increments (time steps).

Typically, such models will have a spatial resolution of about two degrees latitudinally and longitudinally (Kharin et al., 2013) with time steps of 15 minutes or so, although higher resolution models are increasingly being developed. For example, global weather forecasting models, which are closely related to climate models, often have resolutions of about 35 km and there are examples of global climate models with 20 km resolutions.

Comprehensive global climate models (GCMs) simulate the natural internal variability of the climate as well as the response to whatever external forcing may be specified when the model is run. For example, when these models are run with the historical evolution in greenhouse gas concentrations, aerosols, land use change, volcanic eruptions, and changes in solar output since the beginning of the industrial revolution, they simulate changes in global mean temperature similar to those that have been observed; see, e.g., Hegerl et al. (2007) and Jones et al. (2013). The warming response to these forcings is very evident in the global mean because spatial averaging removes a substantial part of the simulated internal variability, but the spatial pattern of warming and how it evolves over time is considerably noisier.

Many detection and attribution studies estimate the response to external forcing by averaging across an ensemble of GCM runs (independent realizations that are obtained by starting the model from slightly different initial conditions each time) to reduce the effects of internal variability on the signal estimates. The signal estimates are also subject to other sources of uncertainty including errors in the formulation of climate models or the omission of some processes that may be climatically important (such as the nitrogen cycle, which is thought to have an important role in mediating the carbon cycle but is not currently represented in all models containing carbon cycle components), forcing uncertainty, possibly missing forcings that are not included in the simulations, and parameter uncertainty (climate models must necessarily parameterize, or approximate, processes that they cannot resolve, and the adjustable parameters in those approximations are often uncertain). Parameter uncertainty in the representation of processes involving clouds and aerosols, in particular, remains large, and one of the implications of this is that the effects of aerosols on the lifetimes and reflectivity of clouds continues to be a very large source of uncertainty in simulating historical and future climate changes. This in turn affects our understanding of a much discussed climate system parameter, the so-called equilibrium climate sensitivity, which is an indicator of the amount of warming that we might eventually experience if carbon dioxide concentrations were to double relative to pre-industrial levels, and remain at that level.

The costs of operating the models are such that it is rarely possible to simulate, with a given model, more than about ten realizations of the evolution of the climate since the end of the pre-industrial era (around 1850) to the present; most modeling centers, in fact, produce smaller ensembles, with five or fewer simulations. Thus estimates of the effects of forcing from individual models are necessarily affected by sampling variability and other important sources of uncertainty as mentioned above.

20.5   A Slightly More Detailed Regression Model

The regression model of Equation (20.1) is often replaced with a slightly more complex errors-in-variables (EIV) model to account for uncertainties in the estimated effects of forcing. This model is fitted with the so-called total least squares algorithm (Allen and Stott, 2003) and has the form

$$Y = (X - \delta)\beta + \epsilon, \tag{20.2}$$

where the columns of random matrix δ are independent of ϵ but have covariance matrices that are proportional to that of ϵ. The term δ is introduced in this model to account for the fact that the signal estimates that make up the columns of X are subject to sampling uncertainty since they are estimated from finite ensembles of climate change simulations. For example, if the signals are estimated using an ensemble of five independent historical climate change simulations from a given climate model, then the columns of δ will have covariance matrices $C_{\delta\delta} = C/5$, where C is the covariance matrix of ϵ that represents the effects of internal variability in the observations Y.

More generally, the signals are estimated by averaging signal estimates from multiple climate models, substantially reducing uncertainty from internal variability in X, but introducing uncertainty due to inter-model differences. Typically this is treated using Equation (20.2), although on rare occasions, studies have used a more complex errors-in-variables model modified to recognize explicitly that X is affected by two sources of sampling variability — from the selection of climate models (discussed below) and from internal climate variability (Huntingford et al., 2006). Philosophically, however, this is difficult since we have no clear way to describe how the "sample" of available models was drawn from the population of conceptually plausible representations of the climate system; see, e.g., Rougier et al. (2010) and von Storch and Zwiers (2013). Also, many current climate models are developed, in part, by using components that are developed elsewhere since no one modeling center has the expertise or capacity that would be required to develop a complete model independently. Thus there is a question of the extent to which simulations from different models are "independent"; see, e.g., Jun et al. (2008). Here the notion of independence again refers to how the population of all plausible climate models is being sampled by the climate modeling community.

The selection of climate models for a given purpose remains a difficult question. Current international climate modeling activities are largely coordinated through the *Coupled Model Inter-comparison Project, Phase 5* or CMIP5 (Taylor et al., 2012), which sets out a number of standard protocols for running climate change simulations in order to facilitate their intercomparison. At last count, about 25 modeling centers around the globe with a combined total of about 60 different model variants were CMIP5 participants; see `http://cmip-pcmdi.llnl.gov/cmip5/availability.htmlforCMIP5statusupdates`. It is well recognized that not all models are of equal quality, but objective measures of model quality are very much dependent upon the climate variable of interest, and overall measures of quality remain somewhat subjective (Gleckler et al., 2008); see the discussion of Knutti et al. (2010).

A very frequently encountered problem in fitting the models (20.1)–(20.2) is that the sample covariance matrices $\hat{C}$ are not of full rank despite filtering that retains only decadal scale temporal variability and continental, or larger scale, spatial variability. Rank refers to a property of matrices such that a matrix has full rank if no column in the matrix can be expressed as a linear combination of its remaining columns; the standard scheme for finding weights for fitting models (20.1) or (20.2) depends upon $\hat{C}$ having full rank.

For example, the space-time surface data vector used by Jones et al. (2013) has dimension 275, but it remains difficult to produce full rank estimates of a covariance matrix with 275 rows and columns from currently available climate simulations. This is because, despite the availability of ever more extensive coordinated international climate model intercomparison exercises such as CMIP5, the total number of century-long control simulation segments that have been run by the different modeling centers remains limited. The implication is that unique weights based on $\hat{C}$, as described in Section 20.3, for fitting regression models (20.1)–(20.2) cannot be obtained; thus regularization, in which constraints are placed on $\hat{C}$, has been proposed.

Most studies use a regularization approach that is based on decomposing $\hat{C}$ according to its so-called eigenvectors, which are called Empirical Orthogonal Functions or EOFs in climate research (von Storch and Zwiers, 1999). The observations and signals can be expressed uniquely as linear combinations of these EOFs (i.e., as a sum of EOFs multiplied by uniquely determined weights). The EOFs are ordered according to their ability to represent the variability that is present in the observations, and generally, a large fraction of the total variability can be represented with only a small number of EOFs. Regularization is thus performed by approximating the observations and the signals using a number of EOFs that is substantially smaller than the length of the data vector Y. In the surface temperature problem, this effectively replaces a data vector Y that is of length 275 with one that is much shorter — typically with 30–40 elements (Jones et al., 2013) — thus making it feasible to calculate a full rank covariance matrix with the available collection of control runs. The columns of the signal matrix X are similarly shortened.

Recently, Ribes et al. (2009, 2013) have proposed an alternative regularization approach which attacks the problem by estimating the covariance matrix differently, using the method of Ledoit and Wolf (2004), which modifies the sample covariance matrix so that it has full rank. This implies that an EOF approximation is not required for either the observations Y or the signals X. While the formalism of Ribes et al. is only beginning to be used in detection and attribution research Ribes and Terray (2013), it has the advantage that all data reduction decisions (such as the decision to use decadal mean temperatures averaged over very large regions) that are taken to reduce dimensionality or increase signal-to-noise ratios are performed prior to fitting any regression model. In contrast, this occurs in two stages in most current detection and attribution studies, with data reduction taking place prior to the analysis and typically also within the analysis as a consequence of the EOF-based approximation of the observations and the signals.

20.6   Interpreting the Results

Regression models such as (20.1) or (20.2) are used in detection and attribution studies to support the interpretation of observed climate changes, such as the warming that has taken place since the beginning of the twentieth century. As explained at the end of Section 20.3, this involves three kinds of statistical inferences.

First, the fit of the regression model to observations is assessed by means of a residual consistency test, which compares the estimated residual variability in $\hat{\epsilon} = Y - X\hat{\beta}$ or $\hat{\epsilon} = Y - (X - \hat{\delta})\hat{\beta}$ with an estimate of internal climate variability. In both cases, the test statistic can be evaluated with the familiar Fisher–Snedecor F distribution. Unexpectedly small or large values of this statistic would flag that the climate model (or models) that have been used to estimate the covariance of the internal variability do not, in fact, simulate that variability correctly. Also, unexpectedly large values of the statistic might indicate that the climate model simulated signals have a different form from those that are actually reflected in the observations. For example, the discrepancy between a climate model simulated response to human induced forcing and observations would be larger than could be explained by internal variability if the climate model simulated the same amount of warming over land as observed, but less warming than observed over oceans. In the case of surface temperature, the available evidence suggests that climate models do simulate internal variability reasonably well on the space and time scales that are of interest in detection and attribution studies; see, e.g., Hegerl et al. (2007) or Jones et al. (2013).

Once we are satisfied that the statistical model fits the observations reasonably well, the next step is to determine whether the climate changes that are expected from external forcing are reflected in the observations. This is accomplished by testing the hypothesis that the regression coefficients (or scaling factors) β are greater than zero. If, for example, global surface temperature observations reflect the warming effect of human influences on the climate system, then the regression coefficient that modifies the amplitude of the climate model's simulated warming effect of that influence should be significantly greater than zero. In contrast, if the scaling factor is indistinguishable from zero, or substantially negative, then we do not have evidence that the corresponding signal is present in the observations.

In Figure 20.1, it is evident that the observed pattern of surface temperature change, in this case for the period 1901–2005 — see Figure 20.1(a) — has a strong resemblance to the pattern of warming that is expected from the combination of human and natural external influences on the climate (Figure 20.1b) and little or no resemblance to the pattern of surface temperature change that is expected from natural external influences (i.e volcanic and solar forcing forcing) alone — see Figure 20.1(c). Numerous studies indicate that the scaling factor β_1 that multiplies the expected pattern of change due to human influence on the climate system is significantly greater than 0 at a very high significance level, while the scaling factor β_2 that multiplies the expected pattern of change due to volcanic and solar forcing is indistinguishable from zero. This then leads to the conclusion that human influence on the climate is clearly detectable; see, e.g., Hegerl et al. (2007).

Assuming that the expected patterns of change are detected (i.e., one or more of the scaling factors is found to be significantly larger than zero, indicating that the corresponding signal is likely reflected in the observations with non-negligible amplitude), then a further question is whether the climate model simulated amplitude of the expected signal corresponds to its amplitude in observations. Our physical understanding of the climate system, which is embodied in the climate models, plus our understanding of the forcing, leads to a physically based estimate of the response to external forcing that is expressed in physical units; see, e.g., °C in Figure 20.1(b). In the case of surface temperature, this is illustrated by Figure 20.1(b). If, for example, we have underestimated the forcing, the climate models would still be expected to produce the same pattern of expected surface temperature change, but it would be more muted than if the forcing had been specified correctly. This in turn would lead to estimates of regression coefficients with values that would tend to be greater than one, indicating a need to scale up the climate model simulated signals, and thus compensating for the error in forcing. Similarly, regression coefficients that are different from one would compensate for a climate model that responds either too vigorously (estimated regression coefficient less than 1), or not vigorously enough (regression coefficient greater than 1) because its sensitivity is not correct; see Section 20.4. Thus the question of the consistency between the signal amplitude simulated by a climate model, and the amplitude of that signal that is reflected in the observations, is an important one. The consistency can be evaluated by examining confidence intervals for the estimated scaling factors, and determining whether those intervals include "1."

Detection and attribution research on the evolution of surface temperature during the twentieth century was assessed for the 4th Assessment Report of the Intergovernmental Panel on Climate Change (IPCC) by Hegerl et al. (2007). They evaluated studies using individual and multiple climate models combined that considered whether three externally forced signals (the effects of greenhouse gas increases, other human influences, and natural external influences) were reflected in twentieth century observations. Figure 20.2(a) shows the regression coefficients that are obtained when signal estimates produced with different climate models or a group of climate models combined are used to evaluate observed large-scale changes in surface temperature over the 10-decade period 1901–99. These analyses cover only the period up to 1999 because the historical climate model simulations available at the time did not extend beyond 1999. The figure shows central estimates of the scaling factors and 90% uncertainty bands. Regardless of the climate model used, the signals associated with greenhouse gas increases (light gray intervals) and other anthropogenic influences (dominantly aerosols, medium gray intervals) are clearly detected. Also, scaling factor estimates are generally consistent with 1, although there is a small tendency for the climate models to warm slightly more due to greenhouse gas forcing than is supported by the observations. The response to natural external forcing (solar and volcanic activity, dark gray intervals), is also detected based on some models and in the multi-model "EIV" analysis, but less consistently. Recall, however, that this signal is weak; see, e.g., Figure 20.1(c).

Having detected the presence of an externally forced signal in observations, and after evaluating whether its amplitude in observations is consistent with the amplitude that is expected from climate model simulations, a further step is to attribute part of the observed change to external forcing. As we have seen, in the case of surface temperature, the patterns of change that are theoretically expected to result from human induced emissions of greenhouse gases and aerosols are found in observations and have about the right amplitude. Other explanations, such as the possible effects of changes in volcanic activity and solar output, do not provide a plausible explanation (e.g., compare Figures 20.1(a) and 20.1(c)), and the pattern of change that is seen is not one that could easily be explained by natural internal climate variations. Thus we must conclude that human induced changes are likely responsible for most of the observed change over the latter half of the twentieth century on all continents except Antarctica (Hegerl et al., 2007).

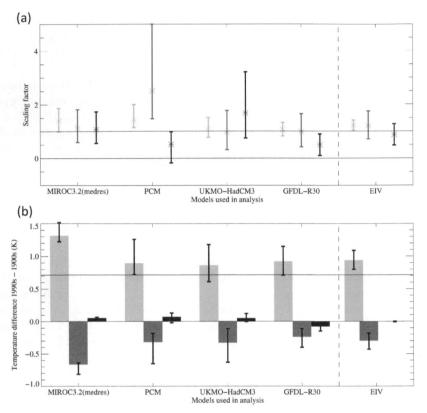

FIGURE 20.2: From Hegerl et al. (2007): (a) 5 to 95% uncertainty limits
on scaling factors (dimensionless) based on an analysis over the twentieth
century, and (b) the estimated contribution of forced changes to temperature changes over the twentieth century, expressed as the difference between
the 1990 to 1999 mean temperature and the 1900 to 1909 mean temperature (°C). The horizontal black line in (b) shows an estimate of the observed
temperature change. Detection and attribution results were obtained using
estimated responses to external forcing from four different climate models (labeled MIROC3.2(medres), PCM, UKMO–HadCM3 and GFDL–R30) in separate total least squares analyses (i.e., using Equation (20.2)). Also shown,
labeled "EIV" are the results of an analysis simultaneously using estimated
responses from three models (PCM, UKMO–HadCM3 and GFDL–R30) that
explicitly considers inter-model uncertainty. See Hegerl et al. (2007) for details
and sources for individual results.

An interval estimate of the size of the detected and attributed effect in
observations over the twentieth century can be obtained by multiplying the
climate model estimates of the effects (from the columns of matrix X) with
the confidence range of the estimated scaling factors. This is shown in Figure

20.2(b), which shows a set of estimates of attributed warming or cooling (bars) with an estimated uncertainty (whiskers) for each of three external forcing agents and several different analyses using either individual climate models or an ensemble of climate models. The bars are obtained by scaling the climate model simulated change in global mean temperature between the first and last decades of the twentieth century by best estimates of the scaling factors for the responses to the forcings represented by the bars.

The estimated uncertainty in the attributed warming or cooling reflects the uncertainty in the corresponding scaling factor, but not other sources of uncertainty such as forcing uncertainty. The figure shows that greenhouse gas forcing, acting on its own, would likely have warmed the planet substantially more than observed during the twentieth century and that this was partially offset by the cooling effects of other anthropogenic influences. It should be noted that, even if a model's sensitivity to greenhouse gas changes was over-estimated, or its sensitivity to aerosols underestimated, the estimated scaling factor would correct for this by scaling the model simulated response to best match the observed changes. Of course, substantially more is required for definitive attribution assessment than simply an evaluation of signal scaling factors; also required is a set of strong physical arguments that would rule out other factors as a plausible explanation of the observed change.

20.7   Discussion

Statistical reasoning and methods are fundamental to the problem of detecting and attributing human influence on the climate. This is, in a sense, the ultimate small sample problem since there is only one realization of historical observed change, and consequently there is a very heavy reliance on physical understanding and climate models. The climate models provide physically based estimates of the expected climatic effects of external forcing and information that is used to describe the climate's natural internal variability. Physical process understanding is required to support statistical inferences about the presence or absence of the effects of external forcing on our climate.

While detection and attribution analyses involve complex physical reasoning and a series of complex data processing decisions, relatively simple statistical models have been used to make inferences about the presence and magnitude of externally forced effects on our climate. A number of assumptions, however, are made in the course of conducting an analysis. These assumptions include:

a) An assumption that the effects of different forcings acting on the climate system are additive. Perturbation analyses suggest that this is reasonable for small amounts of forcing, such as the forcing experienced to date. Nevertheless, there is some evidence that additivity may not hold in all cases, such as for precipitation at smaller than continental scales. This is because the effects on precipitation presumably result from a combination of thermo-dynamic (warming related) and dynamic (atmospheric circulation related) changes.

b) An assumption that forcing does not affect the natural internal variability of the climate system. Again, this should be true for small forcings, although some studies have demonstrated, for example, that the extremes of tem-perature may not change in step with changes in the mean of the surface temperature distribution under projected future forcing in all locations; see, e.g., Hegerl et al. (2004) or Kharin et al. (2013).

c) A strong assumption that modern climate models simulate the natural in-ternal variability of the climate system well. The evidence suggests, over-all, that models do simulate internal variability well on the space and time scales considered in detection and attribution studies; see, e.g., Hegerl et al. (2007) or Jones et al. (2013). Also, detection and attribution studies rou-tinely compare residual variability to model simulated internal variability; see, e.g., Allen and Stott (2003) and Ribes et al. (2013).

d) A further strong assumption is that the collection of climate models par-ticipating in experiments such as CMIP5 can be thought of as a sample that has been drawn from a defined population with a known sampling strategy. While this is not at all an obvious assumption (Rougier et al., 2010; von Storch and Zwiers, 2013), empirical evidence suggests, for exam-ple, that currently available climate models have biases that cluster about zero; see, e.g., Gleckler et al. (2008), Jun et al. (2008) and Sillmann et al. (2013). Designed experiments that control the sources of variation between different climate models, for example, to quantify the uncertainty from un-certain parameters are beginning to be used; see, e.g., Murphy et al. (2004), Stainforth et al. (2005), and subsequent literature. However, these types of approaches to uncertainty quantification have not yet permeated to the large-scale coordinated modeling protocols, such as that of CMIP5.

Each of these assumptions represents an important ongoing area of climate research. Detection and attribution studies acknowledge these limitations, and high-level assessments of the literature, such as those of the Intergovernmental Panel on Climate Change (Hegerl et al., 2007), take them into account when evaluating the available body of science.

Finally, a critical element in detection and attribution is the estimation of the covariance characteristics of the climate's natural internal variability. Up until recently, climate scientists have used sample covariance matrices calculated from samples of climate model output rather than more advanced approaches, e.g., the use of so-called shrinkage estimators such as that of Ledoit and Wolf (2004). While such approaches, as adopted for example by Ribes and Terray (2013), alleviate a technical issue, they bring into clearer focus the question of how observations should be processed prior to conducting a formal detection and attribution analysis. Such pre-processing should isolate, in a transparent way, the spacial and temporal scales at which we expect signals to be evident and also should be designed to retain only scales on which models are judged to be reliable, which is an assessment that remains challenging; see, e.g., Knutti et al. (2010).

Acknowledgments

The authors express their sincere appreciation to the editor, Jerald F. Lawless, for his clear and insightful comments, which helped to greatly improve this chapter.

About the Authors

Francis W. Zwiers is the director of the Pacific Climate Impacts Consortium at the University of Victoria. He specializes in the application of statistical methods to the analysis of observed and simulated climate variability and change. He served as a coordinating lead author of the Fourth Assessment Report of the Intergovernmental Panel on Climate Change, and is an elected member of the IPCC Bureau. He is a fellow of the Royal Society of Canada and the American Meteorological Society. He received the Meteorological Service of Canada Patterson Medal and the 2011 SSC Impact Award.

Gabrielle C. Hegerl is chair of Climate System Science at the School of Geosciences, University of Edinburgh. Previously, she was a research professor at Duke University, and held appointments at Max-Planck-Institut für Meteorologie, the University of Washington, and Texas A&M University. She has a master's degree and a PhD in applied mathematics from Ludwig Maximilians Universität in München. Her main research interests are identifying the causes of observed climate change; she also works on estimation of climate sensitivity and characterization of climate variability. She has been a member of the core writing team and a lead author for the Assessment Reports of the Intergovernmental Panel on Climate Change (IPCC). She is a fellow of the Royal Society of Edinburgh.

Xuebin Zhang is a research scientist with the Climate Research Division, Environment Canada. He also holds adjunct appointments with the Department of Mathematics and Statistics at York University and at the Institut national de la recherche scientifique (Québec). He has a PhD in physics from the Universidade de Lisboa, Portugal. His main research interests are understanding the causes of observed climate change and projecting future climate. He served as lead author of the Special Report on Managing the Risks of Extreme Events and Disasters to Advance Climate Change Adaptation of the Intergovernmental Panel on Climate Change (IPCC) and as lead author of the IPCC 5th Assessment Report.

Qiuzi Wen is an NSERC visiting fellow at Climate Research Division, Environment Canada, and she also holds a position of research assistant professor in the Institute of Atmospheric Physics, Chinese Academy of Sciences, Beijing, China. She obtained her degrees in probability and statistics from Peking University and York University, in Toronto. Her primary research interest is to study changes in Earth's climate system through application and development of advanced statistical methods.

Bibliography

Allen, M. R. and Stott, P. A. (2003). Estimating signal amplitudes in optimal fingerprinting, Part I: Theory. *Climate Dynamics*, 21:477–491.

Allen, M. R. and Tett, S. F. B. (1999). Checking for model consistency in optimal finger printing. *Climate Dynamics*, 15:419–434.

Cressie, N. and Wikle, C. (2011). *Statistics for Spatio-Temporal Data*. Wiley, Hoboken, NJ.

Fourier, J.-B. J. (1824). *Mémoires de l'Académie Royale des Sciences de l'Institut de France VII*.

Gleckler, P. J., Taylor, K. E., and Doutriaux, C. (2008). Performance metrics for climate models. *Journal of Geophysical Research*, 113:D06104, doi:10.1029/2007JD008972.

Guillot, D., Rajaratnam, B., and Émile-Geay, J. (2013). Statistical paleoclimate reconstructions via Markov random fields. arXiv:1309.6702.

Hasselmann, K. (1997). Multi-pattern fingerprint method for detection and attribution of climate change. *Climate Dynamics*, 13:601–612.

Hegerl, G. C., Hasselmann, K., Cubasch, U., Roeckner, E., Voss, R., and Waszkewitz, J. (1997). Multi-fingeprint detection and attribution analysis of greenhouse gas, greenhouse gas-plus-aerosol and solar forced climate change. *Climate Dynamics*, 13:613–634.

Hegerl, G. C., von Storch, H., Hasselmann, K., Santer, B. D., Cubasch, U., and Jones, P. D. (1996). Detecting greenhouse-gas-induced climate change with an optimal fingerprinting method. *Journal of Climate*, 9:2281–2306.

Hegerl, G. C. and Zwiers, F. W. (2011). Use of models in detection and attribution of climate change. *WIREs Climate Change*, 2:570–591.

Hegerl, G. C., Zwiers, F. W., Braconnot, P., Gillett, N. P., Luo, Y., Marengo, J. A., Nicholls, N., Penner, J. E., and Stott, P. A. (2007). Understanding and attributing climate change. In *Contribution of Working Group I to the Fourth Assessment Report of the Intergovernmental Panel on Climate Change*. Cambridge University Press, Cambridge.

Hegerl, G. C., Zwiers, F. W., Stott, P. A., and Kharin, V. V. (2004). Detectability of anthropogenic changes in annual temperature and precipitation extremes. *Journal of Climate*, 17:3683–3700.

Huntingford, C., Stott, P. A., Allen, M. R., and Lambert, F. H. (2006). Incorporating model uncertainty into attribution of observed temperature change. *Geophysical Research Letter*, 33:L05710.

Jones, G. S., Stott, P. A., and Christidis, N. (2013). Attribution of observed historical near-surface temperature variations to anthropogenic and natural causes using CMIP5 simulations. *Journal of Geophysical Research: Atmospheres*, 118:4001–4024.

Jun, M., Knutti, R., and Nychka, D. W. (2008). Spatial analysis to quantify numerical model bias and dependence: How many climate models are there? *Journal of the American Statistical Association*, 108:934–947.

Kharin, V. V., Zwiers, F. W., Zhang, X., and Wehner, M. (2013). Changes in temperature and precipitation extremes in the cmip5 ensemble. *Climatic Change*, 119:345–357.

Knutti, R., Furrer, R., Tebaldi, C., Cermak, J., and Meehl, G. C. (2010). Challenges in combining projections from multiple climate models. *Journal of Climate*, 23:2739–2758.

Ledoit, O. and Wolf, M. (2004). A well-conditioned estimator for large dimensional covariance matrices. *Journal of Multivariate Analysis*, 88:365–411.

Murphy, J. M., Sexton, D. M. H., Barnett, D. N., Jones, G. S., Webb, M. J., and Collins, M. (2004). Quantification of modelling uncertainties in a large ensemble of climate change simulations. *Nature*, 430:768–772.

Polson, D., Hegerl, G. C., and Zhang, X. (2013). Causes of robust seasonal land precipitation changes. *Journal of Climate*, in press, doi:10.1175/JCLI–D–12–00474.1.

Rajaratnam, B., Massam, H., and Carvalho, C. M. (2008). Flexible covariance estimation in graphical Gaussian models. *The Annals of Statistics*, 36:2818–2849.

Ribes, A., Azaïs, J.-M., and Planton, S. (2009). Adaptation of the optimal fingerprint method for climate change detection using a well-conditioned covariance matrix estimate. *Climate Dynamics*, 33:707–722.

Ribes, A., Planton, S., and Terray, L. (2013). Application of regularized optimal fingerprinting to attribution. Part I: Method, properties and idealised analysis. *Climate Dynamics*, 41:2817–2836.

Ribes, A. and Terray, L. (2013). Application of regularised optimal fingerprinting to attribution. Part II: Application to global near-surface temperature. *Climate Dynamics*, 41:2837–2853.

Rougier, J., Goldstein, J. M., and House, L. (2010). Assessing climate uncertainty using evaluations of several different climate simulators. `http://www.maths.bris.ac.uk/?MAZJCR/mme2.pdf`. Accessed 14 December 2011.

Sillmann, J., Kharin, V. V., Zhang, X., Zwiers, F. W., and Bronaugh, D. (2013). Climate extremes indices in the CMIP5 multimodel ensemble: Part I. Model evaluation in the present climate. *Journal of Geophysical Research*, 118:1–18.

Stainforth, D. A., Aina, T., Christensen, C., Collins, M., Faull, N., Frame, D. J., Kettleborough, J. A., Knight, S., Martin, A., Murphy, J. M., Piani, C., Sexton, D., Smith, L. A., Spicer, R. A., Thorpe, A. J., and Allen, M. R. (2005). Uncertainty in predictions of the climate response to rising levels of greenhouse gases. *Nature*, 433:403–406.

Taylor, K. E., Stouffer, R. J., and Meehl, G. A. (2012). An overview of CMIP5 and the experiment design. *Bulletin of the American Meteorological Society*, 93:485–498.

von Storch, H. and Zwiers, F. W. (1999). *Statistical Analysis in Climate Research.* Cambridge University Press, Cambridge.

von Storch, H. and Zwiers, F. W. (2013). Testing ensembles of climate change scenarios for "statistical significance." *Climatic Change*, 119:1–9.

21

Data Hungry Models in a Food Hungry World: An Interdisciplinary Challenge Bridged by Statistics

Louis Kouadio and Nathaniel Newlands

Agriculture and Agri-Food Canada, Lethbridge, AB

Agriculture is an essential component of societal well-being. Feeding an increasing human population, expected to reach 9.3 billion in 2050 (United Nations, 2011), and combating hunger and poverty worldwide remains a tremendous challenge. Producing and distributing food is also increasingly vulnerable to interconnected risks and impacts due to extreme climatic events, harmful environmental pressures and socio-economic volatility. Today's rapid communication and globalization of trade increase the potential for high volatility in commodity pricing from economic disturbances and shocks. This is, in part, due to less constrained, open market-based valuation of major agricultural-based food, feed, fiber, and bioenergy products. There is also, however, a need to better understand crop production risks and commodity price fluctuations. Forecast models of crop yield and production can help to better design risk averse market regulation, trade and policy strategies and instruments that have a higher success of achieving and sustaining global food security.

21.1 Big Challenges for Future Global Crop Production

Statistical-based modeling offers a way to represent and understand real-world agricultural production systems. More accurate and timely crop yield forecasting methods can provide support for local, regional, national and international decision-making when future food supply and demand are highly variable and uncertain. Statistical-based forecasting methods are currently being advanced within Canada, involving strong interdisciplinary collaboration of government scientists, university researchers, and industry technologists and practitioners. The goal of this accelerated, collective effort is to enable a more comprehensive and reliable assessment of socio-economic and environmental-related vulnerabilities, risks and potential impacts affecting global food production and distribution. This chapter showcases how statistics is enabling the integration of data and knowledge to advance operational crop yield forecasting across Canada's agricultural extent.

21.2 Differing Needs for Crop Yield "Outlook" Forecasts

Crop forecasts provide information on the health and maximum attainable yield of a crop across space and time. Their accuracy relies on a sufficient and reliable understanding of specific water, energy and nutrient requirements and their availability to promote crop growth and ensure survival across a series of development stages (i.e., crop phenology). Crops differ in how sensitively they respond to cues and conditions within their environment. Forecasts are most useful if they are timely, consistent, and validated against other available or independent sources of information. They must also meet the needs of a wide variety of different agricultural stakeholders, i.e., agricultural producers/farmers, agri-business, media and provincial/federal government managers and policy-makers (Allen, 1994; Luo et al., 2011). Such requirements may differ by the statistical information presented, different levels of forecast accuracy, and relevant spatial and temporal coverage and scale.

Farmers are the basis of agricultural production and they need to make marketing decisions with financial repercussions before and during the growing season. Agri-business managers and crop exporting-agencies involved in the marketing chain need forecasts in making informed purchasing, routing and warehousing decisions, as well as to determine, in advance, where and how much to export. Agricultural forecasts may also provide users who are involved in the legislation and operational programs that help to protect domestic agriculture and enhance food security — by offering early warning of potential risks and impending impacts and their severity, and providing recommendations for potential mitigation and longer-term agricultural adaptation options.

Models are generally defined as a simplification or abstraction of a real (managed) complex dynamic system. They are considered to be comprised of a vast number of components and processes interacting over a wide range of organizational levels. A crop model can be described as a quantitative scheme for predicting the growth, development, and yield of a crop, given a set of genetic features and relevant environmental variables. In the simplest model representation, a crop grows by absorbing incoming light (i.e., photosynthetic active radiation), with important individual processes of crop growth (e.g., photosynthesis, respiration, assimilate partitioning, phenology), and strong inter-relationships between environmental variables are not taken into account.

More complex crop growth models describe such underlying biophysical processes in great detail, and include: (i) biological modules (e.g., photosynthesis, plant growth and partitioning, pest damage); (ii) environmental modules (e.g., weather, soil water and nitrogen balances, soil temperature); and (iii) management modules (e.g., irrigation, fertilization, planting and harvesting dates); see Boote et al. (1996). These complex models are increasingly being used in theoretical research, yield predictions, and decision-making in agriculture, but in many cases are viewed as "black-boxes," lacking transparency on inter-linkages and on their assumptions.

The wide spectrum of models — from the simplest to most complex can be used to generate crop forecasts. Nonetheless, a detailed consideration of their appropriateness, especially for use in generating operational forecasts is still needed. Forecasting models can be distinguished according to their complexity (number of variables and their inter-dependence), linearity/nonlinearity, assumptions regarding the probabilistic nature of underlying biophysical processes, application scale, and extent to which they integrate diverse sources of available data.

According to their structure and the amount of inputs required, methods for forecasting can be roughly subdivided into three categories: (i) simple methods that are generally empirically based and include descriptive rules and logic; (ii) complex methods based on more structural models and large datasets (they are generally process based); and (iii) integrated approaches that are based on sophisticated statistical methods and use various data sources. The latter appear as intermediate approaches between the first two; see Figure 21.1.

Typically, intermediate approaches are dynamic and multi-variate time-series models with either fixed covariates or a mixture of temporal autocorrelation and spatially dependent covariates involved. Complex methods can involve mechanistic approaches comprising deterministic assumptions linked with system structure and empirical data inputs. Systems of finite-difference equations (based on continuous-time differential equations) are used to model features such as crop growth and yield in relation to soil, carbon, nitrogen, water, solar radiation and other climate variables. Complex models of agroecosystems are also coupled to dynamic global or regional climate scenario models and used in ecological and economic forecasting.

There is a need for a new generation of models for crops and agroecosystems that can make better use of new types of observational data to better capture soil-plant-air interactions (Zhu et al., 2011). A common pitfall of dynamic models of spatial and temporal processes is that they are over-parameterized because their number of parameters increases, with increasing numbers of spatial points or areas. Readers are referred to a recent review of latest advances and approaches in modeling dynamic ecological processes by Leeds and Wilke (2012). In Section 21.3, we discuss the use of remote-

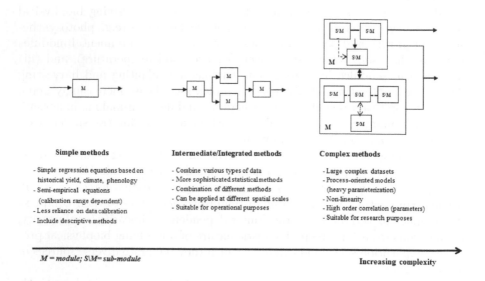

FIGURE 21.1: Summary of the wide range of complexity of statistical models applied in generating real-time, operational crop yield predictions/crop "outlook" forecasts.

sensing data in crop forecasting, and in Section 21.4 we describe a Bayesian statistical model for sequential, regional-scale forecasting.

21.3   Use of Satellite Remote Sensing Data

Regional crop production estimates based on analysis reports are often expensive, prone to large errors, and cannot provide real-time estimates or forecasts of crop condition. The application of crop growth models to a specific spatial region may rely on the selection of representative, long-term sampling locations, followed by spatial aggregation or interpolation. But geospatial analysis is hampered by fine-scale spatial variation in soil properties, management practices or climatic conditions, and the large number of model parameters to be calibrated across spatio-temporal scales. Yet the systematic acquisition and near real-time delivery of high resolution data are needed for critical periods during the growing season. With the development in earth observation data

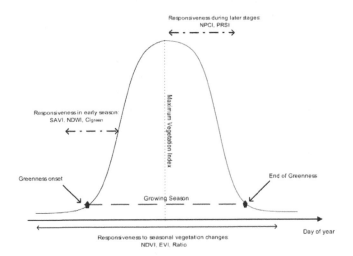

FIGURE 21.2: Selected vegetation indices and their potential use according to the phenological stages. NDVI: normalized difference vegetation index; EVI: enhanced vegetation index; SAVI: soil-adjusted vegetation index; NDWI: normalized difference water index, CIgreen: chlorophyll index green; NPCI: normalized pigment chlorophyll ratio index, and PSRI: plant senescence ratio index.

acquisition, satellite remotely sensed data, owing to their synoptic, timely and repetitive coverage, have been recognized as a valuable tool for crop monitoring at different levels.

Satellite remote sensing (RS) can be defined as the recording by on-board satellite sensors of the electromagnetic radiation that is reflected or emitted from the earth's surface. On the plant canopy surface the electromagnetic energy from the sun can be reflected, absorbed or transmitted (depending on the wavelength of this energy and the characteristics of plants). Vegetation indexes (VIs), that compare reflected light information at different wavelengths, are routinely developed and utilized to assess various plant parameters.

The so-called "normalized difference vegetation index" or NDVI is most commonly used. It is an index that is calculated from the visible and near-infrared incident radiation received by a detector, and compares information on a plant between a time when the chlorophyll it contains causes considerable absorption of incoming sunlight and a time when it develops a leaf structure, thereby increasing the amount of light that it reflects. The evolution of NDVI over time could be used to determine plant development key indicators or seasonal cumulative totals. The onset and the end of greenness in a given crop area determine the length of the growing season. During vegetative growing

TABLE 21.1: Estimation of canopy, leaf or soil biophysical variables as a function of the spectral domain used (Baret, 2000). The current level of accuracy and robustness possible for different parameters is indicated qualitatively by "+" (where "+++" is most accurate and robust and "−" implied no estimates are available). "Albedo" is a measure of how well a surface reflects solar energy, varying between 0 and 1. A value of 0 means the surface is a "perfect absorber" that absorbs all incoming energy, whereas a value of 1 means the surface is a "perfect reflector" that reflects all incoming energy.

	Biophysical variable	Visible Near Infrared	Near Infrared Short Wave Infrared	Thermal Infrared	Active μ-wave (radar)	Passive μ-wave
Canopy structure	Leaf area index (LAI)	+++	+++	+	++	+
	Leaf orientation	+++	+++	+	+	+
	Leaf size and shape	+	+	+	+	+
	Canopy height	−	−	−	++	+++
	Canopy water mass				+++	+++
Leaf characteristics	Chlorophyll content	+++	−	−	−	−
	Water content	−	+++	−	+++	+++
	Temperature	−	−	++++	−	++
Soil characteristics	Surface soil moisture	−	+	+	+++	+++
	Roughness	+	+	−	++	+
	Residues	+++	++	−	−	−
	Organic matter	++	++	−	−	−
	Soil type	++	++	+		
Other variables	Fraction cover (fcover)	++++	++++	++	++	+
	Fraction of absorbed photosynthetically active radiation (fAPAR)	++++	++++			
	Albedo	++++	+++			

phase (from crop emergence to flowering), for example, the SAVI (soil-adjusted vegetation index) can be useful in characterizing the proportion of area covered by the crop. Figure 21.2 depicts crop development reference points and time-intervals when different VIs best track crop growth through the growing season.

Table 21.1 provides a summary of agroecosystem/biophysical variables derived from NDVI and other remotely-sensed indices that measure soil moisture, roughness and other characteristics, crop canopy or leaf surface area. NDVI generated on a weekly basis is currently received using near real-time information collected by NASA and used to generate NDVI maps south of 60°N for assessing and discriminating crops (250 m scale) across the agricultural land area of Canada.

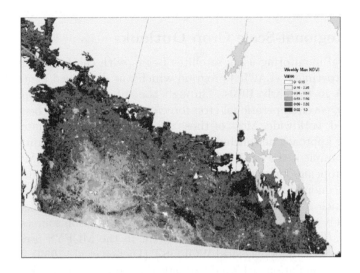

FIGURE 21.3: Example map of Normalized Difference Vegetation Index (NDVI) showing its variability across Western Canada (Canadian Prairie Provinces). This information is used to generate distribution maps of agricultural crops. NDVI indices are derived from bands 1 and 2 of the MODerate Resolution Imaging Spectroradiometer (MODIS) on board NASA's Terra satellite (Source: Agriculture and Agri-Food Canada) (Carroll et al., 2004).

Changes in the NDVI index enable a comparison of current growing conditions to those in a previous week, the same week in previous years or to a historical mean, and provide useful information for forecasting and assessing drought/floods, crop insurance and tax deferral agricultural support programs. Figure 21.3 provides an example crop-type map generated using NDVI observational data. These high resolution data are also now being incorporated into a broad set of process-level and statistical models that explore how agroecosystems respond to a changing climate.

The retrieval of biophysical variables from RS data as input to crop growth models can be used directly to forecast crop yield at the regional scale (Moulin et al., 1998). Most models can use RS data as input in various stages of the modeling process and it is well documented that the performance of the models can be readily improved when RS data are combined with crop models (Mkhabela et al., 2011; Kouadio et al., 2012).

21.4   Regional-Scale Crop Outlooks

Agricultural monitoring using satellites began with the Large Area Crop Inventory Experiment (LACIE) program which was developed jointly by NASA and USDA in the early 1970s. Through the subsequent development of RS techniques, several operational crop forecasting systems have been developed and applied worldwide. Most of those systems consist of an integrated crop forecasting approach and include large and diverse types of data.

One example is the MARS Crop Yield Forecasting system (MCYFS) used by the European Commission for seasonal yield forecasts of European crops of major economic importance. Within the MCYFS different procedures are applied which combine historical yield statistics, climate/weather indicators, and remotely sensed VIs, including additional information sources and expert knowledge. The crop monitoring component in the MCFYS uses a complex annual crop growth model called WOFOST (Diepen et al., 1989). The regional crop growth simulation, achieved through the crop growth monitoring system (CGMS) is mainly driven by climate variability as well as reference sources of statistical information derived from RS data.

A large number of available crop yield forecasts can be generated statistically, and when the accuracy of the predicted yield is deemed to be not sufficient, trends periods and functions are analyzed. Given the uncertainties about the evolution of the season during the forecasting process, agro-meteorological scenarios can be produced and analyzed. Such scenarios are currently based on agro-meteorological similar years as detected by multivariate statistical methods (e.g., clustering, principal component analysis) useful for reducing the number of explanatory variables or "dimensionality" required to forecast crop yield. The reliability of all data sources as well as the accumulated experience and knowledge of the yield forecasting are thus important in selecting the more important variables that explain crop yield and its variability in time and space. In summary, statistical methods are applied in quantitative analyses (e.g., linear regressions, agro-meteorological scenarios/similarities analyses, trend analysis) and qualitative assessments (distribution of errors over time) of yield. Crop yield forecasts are updated in near-real time and able to anticipate most other sources of information.

In Canada, a statistical model with an intermediate level of complexity has recently been developed as a prototype tool for forecasting major crop yields. The model utilizes crop field survey, historical climate and near-real time RS data in generating probabilistic yield forecasts that are sequentially-updated within the growing season as data become available (Chipanshi et al., 2012; Newlands and Zamar, 2012). This statistical model has been embedded within a prototype, operational regional forecasting system (Integrated Canadian Crop Yield Forecaster or ICCYF). This prototype can be used to evaluate regional and local yield variations and give the industry sufficient lead time to

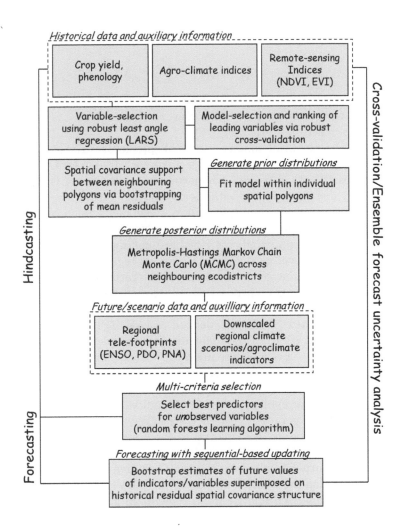

FIGURE 21.4: Overview of the analytical framework and current Canadian integrated regional-scale crop forecasting methodology (ICCYF).

make economic choices, and serve as a cost effective alternative to traditional survey methods. The crop yield simulations will also provide baseline data for projecting future variations in yields under a changed and variable climate.

The integrated forecasting framework for regional-scale forecasting of crop yield brings together a variety of advanced statistical techniques, as outlined in Figure 21.4. A region is divided into individual subregions delineated as

Statistic Canada's Census of Agriculture Regions (CARs). Within each CAR, the variables that best predict crop yield are determined by a method called robust least angle regression based on the following model,

$$Y_{s,t} = \gamma_0 + \gamma_1 t + \alpha y_{t-1} + \sum_{p=1}^{n_p} \beta_{s,t}^p X_{s,t}^p + \varepsilon_{s,t}, \tag{21.1}$$

where $Y_{s,t}$ is the annual crop yield for year, t, where $t \in \{2, \ldots, n_t\}$, within a given CAR, s, where $s \in \{1, \ldots, n_s\}$.

In this model, the total number of possible predictor variables selected in a given year is n_p and $X_{s,t}^p$ denotes the pth predictor variable for CAR s at time t, while $\beta_{s,t}^p$ is the pth regression coefficient. In addition, $\varepsilon_{s,t}$ is independent and Normally distributed error with mean zero and variance σ^2. The regression parameters γ_0 (yield intercept), and γ_1 (technology trend coefficient) are used to detrend the yield data and α is a lag-1 autoregressive term. This model is multivariate (having n_p predictor variables) and has spatially-varying coefficients, $\beta_{s,t}^p$ varying over n_t years and n_s spatial regions.

The model's dynamic covariates or predictor variables comprise a sub-set of possible predictors that include well established biophysical driving variables and indicators or crop growth: PET — potential evapotranspiration, a crop water stress index (WSI) (i.e., defined as (AET/PET−1), where AET is actual evapotranspiration), a thermal-time growth index defined as growing degree day (GDD's) above a base temperature of 5°C, and NDVI. Information from neighboring CARs is also considered, employing an approach described by Bornn and Zidek (2012). Predictor variables can also include other remote-sensing indices, and/or climate variables and indices. A technique called "Random Forests," that has proved very effective in reducing variance and error in high-dimensional datasets, is employed to forecast yield under a wide range of sets or mixtures of the predictor variables.

Some model output results from the new Canadian model are illustrated in Figure 21.5 showing the benefits of integrating NDVI data into a forecasting model and the associated relative improvements in forecast surveys that are possible. These results were generated using the ICCYF model to estimate yield in the past (i.e., so-called "back-testing" or "hindcasting" employing the "leave-one-out" approach to validating model forecasts). This involves dividing the historical data available into two parts, one for "in-sample" estimation and the other for "out-of-sample" validation. This enables one to optimize a model by comparing how well the model can predict yield using previous historical yield and other input data.

Historical data for the years 1988–2010 were used for this procedure. Data for a single selected year within this range of years available (i.e., a rolling window of one year) were sequentially omitted and the model was run to generate a yield forecast. Here, the data for this single year represent the "out of sample" year for which the model forecast accuracy is tested, with all the other data serving as "in-sample" data. An empirical prior distribution is

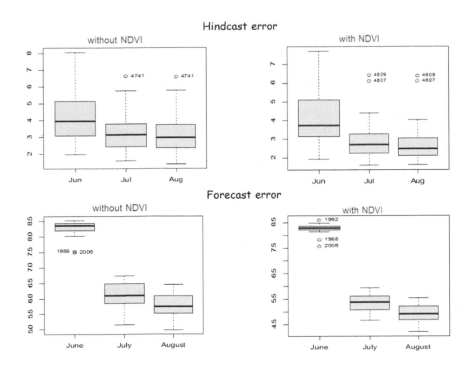

FIGURE 21.5: Relative gain in model accuracy when using the NDVI auxiliary index for forecasting crop yield (spring wheat) within the growing
season (May–October). Top: Hindcast error (Mean Absolute Error, MAE)
(bushels/acre) obtained from leave-one-out cross-validation against historical
crop yield data. Bottom: Forecast error based on forecast margin of error (i.e.,
for a future month).

obtained by residual bootstrapping the regression model in Equation (21.1),
and by pooling the data from neighboring CARs. For each forecast year, the
model was run with the following settings: spring-wheat (crop type), monthly
time-step, scale (40 CARs), 5 as the maximum number of yield predictors or
covariates, 3 as the number of spatial unit neighbors, 500 as the number of
bootstrap samples used to generate a prior yield distribution for each CAR,
and 5000 as the chain size for Markov Chain Monte Carlo (MCMC) sampling
of the joint posterior distribution over the model parameters, agro-climate
and remote-sensing auxiliary variables.

Model output of both historical "hindcasted" and future "forecasted" mean
absolute error (MAE) with associated 95% MCMC distribution-based prediction confidence intervals is shown in Figure 21.5, for each month of the wheat

growing season (Canada), and with/without use of RS data (NDVI). These box plots indicate the median error as the vertical line within the box. Ends of the box are the upper and lower quartiles as a measure of data spread spanning 50% of the dataset and eliminating the influence of outliers. The whiskers are the two lines outside the box that extend to the highest and lowest data values. Here, the CARs 4607 and 4609 are identified as outliers and so warrant further analysis.

To provide additional information and insights regarding how accurate the forecasting model represents past extremes events (i.e., heat waves, drought or persistent flooding conditions), we highlight those years that are outliers (i.e., 1992, 1998, 2006). In this way, we can first focus in on specific CARs and years and improve the historical climate, yield and remote-sensing data inputs and further investigate the selection of predictors within those CARs to help improve the accuracy and reliability of the model's forecasts.

These cross-validation results reveal that there is bias when NDVI data are not used in forecasting yield and that it is reduced when they are included. This indicates that including NDVI data improves the forecasting of yield, particularly in later months of the growing season when the crop (spring wheat) reaches maturity and its maximum attainable yield. For hindcast error, we identify outlying CARs because areas that show marked departure from the mean and lower/upper quartile ranges provide several practical modeling insights: (i) the need or importance to consider and include other predictors of yield to better represent historical trends; (ii) whether more accurate climate data from additional long-term or near-real time monitoring stations is required to more accurately represent the climate and yield variability for some CARs.

Bias in forecasting or so-called "forecast skill degeneracy" can often occur when climate data is involved, and be amplified, when forecasts are verified using the correlation between forecast and actual data resulting from leave-one-out cross validation methods (Barnston and van den Dool, 1993). This can result when a cross-validation procedure uses nearly all of an available sample dataset to develop regression-based forecast equations (i.e., as for historical crop yield). Validation procedural adjustments can help to reduce such bias by leaving out or excluding more than just one year at a time in generating cross-validation forecasts, or using a verification measure other than correlation such as MAE.

Use of remotely-sensed data not only improves crop yield forecasts, but also improves the reliability of soil moisture data related to the amount and efficiency of water use by different crops under different climate conditions. For example, soil water stress is widely recognized as a dominant factor limiting crop growth across the Western Canadian Prairies — a major agricultural region of Canada. Geostatistical validation using soil moisture active/passive satellites (i.e., RADARSAT–1, 2) and wireless sensor-based monitoring networks measuring climate variability, soil conditions and in-situ soil moisture (5–10 cm depth) are currently being tested. It is hoped that these inter-related lines of enquiry will culminate in the future implementation of a national monitoring effort to create a national database of soil moisture to further improve the accuracy of inputs used in forecasting crop yield for major crops.

21.5 More Reliable Forecasting

In summary, previous methods of crop yield forecasting at regional scales relied mainly on survey sampled yield and station-based climate data, but such methods are now being progressively upgraded toward much more integrated approaches that combine wireless sensor network monitoring, satellite and airborne remote-sensing, and higher-resolution climate information (i.e., interpolated historical station-based observational data, as well as data generated from the future scenario output of dynamic regional climate models).

The awareness of the public to environmental concerns related to climate change, environmental pressure exerted by agriculture, and ecosystem sustainability continues to increase. Recurrent problems of dire food shortages and famine in many parts of the world have boosted the demand for more inter-disciplinary studies.

Information about crop production and demand, and decisions for efficient agricultural policies are all shaped by statistical concepts, knowledge and techniques. For both the researcher and a wide range of users of crop forecasts, predicting impacts and evaluating decision options based on forecasting systems requires the ability to anticipate outcomes associated with each decision option under different forecast conditions. Risk analysis is an important element for grounding decision-making linked with scientific recommendations as well as more subjective and incremental decision responses to observed impacts, so that decisions are more consistent with forecast uncertainty and decision maker tolerances to risk (Hansen, 2002).

To reduce human vulnerability to potential impacts of climate variability on agriculture, several methodological approaches exist. In spite of their relative strengths and weaknesses, statistical approaches are widely used to investigate the effects of recent and future climate changes on crop productions (Lobell and Burke, 2010). One of the directions to improve crop forecasting is the introduction of climate forecasts at different time spans. Advances in modeling climate variability, by incorporating information from regional-scale output of dynamic climate models, might aid in generating timely and accurate yield forecasts with the ICCYF model.

Fine scale estimates of future climate and a better understanding of the statistical properties of extreme events will help to provide models for the simulation and for prediction of future crop yields. The numerous prospective analyses and international commitments related to world food security over the coming epochs will require strong interdisciplinary collaboration, so that statistical methodologies can be best applied as operational decision support tools for agriculture. Statistical methods also provide a vital bridge for linking diverse types of complex data and for the development of finely tuned models for forecasting. The challenges are many but the rich variety of data that are now available deserve the best statistical tools we can provide.

Acknowledgments

This work was supported by the Growing Forward Program of Agriculture and Agri-Food Canada. The authors thank David S. Zamar for assistance in generating the example cross-validation output provided from the ICCYF crop forecast model. Thanks are also due to the editor, the associate editor, and Dr. T. A. Porcelli for their helpful feedback and comments that improved this manuscript. The Canadian Crop Yield Forecaster has been coded and tested using the open source software and statistical libraries provided by the R Statistical Software (R Development Core Team, 2008). ArcGIS^{TM}(ESRI^{TM}, Version 10, 2010) was used to visualize model output, processing spatial data and generating crop "outlook" maps.

About the Authors

Louis Kouadio is an NSERC visiting postdoctoral fellow with Agriculture and Agri-Food Canada. He completed his PhD in environmental sciences and management at the Université de Liège, Belgium. His primary research interests are crop forecast modeling, food security, and satellite remote sensing applications in agriculture.

Nathaniel Newlands is a research scientist in science and technology with Agriculture and Agri-Food Canada and an adjunct professor in the Department of Statistics at the University of British Columbia. He completed a BSc in physics and mathematics at the University of Guelph, an MSc in astrophysics at the University of Calgary, and a PhD in resource management and environmental studies at UBC. His research focuses on statistical climatology, crop forecasting, ecosystem risk assessment and sustainable development modeling, as well as resource management decision-support.

Bibliography

Allen, P. G. (1994). Economic forecasting in agriculture. *International Journal of Forecasting*, 10:81–135.

Baret, F. (2000). *ReSeDA: Assimilation of Multisensor and Multitemporal Remote Sensing Data to Monitor Soil and Vegetation Functioning.* Final Report, EC Project ENV4CT960326, INRA, France.

Barnston, A. and van den Dool, H. (1993). A degeneracy in cross-validated skill in regression-based forecasts. *Journal of Climate*, 6:963–977.

Boote, K. J., Jones, J. W., and Pickering, N. B. (1996). Potential uses and limitations of crop models. *Agronomy Journal*, 88:704–716.

Bornn, L. and Zidek, J. V. (2012). Efficient stabilization of crop yield prediction in the Canadian Prairies. *Agricultural and Forest Meteorology*, 152:223–232.

Carroll, M., DiMiceli, C., Sohlberg, R., and Townshend, J. (2004). *250m MODIS Normalized Difference Vegetation Index*, 250ndvi28920033435, Collection 4. University of Maryland, College Park, MD.

Chipanshi, A. C., Zhang, Y., Newlands, N. K., Hill, H., and Zamar, D. S. (2012). Canadian crop yield forecaster (ICCYF): A GIS and statistical integration of agro-climates and remote sensing information. In *Proceedings of the Workshop on the Application of Remote Sensing and GIS Technology on Crops Productivity among APEC Economies*, Beijing, China.

Diepen, C. A., Wolf, J., Keulen, H., and Rappoldt, C. (1989). WOFOST: A simulation model of crop production. *Soil Use and Management*, 5:16–24.

Hansen, J. W. (2002). Realizing the potential benefits of climate prediction to agriculture: Issues, approaches, challenges. *Agricultural Systems*, 74:309–330.

Kouadio, L., Duveiller, G., Djaby, B., El Jarroudi, M., Defourny, P., and Tychon, B. (2012). Estimating regional wheat yield from the shape of decreasing curves of green area index temporal profiles retrieved from MODIS data. *International Journal of Applied Earth Observation and Geoinformation*, 18:111–118.

Leeds, W. and Wilke, C. (2012). Science-based parameterizations for dynamic spatiotemporal models. *Wiley Interdisciplinary Reviews: Computational Statistics*, 4:554–560.

Lobell, D. B. and Burke, M. B. (2010). On the use of statistical models to predict crop yield responses to climate change. *Agricultural and Forest Meteorology*, 150:1443–1452.

Luo, Y., Ogle, K., Tucker, C., Fei, S., Gao, C., LaDeau, S., Clark, J. S., and Schimel, D. S. (2011). Ecological forecasting and data assimilation in a data-rich era. *Ecological Applications*, 21:1429–1442.

Mkhabela, M. S., Bullock, P., Raj, S., Wang, S., and Yang, Y. (2011). Crop yield forecasting on the Canadian Prairies using MODIS NDVI data. *Agricultural and Forest Meteorology*, 151:385–393.

Moulin, S., Bondeau, A., and Delecolle, R. (1998). Combining agricultural crop models and satellite observations: From field to regional scales. *International Journal of Remote Sensing*, 19:1021–1036.

Newlands, N. K. and Zamar, D. S. (2012). In-season probabilistic crop yield fore-casting: Integrating agro-climate, remote-sensing and crop phenology data. In *Proceedings of the Joint Statistical Meetings: Statistics — Growing to Serve a Data Dependent Society*. American Statistical Association, Alexandria, VA.

United Nations (2011). *World Population Prospects: The 2010 Revision, Highlights and Advance Tables*. Population Division, Department of Economic and Social Affairs, United Nations.

Zhu, X. G., Zhang, G. L., Tholen, D., Wang, Y., Xin, C. P., and Song, Q. F. (2011). The next generation models for crops and agro-ecosystems. *Science China, Information Sciences*, 54:589–597.

Index

Abundance estimation
    Animals, 289, 292
    Fish, 275, 306, 307
    Humans, 273, 292
    Species, 283, 305
Aging, 193
Agriculture, 22, 341
Algorithm
    Adaptive MCMC, 103
    Lavallée–Hidiroglou, 24
    Least squares, 329
    MCMC, 95, 157
    Metropolis, 93, 157
    Monte Carlo, 94, 253
    Pool Adjacent Violators, 87
    Tree-based, 136
Association
    Genetic, 129
    Genome-wide, 133, 147
    Measure of, 65
    Statistical, 1

Bayes
    Factor, 152
    Rule, 136
Bayesian
    Analysis, 93, 154, 167
    Method, 119, 276
    Model, 344
Bootstrap, 138, 351
    Parametric, 73

Cancer, 140, 141
    Bone, 177
    Breast, 178
    Diagnosis, 86
    Prostate, 139, 178

    Skin, 93
    Surgery, 81
Censoring
    Dependent, 215
    Time, 187, 194, 212
Chart
    CUSUM, 234
    EWMA, 232
    Monitoring, 228
Climate change, 59, 317, 321, 347
Copula, 64, 256
Correlation
    Analysis, 59
    Linear, 62, 251
    Matrix, 309
    Spatial, 158
    Structure, 119
    Surface, 54
    Temporal, 343
Curve
    Bell-shaped, 59, 243
    Growth, 48
    Registration, 54

Data
    Aggregated, 293
    Auxiliary, 28
    Collection, 25
    Discrete, 50
    Financial, 246
    Functional, 47
    Insurance, 59
    Time series, 37, 244, 327
Dementia, 193
Dependence measure
    Kendall, 65
    Pearson, 61